# Automotive gas turbines

## The dream that faded

**By John Mortimer**

# Automotive gas turbines

Published by John Mortimer

First printing 2016

John Mortimer
37 Stock Lane
Whaddon, Milton Keynes,
Buckinghamshire,
MK17 0LS. UK

Also by John Mortimer

*Zerah Colburn: The Spirit of Darkness*, published by Arima Publishing

*Angel in the house*, published by John Mortimer

*The 'nearly' engine*, published by John Mortimer

*My Mirfield Family*, published by John Mortimer

ISBN 978-1-291-88411-6-9000

Copyright© John Mortimer 2017

All rights reserved. This book or any portion thereof may not be reproduced or used in any manner whatsoever without the express written permission of the publisher except for the use of brief quotation in a book review or scholarly journal.    149125

**COVER. Noel Penny with Rover-BRM gas turbine car.**

# Foreword

This book stems from emails Noel Penny sent from December 2009 to October 2011. They comprised Penny's 'My story' with photographs and diagrams of his various engines. It is a 'warts and all' biography, through boyhood and the night of Thursday 14 November 1940 when the German Luftwaffe launched its then most devastating bombing raid of WW2. Later, for Penny years wrestling with gas turbine engine development followed; then creating new engines for vehicles. Finally, came the period of being an entrepreneur of small aero gas turbines.

Sadly, Penny's story has no happy ending; 'My story' concludes with the eventual demise of his business and the fading of the automotive gas turbine dream.

Without compunction however, I dedicate this book to Noel Penny, the father of the UK's automotive gas turbine; his had unremitting enthusiasm for the engine.

# CONTENTS

Introduction                                    7

One
Growing up                                     11

Two
A world at war                                 41

Three
The apprentice                                 71

Four
Atomic energy                                 107

Five
Rover and gas turbines                        123

Six
The T3 gas turbine car                        151

Seven
Racing gas turbines                           175

Eight
Gas turbines power trucks                     205

Nine
Noel Penny Turbines Ltd                       251

Ten
Downfall                                      279

Eleven
International advisor                         309

Appendices                                    327

# Automotive gas turbines

| | |
|---|---|
| 1 Alvis | 328 |
| 2 Austin Motor Company | 329 |
| 3 Bladon Jets | 332 |
| 4 Ceramics | 344 |
| 5 Chrysler Corporation | 349 |
| 6 Daimler-Benz | 355 |
| 7 Detroit Diesel | 358 |
| 8 Fiat | 359 |
| 9 Ford Motor Company | 361 |
| 10 Ford/Garrett AiResearch | 372 |
| 11 Garrett AiResearch | 376 |
| 12 General Motors | 378 |
| 13 George J. Huebner | 383 |
| 14 Leyland Gas Turbines | 392 |
| 15 Mack Trucks | 395 |
| 16 MTU | 410 |
| 17 Rolls-Royce | 412 |
| 18 Rover Gas Turbines | 413 |
| 19 Volvo Corporation | 422 |
| 20 Williams International | 429 |
| 21 Walmart | 431 |
| 22 Wrightspeed | 433 |
| 23 Miscellaneous | 434 |
| **Illustrations** | 437 |
| **References** | 454 |

Index of names 457
Obituaries 459
Author 464

# Introduction

"IT IS SHOWN that the gas turbine is likely to be superior in all respects to the piston engine as a motive power for road vehicles. It is suggested that a turbine unit on the lines given in this report should be immediately put in hand if we are to enter the car field in the post-war era and successfully compete with other firms who will undoubtedly follow this line of development."

Those words were written by no less than a Rolls-Royce Ltd. engineer in March 1945, just a few months before the end of World War 2. Despite the engineer's positive pronouncement, Rolls-Royce's senior engineers and directors decided against taking his advice and the company never entered the arena of gas turbines for road vehicles, but instead concentrated on aero gas turbines. The company eventually also dismissed the Wankel engine for main battle tank applications.

But other companies, many with famous names, did pursue gas turbine engine development, including Ford Motor Company in the US and the Rover Company in the UK.

Almost a generation after Rolls-Royce's engineer penned the words above, another engineer across the Atlantic, A. F. McLean – Art McLean – of Ford Motor Company in Dearborn, Michigan delivered a paper to the Institution of Mechanical Engineers in London (6 February 1969). In this he wrote:

"Production of gas turbine engines for over-the-road vehicles will begin in the 1970s, when overall operating cost (initial cost, fuel cost, maintenance cost, repair cost, oil cost, etc.) becomes favourable. Initially, the turbine will be introduced in applications with high power utilization in the 300 to 600bhp class. But what of the lower power ranges where the 'carrot' of the big market is ever-present?

"There are at least two major problems which will prevent today's turbine engine from competing in this market. One is fuel economy at part power and the other is initial engine cost.

It is the author's opinion that the first of these will be solved by operating engines with higher pressure ratios and turbine inlet temperatures. The second problem, cost, will have to be solved, not only by exploiting modern manufacturing techniques to develop low-cost turbine components, but also by simplifying the basic engine design so that low-cost ceramic materials and/or low-cost blade cooling techniques can be utilized to facilitate operation with higher turbine inlet temperatures.

"This paper suggests the single-shaft turbine engine as a contender for a lower-cost approach, where fuel economy requirements are not met by complexity of cycle but by operation at higher turbine inlet temperatures.

"Before reviewing gas turbine cycles under consideration for vehicular application, the need for a heat exchanger must first be examined......It is apparent that a heat exchanger is a 'must' in order to get down to the SFC levels of today's diesel or gasoline engines. Whether the heat exchanger is a stationary type recuperator or a rotating type regenerator is not important for this discussion.

"In the vehicular turbine engine, the heat exchanger cannot be considered as a hang-on device any more than the compressor, the turbine, or the combustor; it is a major component of the engine, playing a vital role.

From McLean's perspective in 1969, as in 1945, the gas turbine looked a strong contender to replace the diesel. Yet, up to the present time, the gas turbine has failed to live up to the 'dream' as fuel economy and cost of manufacture serve as the principal reasons for its downfall. But it has not been for the want of trying, as this book shows. Indeed, there were many participants striving to put gas turbines in road vehicles on both sides of the Atlantic. But one man in particular tried hard to bring the gas turbine to life in Britain. His name: Noel Penny.

And so it is that this book is one man's story. It is very much a 'warts and all', very personal cameo of Penny's reflections and experiences of developing an automotive gas turbine – as well as of the politics of the day. Penny was not only a gas turbine

# Automotive gas turbines

pioneer in his own right, but a man of unflinching passion and enthusiasm for the technology; a man with an undying belief the prime mover *did* have a place on the roads of the world.

Such was Penny's determination and belief, that when his state-owned employer – British Leyland – 'chickened-out' from further developing an automotive gas turbine, Penny picked up the reins and formed his company in Coventry to develop small gas turbines for use on land, sea and in the air; no mean achievement. Penny's passions convinced the mighty Caterpillar Inc. of the US to come knocking on his door to harness his technology and enthusiasm. Others came, like Yanmar.

Noel Penny, like Sir Frank Whittle, was a Coventry man. And here in this text, before describing his gas turbine 'adventures', we see Penny recounting his early boyhood in the city and his 'real life' experiences during the bombing of World War 2. He lived through the cataclysmic blitz that almost demolished Coventry in one night – but not its people's spirit – to emerge after the war when Britain began to pick up its feet and step up again to the plate to regain its posture as a great nation. Then, after the war he worked first at the Atomic Energy Authority before turning his hand to gas turbines.

That Penny largely failed in his efforts to widely commercialize gas turbines does not in any way diminish his efforts. He paved the way for miniature gas turbines that have found their way into other areas of society.

So this is Noel Penny's *My Story*, just as he wrote it in his own words, sent by email chapter by chapter... It is a personal and private story, for the most part unedited. Notwithstanding that, I have presumed to expand it through the mechanism of various appendices, largely for my own amusement to record past events – events I observed from the sidelines as editor of *The Engineer* weekly news magazine from 1963 to 1980, and through the pages of my publication, *Auto Industry Newsletter* from 1980 to 2000 – to include most of the other main players in the world of automotive gas turbines.

**John Mortimer, January 2017**

# Automotive gas turbines

# One

# Growing up

**In which I start my journey through life, first as a member of a large family and then when I went to school.**

I was born into a poor family on Christmas Day 1925. From a very early age my eldest sister, Violet, became a mother to my sisters and brothers. I became the fourth brother, together with five sisters. I was a member of a large family. Little did I know that I would to be joined by my youngest brother and yet another sister. With eleven children in the space of just twenty years, it could be said that there was always one baby in the pram at the Penny household.

My father, Charles William, ran away with my mother, Clara Jane, when he was working on my maternal grandfather's estate. He wanted to get away from a rather sheltered life. My maternal grandmother was a school governess. Charles was over 25 years older than my mother, as his first wife had died during the delivery of their seventh child.

I never discovered what happened to my stepbrothers and sisters; the matter became a taboo subject within our family. Later in life, my youngest sister, Norah, was contacted by a Swindon family named Penny. However, at the time it was considered unwise that they should make any contact with them.

I also never knew my grandparents. Cornelius Penny, my paternal grandfather, became a victualler and maintained a village public house. He was also a master carpenter for the village inhabitants.

The days seemed long and sunny during my early childhood. Everyone, it seemed, was poor but very happy. The whole social structure was much different from the present day; people would help one another without counting the cost. To me it seemed so much the normal way of life. The church also played a focal point of our lives. I believed there were so many good things to be thankful for then, and through development and schooling, I was forced almost to appreciate other ways of life.

My family lived in two terraced houses. The second one upstairs was where my brothers slept. The downstairs area was used as my father's workshop for his livelihood. The first house was where we lived. There was a tubular hole in the wall about two inches diameter from my parent's bedroom to the second house where my brothers slept. I never knew what purpose it served, as I was too small to reach it. I remember my father waking up my brothers by calling from one room to the other.

Our two terraced houses were approached from a narrow yard, the entrance of which was in Hertford Place. Hertford Place curved round to both the bottom and the top of Queen's Road. At the bottom end was the Hen & Chickens public house. The Hertford Arms pub existed half way along Hertford Place, on the corner of Junction Street.

Next door to the Hen & Chickens, in Hertford Place, was a group of three shops. The first was Kirk, the hairdresser with a barber's pole above the door. It aimed to be a high-class hairdresser's but did not quite make it as Kirk varied his prices in competition with Hammond's, a hairdresser just round the corner in the Butts. Generally, Hammond's was a penny cheaper than Kirk's for a short back and sides.

Hammond's was well known throughout the district as the 'basin cut'. Small boys who frequented the latter, as I did, would stand facing the back of the barber's chair with their chin resting on the backrest. So they were made to stand throughout and when Hammond cut a straight fringe we knew the 'short back and sides' was nearly at an end.

There was also a newsagent next to Kirk's by the name of Maynard's. The shop sold sweets. Miss Maynard had been there for as long as anyone could remember. She was a spinster and to me she seemed to be very old. Miss Maynard sold halfpenny dips for children; she made these up from old sweets and, sometimes, a chocolate. Her niece occasionally managed the shop; she was also a spinster. A number of boys took the newspapers round the district for a few pence a week.

One of my earliest recollections was that of taking a halfpenny to see what I could get. I must have been about three as I could hardly reach to place my treasured money on the counter. I spotted a tray of broken toffee with a small toffee hammer resting on the toffee. I pointed to the toffee hammer. Miss Maynard's niece seemed to sense what I was trying to purchase as she called her aunt to see whether the old lady might explain to this urchin. As a result I was given one of Miss Maynard's dips and my halfpenny was returned. With a big smile I ran home to announce my good fortune.

Carvell's dairy shop was sited next door to Maynard's. The shop was spotless yet at the same time bare, almost to the point of being austere. Carvell had many mouths to feed, like most people in those days. He was the milkman of the area. He was a large man with a gruff voice. I never remember seeing him in anything other than a cow gown. He was kind to children who followed him round when he delivered his milk.

Children were fascinated with how he would dole out the milk, as he used a half a pint, pint and quart measures in a hand-carried milk churn on his delivery round. At his points of call he would take a measure and empty the milk into jug of the household. The children who followed him on his round enjoyed the musical notes made by the various measures hitting the side of the milk churn. Like a bell being sounded by different sized tuning forks. One of Carvell's sons had an indentation on the side of his forehead, which we observed with great fascination.

Next to the Carvell's was a wide entrance known as the 'big yard'. Here was a general store of the type that sold everything from cooked meat to toothache tincture. The owners of the grocery store changed every few years; the longest running owner of whom was the Owen family. Mrs. Owen had seen better days apart from when she was out on the town and made up. Charles used to say, "mutton dressed as lamb".

On the other side of the 'big yard', past the shops, lived Ma Parnell, a small cantankerous woman always dressed in Victorian black. Mr. Parnell had long ceased this world in protest and her only son Walter had fled to marry. The 'big yard' so called opened out to a very wide space with many terraced houses in a rough circle. The term 'big yard', which really was a court entrance, was far wider than the narrower court entrances in Hertford Place that were sufficient for a good size pram.

I started at infant school at about 1929. Little did I know that I was nearly blind in one eye. This defect was not discovered until I was six or seven. My early schooling was interrupted as a result. I well recall the infant school where Miss Walker was headmistress and it seemed that all the teachers were female spinsters. Miss Walker was a kind lady who understood that most children were nervous of leaving a known environment.

Miss Porter was my favorite teacher of the top class in the infants. Years later she married Mr. Elliot who was one of my teachers in the big school. I recall two names from my infant classes: a lad named 'Ward' whose father was a tailor in the Butts. I remember his dad making a suit of pants with two shoulder straps that were easily fastened at the front with two buttons. Every time he wanted the toilet his teacher only had to unfasten the buttons. The toilet for the infant classes was some distance away, so he had to fumble and his pants dropped once or twice in the classroom.

There was also a girl called Irene Wittrington. She was often chastised because she always laughed outright at Ward's discomfort, setting the whole class in an uproar.

The 1930s came in with mass unemployment. There was a pacifist uprising in the country against anything to do with rearming. The 1930s were a most remarkable decade since unemployment peaked in 1930, yet there was full employment

# Automotive gas turbines

at the end of the 1930s. On looking back, a politician said, "You either starve them or kill them on the battle field".

My brother Roy was born on 25 June 1927, exactly eighteen months to the day of my own birth. He was born during the Godiva carnival. Coventry held a carnival once a year on June 25th and that year was a Godiva carnival. When Roy was old enough we used to pull his leg that a lady rode through the streets of the City stark naked (on his birthday).

The history of the City of Coventry involved the legend of the Danish overlord, the Earl of Leofric. His wife, named 'Godgifu', became so upset with her husband's oppression of the people of Coventry, as a result of the heavy taxes imposed, that she offered to ride through the streets unclothed if he were to relent. The people of Coventry agreed to shutter their windows, so the legend relates. One man however was seen peeping out of the window when she rode through the streets naked. That man gave rise to the name 'peeping Tom' and a statue of 'peeping Tom' could be seen at the bottom of Hertford Street for many years.

Coventry was originally a city of two halves; the military half was in the south and the Christian half in the north. The Earl of Leofric, who became a Christian thanks to his wife, is buried in the north despite his military tendency. Coventry became a walled city early in the Middle Ages, giving rise to the saying "sent to Coventry"; for when Cromwell caught the royalists of Birmingham they were imprisoned within the walls of Coventry.

To the best of my recollection my early years before school were spent with my friend 'Reggie' – Reggie being short for Reginald. Reggie, as he was called, was blond with a mass of curly hair. He had pale skin that complemented his light hair. Reggie was about the same age as myself and he and I became inseparable. He used to call on me every morning crying, "Nole, 'ole" at my door until I emerged. Often we were out for the day, visiting his mother who lived only a few houses from where we lived.

Mrs. Wormell was a tall, thin woman, neatly dressed. I remember thinking that she had one of the longest necks I had even seen. Little did I know that she was not his real mother but an aunt, who was looking after him. Nevertheless she thought the world of Reggie.

The time came when his real father, Mr. Wormell's brother, wanted Reggie back, since he had a new wife. Both Reggie and Mrs. Wormell were heart broken. Reggie's father now lived in Romford, a district of outer London. The day came to collect him; his screams still linger in my ears. Mrs. Wormell never recovered and became an alcoholic. For weeks she played a mournful tune "in the gloaming etc." on an old, scratchy gramophone; I can still hear the mournful tune.

Reggie came back years later to see me. Unfortunately, I was away but my sisters welcomed him. Whenever I hear that tune I think of Reggie, my first real childhood friend, and wonder what he was like in adult years. This illustrated one of my first lessons of life. Just grit your teeth and bear it when there seems nothing you can do. It is another way of saying, "God give strength to stop me worrying over those things I cannot change to concentrate on those issues I can influence for the better."

My eldest sister was due to be married, and, as young as I was, I threatened the 'would be bridegroom', Edward Matthews, that I would get a policeman to prevent him from taking her away. Nevertheless, it was to no avail and eventually she gave birth to a daughter Beryl in 1930. Her father nicknamed the baby 'darky' because of her jet-black hair. My youngest sister, Norah Jane, was born the following year on 7 March 1931, resulting in one of those rare circumstances of an aunt being younger than her niece.

Beryl and Norah became great friends, although of quite different temperaments. Beryl was outspoken to the point of being brash. She hated schooling or any kind of discipline. Norah on the other hand was quiet and well-behaved and became a scholar, winning many school prizes.

The 1930s became a tough time for most working people. The British government sought means for lifting Britain out of the Depression. By the end of the year 1932 a new system was complete in Great Britain. Unemployment reached its all-time peak in 1932.

Unemployment fell in the latter part of 1933 by half a million. Production recovered and international payments were balanced, though more from a decline in cost of imports than from increase in exports.

Charles was due to attend a meeting to discuss a new commission for the refurbishment of Allesley House during January 1934. He had already co-opted Jack, his eldest son (Jack was christened John Francis) into the business of C. W. Penny & Son, Builders & Master Carpenters. Charles and Jack had previously carried out carvings for Coventry's churches. These were works of art and some were quite large. They were well known throughout the city.

Charles was accustomed to travelling about the city by tram. Whenever he was away from a tram route, he would walk. On this particular day he returned by tram to the bottom of Queen's Road. His thoughts were on obtaining work for the best part of a year, not only for Jack and himself but also for many other Coventry workmen. As he passed by the other side of a tram, a 'motorcycle & sidecar' collided with him. His doctor, who just happened to reside nearby in Queen's Road, was called and at first it was thought that he was not badly hurt. However he never quite recovered and Jack had to take over the running of the business in part.

The extent of the injuries as a result of the accident to Charles, both physical and mental, soon became obvious to Clara. He was in his seventy-fourth-year and Clara was anxious that he would achieve what he always strived for. That was, 'to maintain his family in comfort, especially Clara, who had run away from a good life with her own family'. Despite the hardships of the years with Charles, she had no regrets. Clara was worried about Charles and the basic things of life for their large family.

Clara left the children in charge of the eldest girls whilst she walked round the district to think out the way ahead. She passed Saint Thomas's church in the Butts and continued on to Spon-End-Bridge before returning home. Her good friend Cannon James, the vicar of St Thomas's, said, "That anytime she sought counsel, he had an open door." Now was the time to see Cannon James, Clara thought. Cannon James's family had helped the Penny's in many ways.

Clara also belonged to the 'white ribbon' organization, a society run by nuns called the 'Grey ladies' because they all wore Grey habits. The 'white ribbon' was a temperance society of the 1930s. We used to attend the 'white ribbon' as children with Clara wearing a badge of white ribbon.

Clara went to see Cannon James. He said he would visit Charles in order to assess what support he could give and report back to Clara. Although Charles was not a regular church attendee at St Thomas's he had to call there to clean one of his old carvings. Cannon James asked the Churchwarden, whose responsibility it was to attend to Charles, if he would mention that he wanted to see him on a personal matter.

Charles was diffident about discussing the accident and brushed it aside. Eventually, Cannon James was able to give Charles advice that he must slow down and take life more gently. Afterwards, Cannon James reported back to Clara, and as a result she managed to convince Charles that a picnic to the Memorial Park with the children would be a good tonic for the all the family.

It was a summer's day; a blue sky promised another scorching day, so Clara was able to convince Charles without much fuss that the family should visit the park. The children liked to roam the common near to the Memorial Park. A peace had descended on the spot where they decided to hold the picnic; all the children had run off. Charles lay looking up at the sky with Clara sitting at his side. This was one of those rare moments when they could converse together without being interrupted.

They were discussing one of the recent church helpers, Margaret Lea-Fletcher. Margaret was the only daughter of one of the leading families in Coventry whose mother had heard that the church needed helpers to visit some of the poor and destitute in the immediate area. Margaret's mother had seen Cannon James and Clara to urge them to persuade her daughter to use her energies in this social activity of visiting the poor.

In the 1920s and early 1930s people went dancing, often staying out at all night parties. This was particularly so for the children of the wealthy. Mrs. Lea-Fletcher had already rescued Margaret many times from all night and weekend parties popular with the children of the dancing set of the rich. Many of the rich 'play-boys' were killed in fast open top motorcars. The motorcar was transforming social life. In 1920 the number of private cars registered was under 200,000, but by 1930 it had exceeded a million.

Virtually no roads were barred and anyone over the age of 17 could drive without passing a test. Driving tests were introduced only in 1934.

"She seems a nice girl to me," said Charles, referring to Margaret. "Her heart is certainly in the right place," Clara added. "What concerns you?" Charles went on. "Well" said Clara, "I am concerned that Margaret could upset the really poor among the group and drive them away." "Couldn't you give Margaret the responsibility for the funds, so that dealing with people of a very different background would be less of a problem?" asked Charles.

"That could be a good thing to try," said Clara, "but she does have a morbid fascination for visiting their homes".

"It might be good for Margaret to see what life is like for the very poor", said Charles.

"That maybe so," continued Clara, "but she could also turn them away, not intentionally, but because of her lack of experience of poor people. Remember she has had a very expensive education and she is only just out of finishing school."

In the early 1930s, the upper classes were quick to point out the growing economic equality between all classes. This was a sign that the majority of the very rich were embarrassed in light of their comfort in relation to the rest of society. Two-thirds of the nation's wealth was owned by just one percent of the nation.

The strongest potentially were the middle classes who would continue growing. It would become the class that the other two would either fall into or reach.

Few people in the early 1930s believed the First World War had been caused by deliberate German aggression. The League of Nations would prevent mistakes in the future, it was thought. Among the views widely held were those suggesting that private manufacturers of armaments fostered wars. This gave rise to the famous resolution at the Oxford Union in February 1933 that: 'This House will not fight for King and Country'. This resolution was carried by a substantial majority and supported by Dr. C.E.M. Joad, the principal speaker who became well known on television many years later.

Japan invaded Manchuria in September 1931 and it was not until the Spanish civil war in 1936 that the turning point came causing Britain to rearm.

Adolf Hitler became German chancellor in January 1933 with an avowed programme of making Germany once more a great power, to be feared by the rest of the world.

Hitler laid down in *Mein Kampf*, which he wrote while in prison before he became chancellor, the aggressive action he would take when he came to power.

Meanwhile, one day a poor woman was found to be in a terrible state. She had received a black eye, her mouth was swollen and the doctor suspected a broken jaw. The husband involved had been taken away when neighbours called the police. The poor woman had sought the help of the local church group when a kind neighbour called at the church nearby. Clara was already in touch with the doctor who had treated the poor woman shortly after the police arrived to take her husband away.

Clara had been able to stay long enough to get the three young children indoors, bathed and to bed. There was little food in the house and the wife of Cannon James was due to arrive with a basket of groceries.

Mrs. James arrived before Clara left. She sat on the side of the woman's bed. "Are the children alright?" the poor woman mumbled to Mrs. James, her face obviously distorted in pain. "They are all in bed, sound asleep," replied Mrs. James. Mrs. James had already been in contact with the local visiting nurse, as the woman was determined to stay at home, such as it was, and not be sent off to the hospital.

"Where have they taken him? It's that bloody girl with her 'high-and-mighty' ways," said the poor woman. "To whom are you referring?" said Mrs. James. "You know the one at the group, the girl who gives out the money!" uttered the poor woman. Mrs. James was shocked. The woman could only be referring to Margaret, but what on earth did the woman mean?

When the woman was sleeping, Mrs. James crept downstairs into the solitary living room that was both dirty and in great disorder. There was a square table in the middle of the room, and above that hung a gas mantle, infested with flies. Many wooden chairs in need of repair, and a sideboard cluttered with everything imaginable were positioned along one wall. Against the opposite wall was an old settee with one spring showing, covered with all kinds of dirty clothes. The mantle-shelf over the black-leaded grate held an assortment of odds and ends, including an alarm clock, its face stained with a brown discoloration. There was no fire in the grate, itself filled to the top with dusty ashes. Mrs. James found the door to a cramped scullery.

There was no cooker; grease and dirt were everywhere she looked. The floor covering was of worn linoleum. If it ever had a pattern, then wear and dirt concealed all trace. An old galvanised bath, about half the size of a modern bath, hung from the far wall. The brown earthenware sink was crazed all over with many pockmarks. There was no tap inside but Mrs. James peered through the window, thick with grime, and spotted a tap outside dripping into a drain.

She thought back to the days when the vicar must have married them, during one of the husband's few visits from the fighting over in France. As only a boy then, he went off to the 'Great War' and, as she was a pretty bright young girl, the future was theirs, so they thought. He was lucky to return from France but in his absence there was no work for him and he had not been in employment since.

Three quarters of a million men and women from the UK were killed during the First World War. The British Empire lost another two hundred thousand; nearly one-third of them were Indians. In all, the death toll approached one million. The price of victory, though high, was less than that paid by other countries. France, for instance, with a smaller population than the UK, had a death toll nearly twice as great; such is the legacy of war. Lloyd George talked of making 'a country fit for heroes to live in'. His record in social legislation seemed to justify this promise and many people believed in had won the war, but others thought he lost the peace that followed.

The church volunteers moved in and by midday the poor woman's place had been transformed into a clean, if a well-worn home. New floor covering was down; a second-hand handsome settee replaced the worn wreck. Shelves had been installed in the scullery, and the young children were looked after by neighbours.

The doctor came to treat the poor woman of the house. She was now sitting up in bed. The woman related how Margaret came there many times to give her husband cash that he used for drink. She had returned from the group one night to find them in bed together.

"That was several nights ago," she mumbled with difficulty "And when she didn't show up again, he got in such a rage I had to fight him off. If the neighbours hadn't heard, he might have killed me." At this point in the conversation the police arrived and asked if she wanted to make charges. "Of course, I don't. I want him back."

Mrs. James went to see Margaret. She was in a very disturbed mood. "It was only a bit of fun," expressed Margaret. "Well, what you can do is find him some work, otherwise your parents will hear of the mess you have caused," protested the Cannon's wife, with her tongue in her cheek as her ethics would not allow the threat. Margaret found him a job with one her friend's company nearby his home. He was bound over to keep the peace. Cannon James extracted an undertaking from the man that he would not drink to excess. Certainly he would not use any of his wages on drink. Having been found employment and forgiveness from his wife he was pleased to give Cannon James his promise.

Margaret never reappeared at the church group for the poor again, but she did go to see Clara in another tragic way. Through the combined help of neighbours, and the church group, at least one poor family had been given a second chance.

There were still hundreds of others in the area, not all victims of drink, but caught in the poverty trap in many different ways. One of the biggest reasons was that of unemployment.

Clara could see a figure standing in the shadows of the Parish Hall. The rest of the group had left. "Who is it?" she ventured. "It's Margaret, Clara," a refined voice said. "Margaret, what are you doing here?" Clara replied. "I would like to see you please, it is important!"

As Margaret seemed reluctant to show her face, Clara approached. It was dark in the hall but Clara could see from a shaft of light from outside that Margaret was in an awful state.

Her eyes were red and swollen; it was obvious to Clara that she had been crying for some time.

"Let's go to the vestry, I will lock up here!" remarked Clara. "Please no, Cannon James might see us."

"Do you feel like walking home with me?" asked Clara comfortingly. "If you wish, but I won't come in," said Margaret.

The two women walked slowly along the Butts. "I am pregnant, Clara, and I don't know how to go to my parents. It would be the worst kind of shame I could cause them."

"Margaret, come back with me to see Mrs. James. You need the help of someone you can trust."

Cannon James opened the door of their private residence. The Cannon looked worn out. Life for the clergy in those days was worse than that of a busy doctor. He was likely to be called out at any hour. The death rate was high in the early 1930s, as life expectancy showed little change from a century earlier.

Despite his fatigue, the Cannon and Mrs. James listened intently as Clara related the predicament of Margaret, who sobbed throughout. The Jameses decided on an unusual course. His wife put Margaret to bed at the vicarage.

Mrs. James made her way to the home of the Lea-Fletchers wearily. The large house, standing in its own grounds, was in darkness. It was built in 1903 during Edwardian times. It was designed by Mr. Lea-Fletcher who had it built on almost Regency architectural lines. A light came on in answer to the doorbell. A butler answered the door and said that Madam and Sir would not be returning that night, but Madam would be available in the morning. Thankful, Mrs. James explained that Margaret was staying overnight at the Vicarage residence of St Thomas's and that when his Mistress returned he should assure her that their daughter was well and staying with Cannon James and his wife.

The next morning, Mrs. James decided to call her doctor to examine Margaret. To the best of his judgement Margaret was four months pregnant. She had been sick in the night and was in a poor mental state. Following the examination by Mrs. James's doctor, he suggested that Margaret should seek treatment from her own doctor.

"She has rather strange bruising round the lower abdomen which she said was as a result of working in the stables with horses some time ago," said the doctor looking quite pensive. "My best advice is to see her regular doctor who would have her family history. She should be examined regularly. There is something I am unable to access unless she has more tests."

Mrs. James was concerned that Margaret's mental state was not up to further questioning; she had the sense to avoid recriminations. Margaret was now sleeping.

"Her parents must be informed of her condition," said Cannon James sympathetically.

A Rolls-Royce vehicle was parked on the drive of the Lea-Fletcher's residence. Mrs. James was about to ring, when a young chauffeur opened the door. He welcomed Mrs. James and raised his cap as a voice called from somewhere in the hall.

"Who is that, Edward?"

"A lady to see you, ma'am," replied Edward. Mrs. Lea-Fletcher had not met the Cannon's wife, who introduced herself. She was amazed at the beauty of this tall, fine-featured lady.

Mrs. Lea-Fletcher was wearing an expensive light pink suit, beautifully tailored to take every advantage of her perfect figure and poise, despite the age of the wearer. Her fair silken hair was immaculately styled. She smiled with blue eyes, expressing a warm sincerity that Mrs. James found unbelievable.

"Thank you for your kind message," said Mrs. Lea-Fletcher. "Is Margaret behaving herself? Please do come in and I will have the maid prepare some coffee."

The splendour of the furnishings took Mrs. James's breath away. The contrast with the house in Thomas Street was unbelievable. Her mind hovered between a feeling of tremendous restraint for the task she was about to attempt and one of anger for the poor woman of Thomas Street. Mrs. James deplored the task of turning the life of this lovely calm lady of the upper classes into turmoil. She took the approach that first came into her head but one she never believed possible.

"Mrs. Lea-Fletcher, I have to explain that I need your help, immense forbearance and trust in what I am about to explain, perhaps unfairly on so short an acquaintance."

Both women looked into each other's eyes. Instantly they both knew that trust was there. Instinctively they also knew that whatever it was that brought them together for the first time amounted to an everlasting bond. They were about to share in one of the serious problems of life.

"I want to ask you to come with me to our residence where your daughter is resting. Please do not become apprehensive, as your daughter is well but extremely upset. In what I am doing I shall lose her confidence, but I have to do this deed."

For a moment Mrs. Lea-Fletcher turned slightly pale but remained quiet and erect. She said in a soft voice, "Of course I will. I will get Edward to drive us there. Is there anything you would like to tell me; you have my trust? The Cannon's work and tireless efforts undoubtedly reflect your support."

Mrs. James explained in a most gentle way to Mrs. Lea-Fletcher on the way there that Margaret needed all the support and understanding she could give. They soon arrived at the vicarage and were met by Clara and the Cannon. What followed is to be told later.

The summer holidays, although only four weeks in August, were soon over. Poor children roamed the parks and the commons and looked after one another and were no burden to their parents. On returning to school in 1933, I was placed in the big school. My first teacher was Miss Stanley, a typical teacher of the time. She seemed very old to me, but then anyone over teenage was old to me, then at the age of eight.

Miss Stanley was dark with a clean appearance. Despite her long face she had protruding lips and teeth. My best recollection was that she wore glasses but only for reading. I guessed that she was there to ease small boys, like me, into the big school after the gentle ways of the infants. Miss Stanley alternated between severity and kindness. However, I must have performed so well that I jumped a class to class three.

Miss Riley, who rode a big motorcycle, was the teacher over class two. Small boys were fascinated that one of the teachers rode a motorcycle, especially Miss Riley. We used to fight to place the wooden slope over a high step so that Miss Riley could easily push the heavy bike up the step.

Class three, was taught by my favorite, Miss Millerchip. It was the first time my teacher seemed young to me. She played the piano for morning assembly. At that time it was the law that schools must start with a Christian prayer and hymns in a ceremony that was named 'school assembly.' I have often wondered how anyone who was not of the 'Church of England' would fare as most of the religious teaching was decidedly 'Church of England' in nature. All the pupils then were white.

I certainly bloomed in class three as I warmed to Miss Millerchip's teaching. I recall that she promised the class that she would award a prize to the best-behaved boy during one week. The prize was a novelty pencil sharpener with a small man sitting on the square-sectioned pencil sharpener. The small man had protruding eyes and it was one of the smartest novelties I had seen in my young life. I was determined to win it, so at every turn I was on my best behaviour. The time came for Miss Millerchip to announce the winner and she called out my name.

Overjoyed, that night I took the pencil sharpener to bed with me, fast in my closed hand. It was a lesson of life, although it was not obvious to me at the time, namely: That one should strive for the good things we seek in life. If you don't succeed, try, try and try again.

The flags were flying from every building and the bunting stretched over every street. The colour was gloriously mesmerizing. On 6 May 1935 George V was given the unprecedented honour of a jubilee to celebrate the twenty-fifth anniversary of his accession to the throne. This was a deliberate move to revive past glories, echoing Queen Victoria's Jubilee. Also, it was in defiance of the 'King and Country' resolution in the Oxford Union of two years before.

My younger brother Roy and myself were great friends; we were, as most brothers, quarrelsome at times. On the whole though we were close and gave one another good counsel and comfort in bad times.

In 1935, the year of the 'Silver-Jubilee', we were overjoyed with the presents handed out at the schools. I recall receiving a tin with a large bar of Cadbury's milk chocolate inside and a mug. The tin was brightly coloured with Queen Mary and King George V depicted, head and shoulders. So too was the mug.

The street party went on all day with the residents of Queen's Road serving together with the publicans from the Hertford Arms. It was the celebration of a lifetime and went on into the early hours of the next morning. The fireworks display from the Memorial Park, and seen all over the city, left us ore struck. Roy and myself were allowed to stay up throughout, roaming the Memorial Park and visiting the fairground. Most of the amusements were free in loyalty to the King and Queen on the occasion of their 'Silver Jubilee'. English people of all political parties, very sensibly turned the Jubilee into a personal tribute to a King who, in a modest conservative way, had a better record as constitutional sovereign than any monarch since William III.

The Jubilee marked positively the last appearance of the government Prime Minister of the day, Ramsey MacDonald, Labour's Second National Cabinet or 'Coalition Government'. Stanley Baldwin, a strong Conservative, and MacDonald changed places. MacDonald became Lord President of the Council. Baldwin became Prime Minister and so at last accepted responsibility for the power, which, like other 'harlots', the press lord he had previously exercised from behind the scenes.

Baldwin had been a successful Prime Minister in the 1920s but, as young as I was, I suspected Baldwin of undermining MacDonald.

Spain had become a republic in 1931. Early in 1936 a general election gave the majority to the left-wing coalition of 'the Popular front'. In July 1936 Spanish generals, controlling most of the regular army, staged a rebellion and marched on to Madrid. They expected an easy victory and others expected it too. The republican government armed factory workers and to everyone's surprise held the rebels under General Franco at bay. Civil war broke out in Spain.

When Mrs. Lea-Fletcher and Mrs. James reached the parish residence, both Clara and the Cannon were there to usher them in to discuss the situation. Clara, with the help of Cannon James, explained their conversation with Margaret.

Mrs. James took up the dialogue explaining that her own doctor had examined Margaret. Apart from the strange bruising, the doctor had pronounced that Margaret was in good physical condition but she was suffering mentally.

Mrs. Lea-Fletcher asked if she could see Margaret. The Cannon's wife led the way. Margaret was awake but her eyes were red and swollen. "I will leave you together," said Mrs. James, returning downstairs. Clara was talking to Cannon James and Mrs. James joined them.

After some time, Mrs. Lea-Fletcher left Margaret resting. Margaret and her mother proposed to leave Margaret at the vicarage until she was well enough to return home, providing the James's had no objections. Having received Mrs. James approval, Mrs. Lea-Fletcher prior to leaving, thanked them for taking care of Margaret and giving her excellent advice and attention.

Mrs. James confessed to the Cannon that the doctor who had examined Margaret had confided that he thought, in relation to the bruising, that he suspected she had attempted to terminate the pregnancy by self-inflicting injury.

Mrs. Lea-Fletcher explained to her husband Margaret's predicament and begged him to be forgiving and understanding. His first thought was for the child he loved and then he turned angry for the man who had taken this liberty with Margaret. He was determined to find the seducer of his daughter and make him pay. First, he thought he must get Margaret away from the vicarage and, when she was well enough, find out the scoundrel was who has seduced Margaret.

Little did he know that he employed the culprit?

My eldest brothers and sisters were working and earning so that there was a contribution towards the household expense. They had been keen to move to a bigger house for some time. There was an argument in relation to a more spacious residence, particularly as my sisters were out at work excluding Hilda Roslyn and Norah who were still at school. Hilda Roslyn was the next in line or 18 months or so older than I was, but Norah was the baby of the family.

Charles never recovered from his accident and Jack, my eldest brother, had the burden of the business that was proving too much for him. Charles had the expertise in many ways, particularly for hiring the workmen, who worked well for him. He was a craftsman and commanded respect. Not that Jack was any less of a craftsman but he was young, headstrong and wanted to do things his own way.

Jack was working for Mrs. Collins, the eccentric lady who kept an antique shop at the bottom of Queen's Road. He was installing a large covering at the rear of her premises. Tom was working part-time for Edgar Jones, the furrier in Queen's Road, as my brothers Albert and Jack had done before him. When in full time employment, the part-time assignment with the furrier was handed down to next Penny boy of a younger age. So it was my turn next when Tom found full-time employment.

Mr. Edgar Jones was an imposing gentleman. My best recollection of him was he had a fairly large stomach that he carried off well. He was tall and always wore a waistcoat with a gold chain fastened to a gold pocket watch that he often referred to for the time.

Mrs. Edgar Jones was always neatly dressed in a woollen knitted suit with a blouse of fine silk and a cameo broach at her neckline centrally placed. I was a young boy when I went to this impressive house in Queen's Road that had a showroom at the front; the back of the premises came out in Hertford Place. The house was quite large with three stories and a cellar. At the back of the showroom there was a workroom where alterations were carried out. The workroom had a French window that overlooked a huge courtyard at ground level. On one wall there was a large rainwater barrel for catching rainwater from the three-storey guttering. There was a flight of stairs at the back in the courtyard which one climbed to a very large kitchen.

Off the kitchen was a large dining room, then a long hallway to the two sets of stairs, one down and one up to an office at the front of the premises on the second floor. The main bedroom was along the hall on the left before the stairway to the ground and second floors. Prior to the main bedroom on the left was a bathroom and off that there was a toilet. Other bedrooms were located on the second floor and there were also bedrooms on the third floor.

My first duty for the Joneses was to report after school daily, except Tuesdays and Fridays, which were for my choir practice. I also was on duty from 9 to 12 on Saturday morning. My main tasks were to run errands; these also included attending to the post. There were parcels almost daily, each item had to be entered into a parcel book and stamped by the post office 'as proof of posting'. All parcels were registered and the knots sealed with sealing wax, as mostly expensive furs were posted back to the fur companies. Furs were released 'on approval' and, unless sold after the allotted time, were returned.

I recall several names of the companies that Mr. Edgar Jones had dealings with. The names of Hackers & Stein and the Blackburn Fur Company come to mind, but there were many others. Among my duties, after I had been there many months, was the responsibility for taking most of the expensive furs from the showroom to the office of the second floor, which was locked and barred. I used to alarm the showroom and workroom. Looking back I became the general factotum for the Joneses and as time went on I ate there. Mr. Jones died after a short illness and Mrs. Jones came to rely on me more and more. She engaged Mr. & Mrs. Locket to live in on the ground floor and they were a great help.

Mr. Locket was short and rotund; he had been a baker. He and I became good friends and his wife used to feed me with cakes. I became the son they never had. Knowing that my birthday was on Christmas Day they both made the most beautiful large birthday cake for me during their first year and thereafter as long as they remained at 38, Queen's Road. Mrs. Jones had one of those old telephones with the piece that you held to your ear and that so weighed heavily. I always remember the number: Coventry 252.

King George V died on 20 January 1936 after a short illness. He was a model of constitutional rectitude and conservative respectability in his private life. The new King Edward VIII had built up a huge popularity as the Prince of Wales, but the Nation was soon to be thrown into crisis. Baldwin posed a stern choice: renunciation of Mrs. Simpson or abdication. The King wavered. He wanted to have his cake and eat it – somehow to marry Mrs. Simpson and yet to remain on the throne.

Margaret meanwhile became much calmer, following the good care and attention she received at the vicarage that almost amounting to pampering. The time came for the Lea-Fletcher's chauffeur, Edward, to call at the vicarage to take her home. Margaret was no stranger to Edward, the handsome young chauffeur. He had been very reluctant to return her sexual advances, fearing for his job. But Margaret's attraction to Edward placed pressure on him to warm to her. He finally gave in and returned her affections with pent-up feelings. She also had a roving eye, like so many of her rich young friends of the time. Margaret was a regular visitor to the place where he worked, cleaning the Rolls-Royce in readiness for the frequent requests at the Lea-Fletcher's command.

When Edward arrived at the vicarage, Margaret was disappointed and embarrassed that her mother was not present. She obviously wanted to avoid an early encounter with Edward. She remained in silence for most of the journey, but started to sob so that Edward pulled over to one side to console his one-time lover.

"I am pregnant Edward, and I believe you are responsible," sobbed Margaret. The poor chauffeur was astonished. His turmoil was mixed. He was very fond of Margaret but the thought of becoming a father to the baby of the girl of the family who employed him filled poor Edward with immense fear. It was the 'unwritten-rule' that taking advantage of a member of the family for whom you worked was strictly forbidden.

When Edward could find the strength to speak he said, "What are we going to do?" To which Margaret was too overcome to answer. After some time, Edward decided to drive the Rolls round the countryside not returning to the Lea-Fletcher's residence with the hope that Margaret might calm sufficiently to discuss the question. Edward thought, while he was driving, Margaret's days of fun and flirting were over in the foreseeable future. He thought of the outcome and although he was a distant relative of the Lea-Fletchers by marriage, he imagined that he would be unacceptable as a son-in-law. His father had died when he was 10 years old and his mother had struggled, attempting to live in the comfort that they had known. His father gambled on the stock market and had lost his fortune on the London market at the time of the Wall Street crash.

His mother was able to get him a post with Margaret's family and he enjoyed his work; but she too died with a broken heart and he had no other near relatives.

Edward thought that he should make a clean breast of it with Mrs. Lea-Fletcher and he managed to convince Margaret that this was their best course of action. When Margaret returned home, her mother was overjoyed with pleasure and sadness. After a time Margaret settled and was comfortable now she had returned to her own home. Edward approached Mrs. Lea-Fletcher and sought a convenient time to see her, regarding an issue of great importance.

"Please let me have some time with Margaret and I will see you later today," said Mrs. Lea-Fletcher. Margaret preempted the question with her mother and suggested she must discuss with both her parents what was to be done. Mrs. Lea-Fletcher had her suspicions with regard to the subject Edward was about to discuss. She had observed how Margaret had an eye for the handsome lad her husband agreed to hire. Edward had become a great asset to their family, almost like a son.

Edward was nervous of opening the discussion.

"I have cause to know the substance of the issue of great importance, Edward," said Mrs. Lea-Fletcher, in an attempt to put him at ease. "My daughter's welfare concerns me," said she. Edward nodded.

"Margaret and I will discuss the issue with my husband and I will suggest a suitable time for you to join us to decide what can be done in the circumstances."

Edward thanked her and asked to be excused. When Mr. Lea-Fletcher returned home he was relieved to hear that Margaret was now settled as comfortable as possible. He had suffered, as any father would, since he had learned that his daughter was four months pregnant out of wedlock.

In those days, pregnancy out of wedlock was more of a social disgrace than the present day and it resulted in many deaths due to illegal abortions.

The time for discussion between the Lea-Fletchers and Edward was about to happen. Margaret met with her father and mother, owning up that only Edward could be responsible for her pregnancy. Both parents were somewhat relieved but Mr.

Lea-Fletcher put forward an argument to get rid of Edward, as it brought his loyalty into question.

When Margaret, quite rightly, protested that she was as much to blame for the event and that getting rid of Edward, who she loved, would not resolve the issue.

It was decided reluctantly that if Edward refused to marry Margaret then he would be asked to leave the Lea-Fletcher's household. When Edward was called into the discussion he was extremely nervous. He said he wanted to resign his post as he had brought disgrace on the family. Mr. Lea-Fletcher stepped in and asked Edward, "If he loved Margaret?" To which Edward answered, "With all my heart."

An early date was set for the wedding of the two young couple, one an orphan chauffeur in the employ of a rich man, the other his daughter.

Meanwhile, Spain was being transformed into a battleground of rival ideologies. In simple terms, it was fascism versus democracy. The Spanish civil war at last enabled men to wage the fight against fascism in reality.

One of the sons of an outstanding worker of the business of Charles and Jack was a member of the newly-formed communist party. He was anxious to join-up and fight on the side of the workers in Spain. The rebellion was generally supposed to be part of a coordinated fascist conspiracy against democracy, with Franco as Mussolini or Hitler's puppet. Franco, however, was nobody's puppet; he acted without prompting and displayed a remarkable tenacity in asserting Spanish independence. Charles Dunn joined the International Brigade to fight against Franco on the side of Spanish workers. His father went to pieces when he embarked for Spain and, as a direct result, his wife left him, only to return with her support when he was imprisoned for the murder of his wife's lover.

The 'Labour Left', spurred on by the Communists, denounced German fascism from the first day but few of them preached war. Nazi persecution of the Jews did more than anything else to turn English moral feeling against Germany but this moral feeling made English people less reluctant to go to war. Chamberlain shared this feeling but insisted that the best way of beating Germany was appeasement. Britain, at this time, was unprepared for war although two outstanding British inventions were being developed. One of these was radar, which would make a significant impact on the coming World War II. The other, the jet engine, arrived too late to be of use.

Meanwhile, Charles Dunn was soon immersed in the fighting in Spain. He did not know that his father back in England had been jailed.

Three Spanish girls were from the home of a wealthy family whose parents had been slaughtered by the savagery of the brigade to which the Dunn's son belonged. He had saved the three girls from the hands of the factory workers and had fled into France, wounded in an attempt to fight off his own comrades. They travelled to England but the four of them were now turned away from the Dunn's home in Coventry. Clara found them refuge.

Mrs. Dunn grew to dislike the communists and what they stood for. Some saw the Spanish Civil War as an experimental ground for World War II. Spain was a country torn apart as the scapegoat of communists against fascists of the 1930s.

And, while Britain's intellectuals preached pacifism, Germany tested new weapons on the battlefield of a beautiful country savaged by civil war.

The lady who ran the antiques shop at the bottom of Queen's Road kept cats, sold antiques and grew red grapes from vines growing on the wall above her premises.

Every day, the whole shop front was cluttered with sideboards, tables and chairs of all kinds. The window looking out on Queen's Road was filled with vases, china dogs, hanging brasses and warming pans. The front door entrance was narrowed with the obstacle of a grandfather clock. If one gained entrance with difficulty, getting through the first room was a shear miracle but getting out was an even greater miracle.

There was no room to turn; this rotund lady scholar of the area collected every conceivable object. No one witnessed anything having been sold and no one questioned the magic by which she displayed her outdoor wares each day, retrieving them indoors at night. This setting in a district of the City that mixed poverty with riches was so much part of life in the 1930s. It seemed to have been there forever and to remain till eternity.

Clara needed the help of this eccentric lovable lady, who spoke many languages fluently. The lady of antiques asked Clara inside; both were very large. Clara wondered how. Somehow they both struggled to waltz round the grandfather clock, which complained by making resonating an off-key chime.

"Ignore it," said the owner. Clara became wedged in a maze of sofas and sideboards. Eventually they both arrived in the back room where cats were everywhere. Clara was fascinated. An approximate ring of six or eight cats of all kinds, Persian Greys, white ones and a brown furry tabby were sitting or lounging facing in all directions. Some were looking inwards; some were looking outwards, and some at a tangent to the rough circle they described.

"It's a meeting," said the lady of antiques. Clara looked for a place to sit but the only place was a low stool. The prospect of not being able to get up made Clara decide to stand.

"Is Spanish one of your languages?" asked Clara. The abruptness of the remark surprised them both. A cat yawned and stretched, completely oblivious to the two stout ladies. The antiques lady's eyes lit up.

"I can find one or two books on Spanish," remarked Mrs. Collins, the antique lady.

"It's not for me," said Clara.

"Your son Jack is not going over there?" questioned Mrs. Collins.

"Oh no," replied Clara, "I would be against my son going to Spain. I want to help three Spanish girls who seem to be unable to read or speak a word of English".

"That's strange," said the lady of antiques. "I thought English was the second language after French. Bring them along and I will see how well we all get on together."

"I am confident that they would be willing to assist in the business," said Clara.

The arrangement was a good one. For although they doubted knowledge of English the three girls understood every word the antiques lady said. The three Spanish sisters were also fascinated with antiques and the menagerie of Queen's Road. In fact, after the torment of seeing their parents slaughtered in Spain, working alongside Mrs. Collins in antiques was the best therapeutic treatment that could be devised. Not only was their mentor able to speak in the language of their mother tongue, but they all loved antiques and were overjoyed to be kept busy within the business.

With this the lone eccentric lady began a new venture and sales started to improve, much to the mixed feelings of Mrs. Collins. She knew the worth of some antiques and rather than sell them she loved to keep them. Nevertheless there were compensations. The three Spanish sisters started to converse in good English. This Queen's Road oddity became the attraction for the wealthy young men from miles around who suddenly developed a fascination for antiques and the Spanish sisters.

Mussolini had resolved to attack Abyssinia, a member of the League of Nations. The national government of Britain was reluctant to side against Italy. The service chiefs, with the problems of Japan and Germany on their hands, insisted that Italy should not be added to their list of possible enemies. Anthony Eden was chosen to visit Rome and attempted to buy off Mussolini.

Great Britain would secure Abyssinia's agreement by surrendering part of British Somaliland to Italy.

Mussolini rejected the offer. Italy, he insisted, must have a base in Abyssinia, which Great Britain had in Egypt, an awkward analogy for the British government to reject. Italy attacked Abyssinia.

My schooling, after class three, turned out to be an anticlimax. Miss. Holman's class four was a great disappointment. I became disillusioned with school and it was in that year that my left defective eye was detected. The classes were mixed until we reach class five, and from then onwards it was only boys. The girls were taught in other classes.

Mr. Wilson, the teacher of class five, was every inch a character. The headmaster, Mr. Anthony Snell, (nicknamed Tony) had been a captain in the army during the Great War, as was Mr. Torrance (nicknamed Toddy) of class seven, the top boys' class. I always suspected that Wilson did not serve in the forces during the Great War, as he often referred to Captains of the army returning to teaching after the Great War with derision.

I enjoyed Wilson's class, for he was conversational, and I emerged with learning. He read to the class many of Dickens's stories and other favourites such as *Treasure Island* and other adventures that captivated us. He reduced the subject of mathematics to its uttermost simplicity, explaining many of the twists and turns that puzzled many in the class. As a boy who wandered the streets, I used to meet him in passing in Albion Street, two streets away from Hertford Place.

I asked him one day, "Do you live around there, sir?" He replied, "That his mother lived in Albion Street and he went to see her quite often." He was a born teacher.

Apart from my bad left eye, for which I now wore spectacles, I was a healthy, rosy-cheeked boy. People remarked that I was the picture of health. Fortunately my parents were unable to pay for my eye to be operated on, although it was advocated by the eye surgeons that, by operating, the two eyes should be allowed to function separately. This was later found to be the wrong thing to do, and I thanked the fact that my parents were poor. Some of my sisters and brothers had to have hospital treatment; that had to be paid for in the 1930s. I remember my brother Roy needing his tonsils removed, not to mention the drain on my parent's financial resources for his hospital treatment. Fortunately, he was only in hospital for just under a week.

A boy was placed in class five from Hertford Hill, near Warwick. In those days it was an open-air hospital for the treatment of tuberculosis. His name was James Galsworthy. Wilson thought he should be placed next to a healthy boy. Little did I know then who this boy really was.

The pale boy, who was seated next to me, would turn out 50 or so years later to be Joan Alford's uncle (who became my housekeeper); such is fate. At the time, I enjoyed getting James to chew liquorice wood that was on sale at a shop nearby. This was in the hope of eliminating his paleness. In time he actually developed a rosy complexion, whether the liquorice wood was responsible I did not know.

I was already in the choir at Saint Saviours Church, as Saint Thomas's had long abandoned their choir. As James lived near Saint Saviours Church, a small church in Spon Street, I suggested to the choirmaster, Mr. Parker, that he should give James a singing test a with view to James joining the choir. Mr. Parker and I were delighted at the wonderful golden voice of James. James was soon on probation to join the choir, but first he had to pump air into the organ, keeping the weight on a string between two marks – something we all did during the probation prior to joining the choir. In those days we were paid a shilling for the choir and, if we were lucky, two shillings for a wedding, and not very often a funeral. Two shillings in those days was a lot of money and, working for Edgar Jones, I was rich and an ardent saver. I had a moneybox that my elder brothers used to raid whenever they needed cash. One day, Tom persuaded me that he needed new shoes and the price of the new brogues was three and sixpence. I counted carefully seven sixpenny pieces (a sixpenny piece was called a tanner) into his hand. I never remember him returning the debt, but he made up over the years in many other ways.

## Two
# A world at war

**In which I begin to grow up in war-torn Coventry and undergo training at the Naval College at Dartmouth.**

IN 1936, I was 10 and that year, on 11 December, Edward VIII abdicated. He was created Duke of Windsor; the title of Royal Highness was refused to his wife when he later married Mrs. Simpson. By the time George VI was crowned on 12 May 1937, Edward VIII and his abdication were almost forgotten. The sounds of war were apparent and Britain was slow to rearm. In 1936, already the 1930s had been a period to remember. A mood swept the country, almost a dread of pacifism; the nation seemed and unprepared, but some said destiny resided in the ability of the British inventive mind.

The country had already abandoned the gold standard, whereby any notes in circulation could be changed to gold at an old fixed rate. Much of the gold was covered by the return of gold coins to the Bank of England by the year 1918. Leading politicians faced radical changes and moved in all directions, totally opposed to their longstanding beliefs. Winston Churchill remained true to his former views and was steadfast in his beliefs.

In 1937, Jack, my eldest brother, married Betty Slater in her home village of Great Tew. Jack was born in August 1912. A committed Christian, as were his brothers and sisters, he became a choirboy at St Thomas's Church, prior to the disbanding of the choir there. Great Tew was a wonderful setting for the wedding.

The time came for the Penny family to move to a larger house, albeit not far away as the crow flew. A three-storey house became vacant at the top of Rudge Road. Number 29 was the corner house. Violet and Jack, my eldest sister and brother, had flown the nest, as had Alice Maude, who in 1934 married Frank Butler; they went to live in Junction Street.

Number 8 in Junction Street was a residence set apart. It was characterised as the only house in the street with a brass knocker. Alice was quite ambitious and house-proud. This accounted for her getting married before Jack, her eldest brother. She worked on the house until it was wall papered and clean throughout. The residence became a house she was very proud to pass on to other members of the family, when the time came for the Butlers to move on. Frank Butler worked for the railway, as did his father. Alice was earning money at a hosiery company in Queen Victoria Road, so their combined income in those days was more than the average family.

Hilda Roslyn and Lillian Sarah were away in Romford, so the house could accommodate the remaining sisters and brothers with Clara and Charles. There was an air of excitement when the time came to move, especially among younger members of the family.

My part-time work with Jones the furrier continued. I also became quite useful to Mrs. Edgar Jones. I was learning all about the fur trade. I still remember how to tell a good fur, depending on the origin and the number of skins. The late Edgar Jones had a brother, Willie Jones, who kept a shop in Coleshill and traded in musical instruments. He specialized in pianos. Mrs. Jones seldom invited Willie, who was short-sighted and found difficulty in getting round.

Willie visited Coventry on business and was invited to stay for high tea. I was present also, as I usually stayed to tea on completion of my daily tasks after school. Mrs. Jones had a habit of leaving her smelling salts next to her plate. They were the colour of vinegar and contained in a glass cut bottle, looking for every bit like the decanter. Willie was eating with relish and to my dismay was covering his salad with smelling salts. I tried to stop him, but Mrs. Jones hushed, "Do be quiet Noel, speak when you are spoken to." I gave up!

When Willie returned to his shop in Coleshill, it was the custom for him to play cards in the evening with the chemist next door. But that evening he was feeling bilious after his salad high tea. However, he went next door begging the chemist to treat his bout of biliousness. The chemist, who had retired for the evening into his living room, was not about to enter his shop after a busy day. So he suggested that Willie should go into the shop and take a stomach powder from one of the draws. Willie entered the chemist shop. He looked round; it was dark, adding to his difficulty in searching for a stomach relief. The drawers all carried Latin words printed on them above the handles. He ignored these and found a plain oak draw with no label. He felt in the draw and found a box without top. There was a strip of folded paper, like he was used to with his powders contained between the folded papers. Smiling, he went back to the living quarters, but the chemist was upstairs. So Willie emptied the powder into a glass and added water and gulped it down. That night, Willie was taken to the emergency hospital nearby and given a stomach pump to clear the contents he had mistakenly taken, not knowing of his error whilst eating with his sister-in-law.

The following year, 1938, was an eventful one for the family I was just 12 when my father died at the age of 77, on March 26 1938. This followed the move to Rudge Road. Charles was born in 1860 and his father, Cornelius Penny, was also a master carpenter turned victualler. Clara, although heart broken, sighed with relief that he was at rest, because he had suffered since the accident on returning home some years ago.

It was also in June of that year that May Fannie Penny married Ernest (Mac) Adlam to become Mr. & Mrs. Adlam. The wedding was at St Thomas's Church. There were four bridesmaids: Lillian Sarah, Hilda Roslyn, and Norah Jane who were all Penny sisters, and Beryl Matthews a niece and an aunt.

This left the Penny household in Rudge Road with only my mother and baby sister Norah, with Albert, Tom, Roy and I.

I volunteered to distribute gas masks in 1938 before the Munich agreement. On 29 September, Neville Chamberlain flew to Munich once more to meet with Hitler. Chamberlain returned waving a paper saying, "Peace in our time."

Little did we know that we would be at war with Germany in little under a year.

I was confirmed into the Church of England at Coventry Cathedral in November 1938 by the bishop of Coventry. Previous to that, I had attended confirmation classes once a week given by a young curate, K. A. Wells. He gave me a little red 'Communicants Manual', which I still have. My name is written inside the cover and he signed it 'from K. A. Wells and, underneath, Nov: 1938.'

Because of my commitments with the choir and my part-time work with Mrs. Jones, the young curate agreed that I visit him at his lodgings in Chester Street, off the Holyhead Road, during Sunday early evening before the church service. I was a choirboy at St John's on the corner of Corporation Street, but I was soon to be asked to join the Cathedral choir.

My thoughts during my first communion were for my dead father. He was laid out in his coffin in the front room at Rudge Road, as was the custom in those days. We were not forced to see him, but I chose to venture into the room; it smelt of lavender. I well ever remember placing my hand on his cold forehead. He looked very peaceful and Clara had tears in her eyes as she said, "Your father is now at rest."

My elder brothers, particularly Albert and Jack, were over-sensitive about my mother's welfare to an extent I thought almost obsessive at the time. Tom, Roy and me, to a lesser extent, were not exposed to Clara's early struggles and the subsequent toll they had on her health, especially in contrast to her former life. In later life I grew to appreciate Albert's behaviour, particularly towards the opposite sex. He treated every woman as if she was a 'Queen'. We all had great love for our mother, but particularly Jack and Albert who, to a fault, would do anything for her. Not that the others thought any less of her, as we all had tremendous respect for her intellect; she was truly a lady of immense forbearance and wisdom.

I recall that one of her relatives was a jockey; I think his name was K. Humphreys. She came from a part of the country that was famous for flat horse racing. She used to confide in me that she would like a small bet on the horses. We selected together the race and the horse that was ridden by a famous jockey. I agreed to go to the corner grocery shop where I knew bets were placed. Whether it was legal or not did not enter my mind. We won, and Clara was overjoyed that we picked the winner. The jockey was Gordon Richards who rode the favourite for the Derby that year, so the returns were small.

As 1938 came to a close the lesson to be drawn from the crisis over Czechoslovakia was that Britain should rearm.

However, there was a mixture of reactions. Most ordinary people in the street had a fear of war and consequently considered the outcome of the Munich agreement at least a chance for the British to rearm, despite the lack of support for the people of Czechoslovakia. The service chiefs certainly treated war as inevitable. Some people claimed that Chamberlain was playing for time and that he mistrusted Hitler. On the other hand, Hitler himself tried to wreck the Munich agreement and the policy of appeasement. He had his map in the form of '*Mein Kampf*' for carrying out his future plan and was patient enough to appease the British; but he would not be deterred in the long term. The Munich alarm resulted in the acceleration of plans already in train. More shadow factories were built and production of aircraft increased to over 600 a month at the start of 1939. Radar developments benefitted. Designs drawn up earlier gave the R.A.F. more squadrons of fighter aircraft.

One direct impact of Munich was the speeding up of air raid precautions (A.R.P.). A firm decision was to move from the cities all those who had no need to stay. All schoolchildren and mothers with children under five were to be moved from cites in the event of war. A few people claimed that Munich was the first step towards a new system of peace; they pointed to the good that Hitler had achieved in Germany and even mentioned that Mussolini had made the trains run on time in Italy.

In 1938 I had joined class six and was taught by Mr. Elliot who eventually married the teacher from my infant class. Mr. Elliot was quite slender and tall. He often wore an immaculate dark blue suit. His fair hair dropped down over his eyes at one side. He was frequently brushing it back with his right hand.

Mr. Wilson, of class five, on being told of the elevation to class six indicated that he considered Mr. Elliot to hold certain views against war. Later it was confirmed that he was a 'conscientious objector' and therefore would be unable to fight on the battlefield. My recollection of Mr. Elliot's class was pleasant but I was unable to draw great inspiration from his teaching, as I had done with Mr. Wilson and, later, the top class under Mr. Torrance

Most of the shop premises which fronted out into Queen's Road had back entrances in Hertford Place. One of the neighbours of Jones the Furriers was Kennedy, the bicycle shop. This shop had a front in Queen's Road with a display of Rudge bicycles of all kinds. Kennedy's had an extensive covered garage space that could be entered from Hertford Place, and the few people of the area who had cars in those days rented their vehicle space in Kennedy's garage. Another of Jones's neighbour was Dale's, the veterinary surgeon. We boys used to watch with awe the animals brought to Dale's.

In the mid-1930s, someone made me a gift of an old 'sit-up-and-beg' bicycle, which I treasured for a very long time. Now came the time to throw it on the scrap heap. During the latter part of 1938 Roy and I had our eyes on new bicycles, but paying for them was well out of the question. Mrs. Kennedy, of the bicycle shop, observed that Roy and I were frequent visitors to the shop front, admiring the display of new Rudge cycles. She was soon in conversation with us suggesting that we could pay a deposit and pay a small sum weekly, until the debt was paid off for two new bicycles. She was a trusting lady.

I raided my savings and paid both Roy's and my deposits for new 'Rudge' cycles. I was keen to establish that we could pay the weekly amounts. Together with my choir money, and the money I anticipated from Mrs. Jones, I estimated that if I had to I could easily pay off both sums by the end of 1940. Roy's bicycle had a smaller frame than mine, which was full size. We were both proud of our new acquisitions and made many excursions into the countryside. But I perceived that Roy had to peddle a lot faster to keep up with my full frame bike.

Most schoolboys of similar age had bicycles. Families were earning more money as a result of a significant improvement in employment. Britain was on a war footing as 1939 came into being. In early 1939 the British government recognized Franco as the rightful ruler of Spain, though the Spanish Republicans went on fighting until the end of March. I had now entered the final class of my school prior to further education at technical college.

Tragic news came into the class that James, the boy I had fed with liquorice wood, had been in an accident on his bicycle and was in hospital. Marriot, who lived near James, was sent to the hospital to see how James was fairing.

Marriot, as he was called, returned with tears in his eyes; he could hardly speak, "James is dead." The class and teacher went silent for what seemed a long time.

An account of the incident was given in the Coventry evening paper. It said: 'James had been asked by his teacher, Mr. Elliot, to visit a school friend in hospital. While on the way his bicycle became trapped in the tramlines. He had struggled to free the bicycle but went over. A lorry following behind ran over him. The lorry driver stopped and, in the turmoil, went over to James who looked up and said, "It wasn't your fault Mister, it was the tramlines." These were the last words James uttered. James died before he could be taken into the ambulance.'

I was broken-hearted at the time, but I had to wait 50 years or so to find that James was Joan Alford's uncle, for he had the Galsworthy name of his sister, Joan's mother.

Preparations for war continued in 1939. Great Britain and France had an alliance with Poland that was a complex issue. Poland, Russia, Germany, France and England were the main countries involved in the contention that was arising. The map of Germany had been redrawn in 1919. An area of the fatherland of Germany was isolated from the rest of Germany by the Danzig corridor, which was Polish. Danzig was declared a free city.

Most British people saw things straightforwardly as ministers, including Chamberlain, had told them that Munich was the final settlement. Hitler himself said that Munich, (which allowed Germany to annex Czechoslovakia) was his last territorial demand in Europe. It was not as simple as Danzig, a free city, or the Danzig corridor that isolated Germans from their fellow Germans. There were two dictators: one in Russia and a madman in Germany, both with their eyes on Poland.

It was announced that Ribbentrop, the German foreign minister, had been invited to Moscow. On 23 August 1939, he and Molotov signed the Nazi-Soviet pact, thus allowing Russia to invade Poland if Germany attacked. Hitler blundered in thinking that he could attack Poland without provoking Britain and France. On 25 August, the Anglo-Polish treaty of mutual assistance was at last signed. On 31 August, Hitler gave the order to attack Poland. At 4.45 a.m. on 1 September 1939 German troops crossed the Polish frontier. At 6 a.m. German aircraft bombed Warsaw.

The Poles appealed to their ally. They met with cool response. First, just a warning was sent to Hitler. This was not an ultimatum. Prior to the British ultimatum being delivered to the Germans, there was a meeting in the House of Commons at 7:30 p.m. the day before. Members expected to be told of the ultimatum, but instead Chamberlain entertained them. He sat down without a cheer. Amery shouted from the conservative 'back-benches' to the Labour leader, "Speak for England, Arthur!" The house broke up in confusion. Arthur Greenwood went to Chamberlain, the Conservative leader, and told him that unless war was declared next morning, it would be impossible to hold the House.

The British ultimatum was delivered to the Germans at 9 a.m. on 3 September 1939. The Germans made no reply and the ultimatum expired at 11 a.m. The French declared war at 5 p.m. The British declaration of war automatically brought in India and the colonies. The Dominions were free to decide, but Australia and New Zealand followed the British example. The Canadian government declared war on 10 September. South Africa remained neutral for a short time until Field Marshal Smuts became Prime Minister and he declared war in September.

The British people accepted with quiet resolve that the House of Commons had forced war onto a reluctant government. The French government were even more reluctant but followed the British decision later on.

Clara and Mrs. Lea-Fletcher were weary. They had been working for the best part of a day organising mothers and children for a journey to the pre-arranged evacuation areas. Only a few mothers accompanied the under-fives into the unknown. The last of the mothers had left; most broke down and cried. Emotions had brought out strange questions.

The two women were in a reflective mood, their childhood upbringing was not so different but their living had become quite dissimilar. Both now had lost their husbands, but found solace in working for the good of others, as they always did. The last 10 years had taken an enormous toll. That so much had taken place in the last decade would be almost too much for history to recount. Tears came into the eyes of Mrs. Lea-Fletcher.

"Sometimes Clara, I think that we have lived through a turning point in history, equal to hundreds of years." Clara knew the fear of her hidden secret, even too much to share with this woman so close. They were one and the same ages she supposed. It was dark as they walked through the churchyard; the poor and the rich ladies joined arms together. That night Mrs. Lea-Fletcher died.

It seemed to me, after the initial shock of the war being declared, that there was a general feeling of excitement, especially amongst those of my age group and younger. We had no idea what might happen, but it came somewhat as a relief to know that the last two years of uncertainty were at an end. The decision now was to stop Hitler and his plan that was documented in *Mein Kamp*.

My best recollection was that things went on the same way, almost as if there was a lull in the fighting on land. The R.A.F. made some attacks on the German fleet and then dropped propaganda leaflets over the German lines. The British Expeditionary Force (B.E.F.) was placed under the French supreme commander. A few British soldiers took their turn in attacking the German lines but the first casualties were not reported until the end of 1939. It was in December that war at sea became front-page news. The German pocket battleship *Graf Spee* raided the South Atlantic and sank several cargo ships, but a British Naval force found her. The British ships, though less heavily armed, damaged her in using daring attacks and forced her to seek refuge in Montevideo. On Hitler's orders, the *Graf Spee* was scuttled. This recalled incidents when a German U-boat sank the aircraft carrier *Courageous* in September. In October, the defences off Scapa Flow were penetrated and the *Royal Oak* battleship was sunk.

In October 1939, the Soviet Forces occupied eastern Poland; Hitler's path to the east was closed for the time being.

I continued with my work at Mrs. Jones, who had a niece, Catherine Scott. Another of Mrs. Jones's relatives had jilted Catherine, because he was a first cousin. Catherine had been despondent ever since.

One of my weekly tasks over many years was to take groceries to Miss Scott to her flat in Earlsdon. Miss Scott became pregnant and she had to move to Tile Hill to one of the newly built houses in Hathaway Road. There was some mystery in relation to how she became pregnant, but I suspected that Mrs. Jones's relative was the father, as Catherine became a new person after the event.

I continued to deliver groceries to Miss Scott on behalf of Mr. Jones until she sold up and moved to Market Harborough in the summer of 1940. Mrs. Jones was a good friend to me and taught me a great deal in decorum and good manners. She and I were conspirators, for she had one weakness: Red label Johnny Walker Scotch, which I used buy regularly from the outdoor of Hertford Arms for twelve and sixpence. It was of course illegal.

Of course, I was under age to buy liquor from any part of a public house.

The official evacuation operation in September 1939 was a disappointment. Only a million and a half were evacuated compared with the four million planned on paper. Difficulties arose. City children were ill equipped and foster parents were often poorer than those children billeted to live with them. The city children were often verminous and, far from home, relapsed into bed-wetting. In other words, in general the poor housed the poor. The wealthier classes generally evaded their responsibilities, with the secondary and public schools arranging their own schemes. When by January 1940 no bombs had fallen, nearly one million had returned home.

The Ministry of Food called for rationing solely as a means of distributing supplies more efficiently. Some people thought that the country was ill prepared for war. Generally, we ate well, however long queues would form for anything off rationing, like fruit, sweets, eggs and cigarettes. An exchange and mart arose which encouraged a black market. Dubious men could get anything from sweets to butter at a price.

During the early part of 1940, I joined the sea cadets and later became a member of the 'Y' scheme which trained boys at Dartmouth for entrance to the college and, if they made the grade, to enter the service as mid-shipman.

I still had choir practice two nights a week, with church on Sunday, my work with Mrs. Jones and schooling.

In addition, we were asked to remove any notice with the name of the place stated, or to inform the right authority of any signpost that was still standing. This proved quite a difficult task, as the city name appeared in most unlikely places. The first Christmas of the war was a good one as most of the family, wives, husbands and sweethearts came for Christmas lunch and stayed on for the celebrations that ensued. Clara was sad because of the absence of those near and dear, but drank to their health.

She was particularly sad for the James family. Cannon James of St Thomas's church had travelled to Paris in June of 1939 and was shot. The papers of the day reported he was burdened with the plight of poor people and, in a disturbed mind, shot himself. Another newspaper reported that he had a terminal illness. However, the actual truth will remain unknown. Clara suspected that he was killed; she knew that he would not take his own life.

Norah, my baby sister, and Beryl, her aunt, were evacuated to Great Tew the day war was declared. When they arrived, there were two children from the East End of London already billeted with Aunt Min, the mother of Jack's wife Betty. Aunt Min already had one of her daughters at home so, with her husband, Bill Slater the cottage was quite a squeeze.

Roy and I comforted Clara. She was losing her children one by one. Tom, who was working for the Rover Company, had announced that he was required to consider moving to the North of England. He was already a member of the Rover Home Guard as well as an air force cadet in the Air Training Corp. Albert moved out when he married Maud Smith, his first wife, in January 1940. Lillian Sarah and Hilda Roslyn were bridesmaids, along with Maud's sister. Jack and Betty were already living with Alice and Frank in Junction Street. May and Mac had taken a house in Coronation Street in the Hillfields district of Coventry. Only five children still remained within the

Penny household, and three of these expected to leave in the foreseeable future. Clara confided in me that during the last few years she had lived a thousand deaths.

At the beginning of 1940, the lull in the fighting in France continued. The French felt secure behind the 'Maginot' Line fortification and Great Britain equally so behind the shield of sea power. Things seemed to warm up as March came to an end. Hitler planned a total occupation of Denmark and Norway.

The Germans invaded Denmark with practically no opposition and seized every important Norwegian port from Oslo to Narvik. The British imagined, mistakenly, that they controlled the North Sea. The British had tried their best in Norway but by then things started to go wrong elsewhere. On 7 May, Chamberlain opened a two-day debate on the Norwegian fiasco. Amery, sensing the mood of the House, ended with Cromwell's words to the Rump, "Depart, I say and let us have done with you. In the name of God, go!"

Chamberlain tried to win over Amery, but in vain and eventually resigned. Churchill was the obvious man to take over.

My schooling had taken a knock, so I joined technical college and enrolled for a course leading to a degree. But first I had to gain matriculation and that was a two-year solid slog from my then present standard. I argued, quite rightly, that I was self-taught but my mathematics teacher, Leonard Turner, said that I had to pass the examinations to be allowed to eventually work for a degree.

Roy and I could but only go along with the events of the war events that started to accelerate faster than could be assimilated in 1940. On 10 May, Germany invaded Holland and Belgium without warning. By this time British people knew that Hitler had a mania for invading without excuse. Hitler now placed his main weight where the Maginot line ran out. There his armies would break thorough the French defences and go straight for the sea. On 14 May, the Germans broke through at Sedan and

# Automotive gas turbines

within five days they took Amiens and reached the sea at Abbeville. The allied forces in Belgium were cut off. Gamelin, the supreme commander made many mistakes and could only look on with dismay.

On 16 May, Churchill went to Paris; the Germans would be in Paris within a few days. Field Marshall Gort initiated the end of the Anglo-French alliance and saved the British army at Dunkirk. The Belgium forces capitulated in the early hours of 28 May. 'Operation-Dynmo' - the evacuation from Dunkirk began on 27 May. Gort's instruction was 'to evacuate the maximum British force possible.' He succeeded beyond all expectation.

On 3 June, the last men were moved. In all 338,226 men were brought to these shores and of these, over 139.000 were French. Almost the entire B.E.F. was saved. All guns tanks and other heavy equipment was lost.

In France, Marshall Petain formed a government and at once asked the Germans for an armistice; that was concluded on 22 June. Northern France and the coastline to the Pyrenees fell under German occupation. Britain faced a hostile Europe. Italy declared war on 11 June, fearful of being too late for the spoils.

The British people gritted their teeth and remained unruffled. The weather was glorious that year and Churchill rallied the nation with his speeches, the main gist of which was that 'we shall fight on'. Hitler offered terms of peace and on 19 July 1940 repeated the offer in a public speech. Churchill rubbed it in by defining British war aims until the end of the war. There must be total victory or, unconditional surrender of Germany and all territorial gains must be returned to neutral countries. Hitler never repeated his peace overtures again.

Clara had been diagnosed with cancer when war started – that was her secret. The family had a meeting and decided that the house at the top of Rudge Road should be relinquished and Clara would be looked after by May, at Coronation Street. Clara never counted the cost to her own life.

Clara had made the supreme sacrifice for others, particularly her own family. It was Clara who needed care and loving attention now. It was decided that Hilda Roslyn would stay with May and Mac to care for her at Coronation Street, while May and Mac went out to work.

Violet and Ted had taken over 8 Junction Street from Alice and Frank, who had found a house at 49 Rudge Road. This suited Ted, who trained as a gardener; the house had a large cultivated area at the back that was ideal for growing vegetables. Also, there was a long workshop upstairs, much to Ted's delight. Apart from the wartime, Ted was in his second heaven at Junction Street. The Hertford Arms was only a few steps away at the top of the road. He was fond of playing darts and, with Roy and I due to stay there, he had two potential competitors. I liked Ted because he was forthright; he was often mistaken for being tactless, but it was just his way. You knew exactly where you were with him. To the great amusement of Roy and myself, he was the first man we encountered who could swear in the middle of a word. He would often say ali-bloody-minium for aluminium. He was an air raid warden and during one all-night blitz he came home the following morning and cried and sobbed for quite some time.

When he had rested, Violet questioned him and he referred to a bombed house in Moat Street where they had to look for any survivors. In the debris, he came across little severed hands clasping small forks. It was obviously a young family having a meal when a bomb dropped on their house. Ted said, "It was a blessing that not one of the young family would have known anything about what was to happen."

Jack and Betty had a new house in the Cheylesmore district of Coventry, near Jack's work at Armstrong Siddeley. This was in 'Hermit's Croft,' a new road. The next door's family, during one of the Coventry blitzes, and with whom they were on very good terms, received a direct hit on the garden shelter and were all killed. Jack's house was badly damaged but fortunately they were staying away at the time. Albert and Maud had a house in Hillfields, near the city football ground. Tom, prior to going to the North of England with Rover Company, would stay at the house of the Dryburgh's house also at the bottom of Rudge Road. Elsie and Joe Dryburgh had a sister named Olga to whom Tom was engaged. Lillian Sarah would stay with Alice until she married. Clara remarked, "Most of my children are due to leave me. I fear that I may never see some of them again."

Clara's children were leaving the homestead to live mainly elsewhere in Coventry. Only Norah was some distance away, having been evacuated to Great Tew with Aunt Min.

Far away, as the crow flies, the French fleet had gathered to capitulate to the Germans. Its powerful battleships would turn the balance at sea against the Royal Navy if used on the enemy's side. The terms of the armistice stated that the French ships were to be disarmed under German and Italian control. The British government had long given up with regard to Hitler's promises and the war cabinet resolved that the French fleet must be put out of action.

Two French battleships and a battle cruiser were destroyed by the Royal Navy with considerable loss of life to the French crews. The French duly sank their remaining ships.

The start of the Battle of Britain, according to the Germans, was on 13 August 1940. On 25 August, the British began night-bombing German towns, including Berlin and Hitler felt he must retaliate. The German bombing of London saved the Kent fighter airfields. The Germans had set out to destroy the fighter bases in Kent. The British believed that the Germans were preparing to invade and hundreds of barges had been accumulated along the hostile coasts.

The Battle of Britain was all about seeking mastery of the air over Britain; without it Hitler thought the Germans would be unable to achieve a successful foothold in the event of the German invasion of Britain. The German invaders never came and on 17 September Hitler postponed the invasion until further notice. The Battle of Britain had been won partly by the daring and skill of British fighter pilots in Spitfires and

Hurricanes. Churchill paid his respects to those pilots who lost their lives in the Battle of Britain.

The other significant factor of success in the flying battle was the radar system of directing pilots to the core of enemy aircraft and the girls of the WAAF who operated in a most skilled way, often under attack by German aircraft.

Air Chief Marshall Dowding, the head of fighter command during the Battle of Britain, was the only chief to oppose Churchill. Dowding was a devout man with a strong belief in God. Churchill accused him of being a religious fanatic; but Dowding was responsible for winning the Battle of Britain that ultimately saved the nation. Air Chief Marshall Dowding was never honoured, as were his contemporaries; he died a man of God, true to his beliefs and at peace with the world. Churchill did not.

The Rex cinema was built in Coventry just before the war and in 1940 it was a splendid example of the latest cinema of its kind. On the night of 25 Sunday August 1940, the Rex was completely wrecked by a German high explosive bomb. The film *Gone with the Wind* was due to be shown the following day.

Coventry escaped the early part of war without substantial damage or loss of life. The first air raid on Coventry took place on 25 June 1940 on Roy's birthday. Five bombs dropped on Ansty Aerodrome, a few miles east of the city centre. Between 25 August and 31 October, Coventry was attacked some 17 times in air raids. Casualties numbered 176 dead, hundreds of people were injured, of which over 200 seriously. During 1 November and the main blitz on the night of 14 November there were eight air raids on Coventry. Days before the main blitz on the night of 14 November, Roy and I observed hasty preparations going on within the City. Anti-aircraft guns had been brought along the railway line on carriages and placed just behind the technical college, there to be manned on the railway line. Guns, it was reported, had been installed in other parts of the City.

The Coventry and Warwickshire hospital was partly evacuated and the schools closed. Children and their mothers were offered rides out of the city into the surrounding countryside to social centres manned by the Women's Institute (WI). Speculation to this day was that the war cabinet had broken German codes and Churchill knew about the target beforehand. Whether this was so or just the heightening of air raid precautions still remains to be revealed.

On Thursday 14 November, there was not a cloud in the sky; it was a cold but a glorious day. Ted was on duty behind the technical college. A full hunter's moon was lighting up the 'blackout', illuminating the city. You could nearly read a paper in the streets. Violet, Roy and I had already taken up our places in the underground shelters behind the college, nearest to Albany Road. The siren sounded a warning just before 7 p.m. The characteristic throb of German bombers, with their twin-engines and propellers out of phase, brought dread to people night after night. The whistle of incendiary bombs followed almost at once. We were sitting with crowds of people and animals from all over the area. The lady from Thomas Street, who sat next to where we were sitting on bare wooden slated seats, went on knitting faster and faster.

An incendiary bomb whistled down the sloping entrance of the shelter, exploding into an intense bright light that was the result of a magnesium fire. The incendiary continued to burn with such intensity that its acrid fumes were soon filling the shelter with a white fog. We were coughing and holding anything available to our mouths and noses. A white mist was causing people's faces to take on a funny appearance in the dim light. The air was filled with 'bits' from the bomb.

The wardens from the back of the technical college were armed with a stirrup pump, buckets of water and sand. They soon smothered the bomb with sand. By so doing, the bomb continued to burn because it contained its own oxygen, but the burning would slow up and fumes were no longer given off.

There was a debate among experts during the war in relation to the use of sand or water to extinguish this type of German incendiary bomb. Use of water only served to speed up the combustion, so in cases where it was desirable to use water and the resultant fumes could be tolerated it was an alternative to sand. Ted, who was the senior warden, decided the shelter must be evacuated and people sent to the other underground shelters on the site. There was no panic, but one or two complained that they would take their luck in staying put so as not be exposed to the shrapnel that was now beginning to rain down.

Someone asked what time it was and the reply came, "Time it was all over." But the humour was lost on the people nearby and sensibly an answer was given of "just gone eight".

I recall thinking that this would be a long night as it seemed that many hours had already passed since the sirens sounded the alarm; in fact it was about an hour. Ted called for the women to accompany the children and run over to the other shelters, situated at the back of the college away from the road.

Violet, Roy and I were some of the last to leave our shelter, apart from those who wished to stay. As we ran over, the sky was alight; it seemed to be on fire. There was an eerie feeling of noise in the distance, thuds and fire engine bells. As I looked towards the railway line, the gun barrels glowed red hot in contrast to the dark; they were silent. It seemed to me, as we three ran, that life had stood still. We reached the other shelter with its sloping ramp to the underground entrance. There were two thick concrete walls at the two sides of the ramp. This appeared to me to be a bigger shelter, so Roy and I explored the many long passages, each about eight feet wide with bare wooden slated seats on either side of the passageways. Despite the influx from the other shelter we had vacated, there were only a few families and dogs occupying this one. Most were sleeping.

By divine providence we had selected a place well away from the entrance, along one of the passages with an escape hatch at the end. We were tired and sleepy, despite the noise of bombs being dropped. Some of these were near, some far and some delayed action. It was during this air raid on Coventry that the first land mines of the war were dropped by parachute. There was a very loud crunching noise followed by a shock wave that moved along the passage and our ears turned stone deaf. Much to my fascination, a nearby man's trousers set up a continuous flapping motion. We could only but lip-read. I moved further down the passage where people were starting to panic and scream but with little sound.

Then Ted appeared. By this time our hearing was returning, albeit slowly. Ted indicated that a high explosive (H.E.) bomb had been dropped near the entrance, just after he entered. The thick concrete walls had caved in and we could no longer get out through the way that we entered. The mood in the shelter became more subdued with the exception of some people who kept saying "They wanted to get out!" They became almost hysterical. Ted tried to calm them down, and he succeeded until they were able to emerge. When dawn came up, the sirens sounded the all clear. It was 6:45 a.m. on Friday 15 November. We had to leave by the escape hatch.

A ladder close to the end wall was deployed and people started to climb out of the hatch into a smoky atmosphere. The sky was still red; I remember trying to puzzle over whether it was a red dawn or as a result of the blitz. As we emerged from the shelter, the sight that hit us stunned everyone with profound shock. The area behind the college looked like a battlefield. There were several bomb craters and small pieces of shrapnel scattered throughout the site. The anti-aircraft guns, which were on the railway line, were not in evidence.

Outside, in Albany Road there was a bomb crater near to the back of the technical college. We walked home expecting to see the house bombed but apart from an incendiary bomb that dropped in Junction Street there was no damage, other than to the roofs from shrapnel. We learned later that the onslaught was widespread, with whole families wiped out. Lone children were left to starve, bewildered at the wreck of their homes and their families buried beneath the debris. There was no water, gas or electricity. The hospitals were overcrowded with casualties screaming for relief. The shops were in ruins, as were the bakeries and food warehouses.

St Michael's, the cathedral, where I sang in the choir, was a smoldering shell and the whole city centre was a complete smoking ruin. Shelters were hit, burying people alive. The dead lay piled high in temporary morgues or under the miles of rubble, some of it still burning. The scene was a horror and was one of the worst atrocities of the war.

On Friday 15 November, Roy and I met Tom on Spencer ,Park. He had been on duty all that night with the Rover Company in temporary offices at Chesford Grange, between Kenilworth and Leamington Spa. He was thankful to see us. We walked to Spencer Avenue, where a Women's Voluntary Service (WVS) mobile van was serving hot tea and sandwiches. None of us had eaten for several hours, so we ate ravenously.

On the night of 15 November, Violet and Ted decided to move away from Coventry and travel to Kenilworth. Roy and myself slept on the front room floor of a nearby terraced house of Violet's close friend 'Win'. We were all very tired, despite the hard floor. Violet and Ted went to sleep as soon as we arrived; however, we remained awake, comfortable in our sleeping bags. As we had cycled there Roy and I did not go to sleep immediately but talked about the events of the past 24 hours. We had heard that Clara and other members of the family were safe and uninjured. We thought of our school friends who lived in the city. There was no way at the time of knowing whether our close friends had survived the blitz.

In the middle of the night there was an immense explosion, rattling the windows of the front room where we slept. We all woke-up in terror. Surely the Germans were not bombing Kenilworth? No siren warning was apparent. If the sirens had gone off, we were in such deep sleep as not to have heard them. Win and her husband got up and made tea for us, which was very acceptable. In the morning we explored the area and in the main street a land mine had completely demolished the Globe

Public house and several people were killed. The army cordoned off the whole street. Later we were told that a stray German aircraft had let the land mine with its parachute loose over Kenilworth by mistake.

King George VI and Queen Elizabeth visited the city on Saturday 16 November and stood within the cathedral ruins. The headlines in the *'Midland Daily Telegraph'* on Friday 15 November 1940 stated, "The enemy were heavily attacked by intense AA fire which kept them at a great height and hindered accurate bombing of industrial targets, but the city itself suffered very seriously. Preliminary reports indicate that the number of casualties may be in the neighbourhood of a thousand."

We continued to ride back and forth to Win's house in Kenilworth for several days until Ted thought he must report back to the Air Raid Warden's headquarters.

An organisation under the chairmanship of Billy Rootes was placed in train to arrange the restoration of the gas, water, electricity and transport services so that industry and homes would be able to function again. Water standpipes were installed in most streets and notices posted everywhere warned people to boil every drop of water for drinking. Much food had been transported into the city. Often, if you were lucky, it was freely distributed.

The sirens sounded little for the remainder of November but the disarming of delayed action bombs continued, as many would remain hidden for years to come. The army demolition corps lost many trained experts as a result of the November 1940 Coventry blitz as they removed delayed-action bombs. The district of Hillfields was the worst hit, in relation to the number of houses demolished and people killed. Mass graves were dug.

Between December 1940 and January 1941, the population of the city, estimated at 260,000 prior to the blitz, fell to 210,000, though it had risen rapidly by February 1941. Out of the ashes of terrible destruction of the blitz a new and proud city was awakened.

Lillian Sarah was married early in 1941. Her husband, Herbert (Bert) Argyle was soon to be called into the Navy. Also, in the early part of 1941, Clara, who was suffering with cancer, left for Great Tew to join Norah and Beryl. Aunt Min who was having a bad time with her husband, Bill Slater, was unkind and spiteful to Clara.

Roy and I were very worried about Clara's treatment at the hands of Aunt Min, as she was called. News from Great Tew, where Clara, Norah and Beryl were living with the Slater's, was not good. The cottage where they were staying was crowded with evacuees from the East End of London; one of the evacuee's mothers was billeted there also.

Min was on the make as she was paid for every evacuee staying at the cottage, even for Norah and Beryl. Clara and Norah had to climb three flights of stairs – their room was in a small attic at the top of the cottage. If they had to get up, to answer the call of nature, to go downstairs, was one thing without waking the whole house, but there was also a walk to the bottom of a long garden to the lavatory in the dark and cold of the night.

Bill Slater was fond of the ladies and was having an affair with the evacuated woman who was staying at their cottage. Every evening the woman would take a walk and a while later Bill Slater would leave for the public house in the village. They would meet on the high road near the church, and kiss, cuddle and embrace. Aunt Min, in her crafty way, would send Norah and Beryl out to spy on them. The children, thinking it was a game, would return and tell Aunt Min what they had observed. Night after night this took place until Aunt Min confronted them both with what she knew. The woman and her children had to find another place to billet. Whether the affair continued was unknown, but it was only one episode for Bill Slater as he was a born philanderer.

Clara had been admitted to the Radcliff Hospital at Oxford for radium treatment. Little did we know that both Clara and Norah were expected to go by bus once a week into Oxford, and find their own way to the hospital. Also, they had little or no money for their fares and no help from either Bill or Min Slater. In time, Violet was made aware of Clara's plight and made arrangements for her to return.

There was another big blitz on Coventry in April 1941. This raid of a devastating nature was concentrated on Hillfields.

A few days earlier, Lillian Sarah's marriage had taken place and Norah and Beryl had returned from Great Tew to take part in the wedding that Saturday. They were to return next day by bus, accompanied by Betty, Violet, Jack and Ted. Regretfully, it was a Sunday and, being in wartime, the journey had to be done in stages; first by bus to Leamington from Pool Meadow, then by bus to Banbury. To their horror, there was no bus to Oxford passing the high road, about a mile from the village. They started to walk.

An old lorry stopped and gave them a lift to the high road. The two women were in the cab and the men were holding on to the children on the back of the open truck. It was a rough ride and it was late when they arrived in the village. The adults stayed overnight and returned on Monday morning.

My schooling had taken a further knock but everyone I met was in the same boat. I had been selected to go to the Royal Naval College at Dartmouth in August 1941 and banked on the August holidays to catch up.

There was little realisation of the implications of Pearl Harbor on 7 December 1940, other than that of Japanese treachery, until Hitler declared war on the USA; Italy followed suit. Britain had already declared war against Japan on the Japanese attack on the Americans at Pearl Harbour, which sank most of the fleet without warning.

I remember thinking that the world was now at war and if I had any doubt about the outcome of the conflict I was more than confident that eventually we would be victorious. In other words, I could see daylight at the end of the tunnel. I also recall the words of the Japanese commander of the fleet from which the aircraft took off to attack Pearl Harbour. He said, "What we have done today is to awaken a sleeping giant." He proved right!

On 15 June 1941, General Wavell launched an offensive against Rommel in the North African desert. It was a total failure and Churchill replaced him with General Claude Auchinleck. Wavell was sent to India as commander-in-chief. On 22 June 1941, the Germans invaded Russia. The Germans code name was 'Operation Barbarossa'. It was the same time of year that Napoleon invaded Russia.

There was no evidence that Hitler used this date to invade, other than that he studied Napoleon's achievements. I well recall that I thought that this was Hitler's big mistake.

From the declaration of war until Pearl Harbour, Great Britain had virtually conducted the war against Germany and Italy alone, apart from the hostilities up to the fall of France. Now that America had entered the war, only the formal agreement had to be delayed until June 1942. The USA was critical of the North African Campaign and the only point of agreement was sending supplies to Russia. America thought Britain's forces were too spread thinly and lacked concentrated action. Churchill regarded it as an observation by a great nation from the sidelines and did not take it directly as criticism.

In the August of 1941, I set off with two others by train to Dartmouth. The journey was tortuous; first of all from Coventry to Birmingham and then by stopping train to Torquay. Fortunately, there were only six stops but the train arrived in Torquay over two hours' late, and we had missed our connection to Kingswear. Later that day, we were able to catch a train only then to miss the last ferry across the River Dart as we had been hanging around Torquay station for too long.

I had never been to Dartmouth before; neither had my two colleagues. Dartmouth echoed the days of British maritime greatness. Crusaders, explorers Raleigh and the Pilgrim Fathers in the Mayflower had all used the harbour on the Dart estuary that was famous for the naval college that trained former and future royalty. The eldest cadet had the responsibility for calling Dartmouth College to account for our movements.

Kingswear is situated on one side of the River Dart and directly opposite is Dartmouth, at the river mouth which is to the open sea. The train only ran from Torquay to Kingswear; however, there is a long journey by road via Totnes to Dartmouth.

A war-torn and austere appearance faced us from the countryside. What had started off as an adventure was now a prolonged ordeal. I craved for sleep, as did my other two companions. We had lost much sleep during the sirens, which went off virtually night after night. The sirens alerted the path of enemy aircraft even, if an air raid was away from your area.

After what seemed like an age, the stationmaster, who was knocking off for the night, took pity on three young boys dressed in uniforms. He said that a naval boat ferried personnel across the river and he would lead us to the point of docking. A growling petty office manned a small ferry pontoon with an outboard motor. He looked down his nose and in a superior voice said, "Papers!" The other two became nervous and speechless. So I addressed him, "Sir, could we get a ride to the college?" He must have softened with the forlorn look on my colleagues' faces.

"Get aboard and at the double," he remarked. We almost fell into the pontoon with our standard issue knapsacks. When the pontoon reached the other side we were relieved to see a Royal Naval College vehicle waiting to transport any newcomers. We climbed aboard the vehicle alongside a naval officer with three rings; these allowed him to board first, after saluting. "Are you cadets going to Seafield House?" he queried. "Yes sir," said the eldest lad, at last finding his tongue. "Well, the college is a naval barracks now and Seafield House is up river, near Dittisham." the officer replied, "You must sleep at the barracks tonight."

The darkness was closing in as we were ferried across the River Dart but we were in no fit state to notice the men-of-war at anchor in the shelter of the river. We were shown by a petty officer to where we could bed down for the night, after reporting to the guardhouse, helped by the commander with whom we had travelled to the college.

We slept on the floor of a large barrack room that was in the course of being prepared for habitation. We were overwhelmed with the size of the barracks. It was empty apart from the petty officer and us three. The petty officer indicated a washroom and toilets at the end of the room.

"If you hear gunfire in the night, please stay put. There are several destroyers at anchor in the river awaiting convoy duty and the Hun has eyes in his backside, so there could be a raid," he said. "Also, when you hear reveille it's a signal to wake-up and wash before breakfast. You will see other ratings making their way to the cookhouse." He pointed to blankets on the floor. "Make your beds out of those and use the blankets for pillows, you have all you need to make yourselves beds, sleep tight," he said as he walked off.

I was awakened by a 'pom-pom' somewhere outside. It made a deafening din, but the other two hardly stirred. I was in a sleepy daze and I remember wondering: Who was the inventor of the 'pom-pom'? Was it a 'Browning' or a 'Lewis' or an 'Oerlikon' cannon? BSA in Birmingham manufactured all three. The bugle sounded reveille, as dawn came up. We were all in some kind of stupor. How we washed was a mystery? We dressed in full uniforms the best way we could and followed one or two sailors wearing only bell-bottoms and white under tunics. We entered the cookhouse and ate ravenously. We had no mugs for tea so the servers found three old enamel mugs. We felt well out of place in full dress, as all the ratings wore just their tops or overall fatigues. The vehicle we used the day before came to transport us to Seafield House.

We were amazed at the number of men-of-war at anchor on the river. All were flying small rigid type airships from the bows;

this was an attempt to keep enemy aircraft at a height, and as a form of protection against machine gun fire from low-flying aircraft. A warrant office of the wavy navy met us at Seafield House and questioned us regarding our late arrival. We tried to explain that the train was late but he would take no excuses. It was a salutary lesson for me – taking on the responsibility for others, 'come what may'!

During World War II there was a group called 'combined services'. They were characterised by the wearing different garments of each of the three forces, sea, land and air. The combined services operated Seafield House so it was not unusual to see a naval officer's cap, an RAF jacket and army trousers. They were there to train us like commandos.

We also had our own naval officer from the wavy navy and an old leading hand 'Jack Tar', who was there to guide us. He was quite friendly and we called him Jack after 'Jack Tar'.

We were soon thrown in at the deep end. On our first day, we were expected to complete the beginner's assault obstacle course, gravitating to the more hazardous one. The beginner's course started off with simple obstacles, such as old lorry tyres hanging from ropes tied to what seemed like a horizontal ladder; they became more difficult as you progressed. There was a sort of telegraph pole across a ditch of deep water. The group I was with managed to complete the simple course without mishap but the more difficult one was quite another matter. With mud and dirt up to our eyeballs we limped back to the starting point. Our instructor mentioned that we would have to complete the two courses every day before breakfast.

After lunch, we were given an outline of the week's events. Starting with the first day, routine training involved the obstacle courses, which we had just performed, and then dismantling a defensive weapon such as a 'Sten gun'. This had to be done until it could be accomplished in the dark. The other obligatory training involved us walking five miles with full kit in less than one hour, running 100 yards inside thirteen seconds, completing one mile in at least five minutes and rowing a double-bank whaler for eight miles in company with the others in the boat. During the day, there were several organised trips included; one night we had free time.

During the week I hardly saw my other two Coventry cadets with whom I had travelled. There were two events that stay with me today, quite apart from completing the training satisfactorily.

The first occurred while rowing towards Dartmouth in the double-bank whaler. This involved rowing four miles each way on the river. When we reached the men-of-war at anchor, a German Messerschmitt 109 dived out of the clouds, just above the naval airships. The enemy aircraft was machine-gunning the destroyers, but the whole episode was over inside a few seconds and well before the men-of-war had time to open up. We saw the German pilot clearly in the cockpit of his fighter. Seconds later, a Spitfire flew over in pursuit and shot him down over the sea. His body was never found. We realised why the destroyers remained silent. They could have known that the Spitfire was in pursuit otherwise the guns of the men-of-war would have opened-up.

The second event that I well remember took place during the training session was when I was swinging on a rope that bridged deep water. I found to my dismay that the rope was covered with grease and I dropped into the deep water below. After wading from the ditch I recovered sufficiently to continue. The instructor referred to the rope being greased, in order to add to the difficulty.

I well recall leaving the training at Dartmouth feeling on top of the world; I was never fitter. I still have the photograph of the group in 1941 among my treasured possessions

## Three
# The apprentice

**In which I become an apprentice and meet Joan, my first girlfriend.**

WHEN I returned, Ted said that he had found work for me with a company near to Meriden. He was working as a charge-hand on the milling section and I could become an apprentice, if I liked the firm. I had completed my school certificate course at the technical college and I was ready to sit the examination. Ted was trained as a gardener, as was his father; he had 'green fingers' and excelled at his vocation. However, during the war all manpower was directed to essential industrial work and Ted was sent to the Keelavite Hydraulics Company when war began.

Keelavite made hydraulic transmissions; it had been started by two men: one was Mr. Keeling, an ex-miner, and the other an Oxford Don, Mr. Laviteous. The two met at Oxford where Mr. Keeling had been a very late student from the miners' union. Both men were brainy I suspected that Laviteous was the innovator responsible for many of the outstanding inventions. Mr. Laviteous was a big man, tall but broad; he had unruly hair and spoke loudly with an Oxford accent.

There were two others in senior management: Mr. Downs, who was general manager and had served an apprenticeship at Alfred Herbert, famous for its machine tools, and Mr. Riley a designer of note who headed the design department.

Mr. Keeling was chairman and lived nearby. He used to recall his humble background and had a paternal way of running the company. He had a loyal following. I suspected that he was of foreign extraction, maybe of the Jewish faith, for he had the appearance of Telly Savalas. He had dark pale skin and a Roman nose to the best of my recollection. I joined Keelavite and my brother Roy did some time later. My first experience was a good one and I soon became skilled on most machines.

Mr. Keeling, the chairman, took a keen interest in my welfare and he had an indentured apprenticeship agreement prepared, I recall, that Ted, as my guardian signed it. Keelavite allowed me one day a week at Coventry Technical College to take the Higher National Certificate (HNC) course. At the time, the HNC could lead to a degree in engineering after three years. My views of education were quite simplistic in those days. The key, I thought, was 'the ability to think'.

Clara was very ill and in hospital at Kineton. Roy and I set out to cycle to see her from Coventry. Roy, on his small frame bike, seemed to be taking many more turns of the crank. He was very tired by the time we reached the outskirts of Kineton. Our first call was to see if the hospital would allow us to visit, but the sister in charge of the ward taking care of our mother said: "She was too ill to receive any visitors. Mrs. Penny could be well enough to see visitors tomorrow."

We called on the Brown's in Kineton and Mrs. Brown thought that Roy looked so tired she would call over the road to ask her neighbour, as whether she could put us up for the night. The Browns were related to Frank Butler, who had married our sister Alice. Mrs. Brown had a large family of small children; her husband worked as a porter at the station in Kineton. He seemed very old to us and reminded Roy and me of Moore Marriat, who appeared in a comedy act with Will Hay at the time. Roy and I slept until midday. Our intention was to call at the hospital on returning to Coventry. However, we had the same response as the previous day, namely that 'Mrs. Penny was too ill to see visitors'. No appeal would weaken the nursing sister's resolve. We never saw our mother again.

On 30 November 1941, my sister Alice had a son David and I became his godfather. Frank Butler, who was called into the army, was given a short leave before being sent to Burma. About two weeks later, in December, John Penny was born in Leamington Spa to Betty and Jack.

Their newly built house in Cheylesmore, of which they had been truly proud, was demolished in the November blitz of 1940. Clara, who was very ill, had been told of her two grandsons. She said to May that Christmas: "That makes three grandchildren, look after them and make sure they come to no harm in this terrible conflict." She then faded into a deep sleep, dreaming of happier years. A wonderful lady!

On 10 December 1941, Japanese aircraft sank both the *Prince of Wales* and the *Repulse* battleships. Two days before, a Japanese force landed in Malaya and the fate of Singapore was sealed. Singapore had taken a back seat, as most people expected the Japanese would remain neutral. Once more, Churchill could be criticised for attempting too much with inadequate resources. The Far East at that time was an area of the world left impoverished for the European theatre of war. Even the American forces were thinly spread over Asia. The USA had been forced into a world war for which they were ill prepared, all due to Japanese treachery.

At the very young age of 56, Clara died on 27 February 1942. She proved to be a loving mother of great intellect who made a supreme sacrifice for her large family. No words can express how we all felt. It must have had a profound effect on many of the older members of the family. Clara was laid to rest in the same grave as Charles, in the London Road Cemetery. There are many delightful moments I could relate but one thing stands out more than any other: my mother always gave me two gifts at Christmas, one for my birthday and one for Christmas. She had good course to regret that Christmas day, as I was born at noon. Instead, she was always proud of me, as she was of her other children.

The Japanese forces overran Singapore and on February 15 General Perceval, together with 60,000 of British forces, surrendered; turned out to be the greatest capitulation in the history of Britain. The ironic thing was that the guns of Singapore were pointing out to sea and not to the mainland.

In July 1942 Roy and I planned to go to Aberystwyth with Lillian Sarah, who was pregnant, and her husband Bert Argyle. We had a wonderful time at a campsite. We made a bad error on calling at a Welsh public house on a Sunday only to find it was closed, as indeed were all the pubs in Wales at that time. The highlight of the holiday was when we went to the local theatre; I recall that the anniversary waltz was played.

As we returned by train via Birmingham there had been an air raid. The raid was repeated and over 900 were killed or injured, many because they failed to take shelter. Apart from a single raid in April 1943, Birmingham remained free of air raids for the rest of the war. My sister Hilda Roslyn married Harold Shaylor in August 1942. Roy and I often visited the Shaylor's home in Widdrington Road, Radford. Mrs. Shaylor made us particularly welcome, as did the whole family. Harold had a sister Eileen, and a brother Len. Eileen was a bonny, tall girl and well built, but not fat, while Len was of average height and of slight build. Both Roy and I looked up to Harold, who was quite handsome. He went into the army and served in India where he was a sergeant in the military police. We were unable to imagine Harold in that role as he was such a happy-go-lucky individual. I would certainly have said that he was not cut out to man-handle drunken soldiers. In later life, my sister Hilda Roslyn, although she had a strong love for Harold, thought he could be more assertive.

Christmas 1942 turned out to be a bad time for what was left of the Penny family; luckily all had survived the war so far. Albert, Frank and now Harold had been called into the army, and Mac, who married May Fanny, was to be called into the Royal Navy, as too was Bert Argyle. Tom had left for the North of England and was about to marry Olga in Skipton.

I am reminded of my school days when I was selected to take the lead in a play at Christmas time. I would be about nine; the play was the *Christmas Carol* by Charles Dickens. I was chosen to be 'Scrooge'. I wore an old white beard that was wired round my ears, like spectacles. I well remember the words at the end, "Before you dot another 'i' Bob Crachit." In 1942,

Christmas was as frugal as the *Christmas Carol* but the ending was not as pleasant.

The North African campaign was a thorn in Churchill's side. America was critical of the lack of focus. Auchinleck had replaced Wavell but was facing Rommel, who seemed invincible. Churchill realised that there were many factors involved, not just the general in charge or the leader of the desert forces. The supply line was also a large part of the equation, particularly the supply of fuel for tanks and other military vehicles that are so significant in desert warfare. Ribbentrop, one of Hitler's trusted ministers, emphasised over and over again that, "The Fuhrer is absolutely right to attack Russia now". He kept repeating it as if trying to convince himself.

General Alexander was summoned to replace General Auchinleck and General Montgomery was called from England to command the Eighth Army at a time when Germany and Italy were becoming hard-pressed to supply the ever-growing needs of the Russian front. Montgomery had a natural instinct to inspire men and would not move too prematurely.

In January 1943, a baby girl was born to my sister Lillian Sarah Argyle; they christened her Dianne. Her husband Bert Argyle was away in the navy.

In February, two German battle cruisers, the *Scharnhorst* and the *Gneisenau* sailed through the mist of the English Channel in defiance of both services, which were alerted. They were withdrawing from Brest in France to Germany and were no longer a threat to British shipping in the Atlantic.

Little did I know that in August of that year, as a cadet, I would be sent to a naval airfield at Lee-on-Solent; there I would receive my first flight in a Swordfish, a torpedo-carrying aircraft. The successful Swordfish aircraft was now quite old and should have been replaced by the Barracuda aircraft, but the latter had problems with their undercarriages and were grounded. Two squadrons of Swordfish had taken off from Lee-on-Solent to attack the German battle cruisers but had to turn back due to bad weather.

Meanwhile, trouble in the form of civil disobedience against the British in India was flaring up. The root of it was the old insoluble conflict between Moslems and Hindus. Gandhi and Nehru, and many other Congress leaders who were imprisoned, also joined the opposition to British rule.

I continued to sing in the choir long after my voice should have been rested. So I did not have to pay the penalty of neither one thing nor the other. As my voice kept breaking up so reluctantly I had to retire from my choir days. I became immensely sad as I had drawn a great deal of comfort and stability from my time in the Church.

I made good progress with the Higher National Certificate course. The books I needed could be bought new from the college bookshop in the Butts. The college bookshop, it was rumoured, was owned by the wife of the head of the mathematics department, Leonard Turner. Mr. Turner had published many good books on applied mathematics, which were mandatory in the Higher National Certificate classes.

I sailed through the first year with distinctions in all subjects but, like most students, I was hard-up. So I had to resort to buying secondhand books rather than forking out for new ones each year. By so doing I was able to buy the yearly expendables from the college bookshop, such as erasers, pencils, lined foolscap and new folders. In the second year, it became more difficult, as we had lots of homework as well as laboratory experiments to write up.

Roy and I were avid followers of the Russian war. We kept maps of the German progress and Russian counter-attacks. One of the outstanding aspects of the Russian resistance was the battle of Stalingrad. Stalingrad formed a special relationship with Coventry and, later in the war, became a sister city. Marshall G. K. Zhukov, of the Red Army, set a trap that surrounded the German army in the battle of Stalingrad; it was aided by Hitler's order of no surrender. The German general of the Panzer Troops, Field Marshall Frederick Paulus, fought gallantly but with the Red Army fighting on its own ground proved overwhelming. Hitler simply refused to acknowledge the possibility of defeat, but his Sixth Army, commanded by Field Marshall Frederick Paulus, was surrounded.

The German Sixth Army's strength in total was of 275,000 combatants. Inside the Zhukov trap it was estimated that approximately 250,000 men, including Germans, Romanians, the 'Hiwis' (short for Hilfsfreiwillige or volunteer helpers) and the Italians were trapped. Before the battle of Stalingrad, German doctors had observed the disturbing phenomenon of the increasing number of soldiers who died suddenly 'without having received a wound or suffering from a diagnosable sickness'. Supplies were indeed severely reduced but it appeared to be far too early for cases of death by starvation. "The suspected causes," wrote the pathologist with the inquiry, "included exposure, exhaustion and hitherto unidentified disease. Many of those captured died after years of imprisonment in the cold of Siberia."

One interesting factor of British life was the way everyone warmed to the Russian people. The Royal Navy risked numerous convoys of supplies to help the Russian people. Supplies of food and war materials that were in great need to the British war effort were sent, but with the support of 'every man in the street'. Russian culture, the arts and music became popular in Britain. One song that was on everybody's lips told of a Russian girl: 'My little Russian Rose, break the chains that bind me etc."

Selected Russian soldiers, mostly female, visited Britain to address the people, sometimes expressing the ways of communistic life in Russia. The King, in his radio broadcast message to the nation at Christmastime 1942, praised the Russians and recounted the story of the Russian soldier on the Stalingrad front. He walked for 20 miles or so with his comrade on his back to a first aid post. When asked, "If the load was a burden?" he replied, "No, it is my brother!"

Using my holiday to go to camp in August, before the new term started, was fortunate. There was a plant shut down to install new machines that year and one of the two weeks' holidays coincided with the training at Lee-on-Solent Naval Airfield. The journey was straightforward compared with the last train tour. Another 'Y' scheme cadet from Coventry travelled with me. His first name was James and I became his close friend. He preferred the second name, which happened to be my first name, Robert, so I called him Bob.

There were no obstacle courses at the airfield and we were treated with a fair amount of respect. Most of the naval ratings were aircrew or mechanics with who the officers were on good terms, as they came to rely on good mechanics for their well-being when flying.

As soon as we arrived, we were requested to report to the stores to be equipped with full uniforms and other items. The leading-hand that gave out the equipment said, "It must be returned in good order when we were due to leave." The first day was taken up with a tour of coastal defences. We were amazed at the heavy fortifications all along the coast. A sandy beach with skull and cross bone signs indicated that the beech down to the sea was mined. The signs were arranged in such a way that an enemy approaching from the sea would be unable to read the boards. It seemed that every part of the coast was covered with gun emplacements, while the beaches were either mined or covered with death traps against invasion. When we returned to the naval camp we were led into a large briefing room to have the week's intensive training programme outlined, we guessed. We were about 35 in number from all over Britain; the whole atmosphere was surprisingly informal, just as if outlining another flying mission.

We were separated into groups, as during the flights each aircraft could only take two cadets. The 'Swordfish' is a biplane with a crew of three: a pilot, a rear gunner and an observer whose responsibility it was to release the torpedo. The slowness of the aircraft indicated its age. My pilot said, "It could be overtaken by a cow in a field."

# Automotive gas turbines

We were due to visit 'Whale Island', the Naval Gunnery School at Portsmouth and go aboard a submarine of the 'S' class. Of course, the flight was the prize of the whole training. In the afternoon, we were shown around the parachute room. Here there were long tables on which parachutes were packed by 'Wrens', members of the Women's Royal Naval Service, now abbreviated WRNS. There was some superstition in relation to the packing of parachutes. Certain aircrews would insist that a particular 'Wren' should pack their 'chutes.

My best recollection of Lee-on-Solent was that of two squadrons of 'Swordfish' that took off from the airfield to attack the German battle cruisers as they fled through the English Channel; unfortunately few returned.

The day arrived for my flight. I was due to fly in the 'observer's' place and Bob, my friend from Coventry, in the rear gunner's seat. We both had parachutes that were different from our young pilot's 'chute. There were three seats: one for the pilot and, behind the pilot but separated in another sort of cockpit, the 'observer' and rear gunner. Bob and I had parachutes at the front with metal plates.

The pilot, during pre-flight instruction, said that in the rare event that we might need them, before pulling the ripcord we should place our heads backwards so that the metal plate came away without hitting the chin. Apparently several people had been injured and one had a broken neck when deploying the front-mounted parachute with a metal plate. We asked the pilot whether there was any likelihood that an enemy aircraft could appear. He answered, "It was most unlikely and, in any case, Jerry had fast fighters which would require an immense space to maneuver compared with the low speed of the 'Swordfish'.

The 'Swordfish' would virtually hover; the pilot would also stay at a low height. The flight was thrilling for its duration of over 30 minutes or so. We skirted round the Portsmouth area, keeping well away from the balloons. He flew low over cattle in a field indicating the slowness of the aircraft. The pilot made an excellent landing and, as we walked from the aircraft together

# Automotive gas turbines

with the pilot dressed in our flying garb and parachutes, we had a feeling of being old hands.

I was astonished during our visit to the submarine docked at Portsmouth how small the vessel appeared.

We had exchanged places with the other groups that were at Whale Island gunnery school. We were shown round by a petty officer from the crew of the submarine. One cadet in our group appeared to man the gun, lining up the sight on a moving river barge. The petty officer knocked the cadet from the seat. We were all amazed at his swift action. He said it was forbidden to line up a gun on a river barge, especially one flying the white ensign with a red ball in the upper inner canton. It was the rear admiral's barge. We found the visit to the submarine fascinating, apart from the cadet's disgrace in relation to the gun. We were informed later that the rear admiral was aboard at the time and a stern signal was transmitted to the commander of the submarine.

On our return to the airfield, our group had been 'placed on the mat' as a result of the gun incident. The duty officer's explanation from Lee-on-Solent was sufficient to get permission for further visits to the submarine reinstated, on condition that cadets were trained with respect to rules of naval conduct.

The visit to Whale Island was one way of redeeming the medals of our group; we played it by the letter and there was no informality. Whale Island had been a gunnery school from way back in time. It was a fortification, looking out to sea over the whole of Portsmouth Estuary. We found every imaginable gun there. The expertise of the instructors was excellent and turned out to be as equally fascinating as the flight, and that of the visit to the submarine, albeit somewhat marred.

The history of naval projectiles is interesting; from the solid cannonball to the exploding cannonball and then up to present-day nautical high-explosive shells that demand thick armour-plating for men-of-war. We were taught the proper way to use our lanyards when operating naval guns and how to man the various gun platforms, as well as basic issues of locating accurately the target.

Under pressure from Churchill, Montgomery insisted on having adequate tanks, guns and other supplies that were to be available prior to attacking Rommel. On the eve of the battle of El Alamein, Montgomery assured the Eighth Army that their advance could end only in Italy. He was right! Rommel, the German counterpart and a great Field Marshall, defied Hitler's orders and retreated as soon as the Eighth Army attacked. The Germans were short of fuel for tanks and military vehicles. The German forces were virtually driven into the sea.

When Rommel was driven from Libya, he took up a position on the borders of Tunisia. Montgomery attacked across the desert and German and Italian resistance ended. The attack on Italy was just a matter of time, but Sicily would be invaded first.

I was informed that should I apply, I would be accepted into the 'Y' scheme. So I had a medical and passed 'A1' for the Navy. After expecting to report to a naval establishment I was notified that, as I was an apprentice studying engineering, I was in a reserved occupation. Keelavite supported the reserved occupation order so, if I opposed the order by appealing, it was said I had little chance of winning. I often wonder what would have resulted if I had taken the naval career that was on offer. The thought reminded me of the poem by Tennyson: 'The stately ships go on.'

Germany's only remaining battleship, the *Tirpitz*, was severely damaged and finally sunk at the latter part of 1944. The *Duke of York* sank the *Scharnhorst*, the *Gneisenau* had been damaged beyond repair; thus ended the war for Germany's capital ships. The British bombed the Ruhr Dams and two were breached. Bomber Harris, as he was named, had a rift with Churchill and on 13 April was told, much to his indignation, that bomber command had been placed under Eisenhower, the supreme commander of Overlord.

# Automotive gas turbines

On 9 July, 1943 allied forces landed in Sicily and on 25 July Mussolini was overthrown. Italy signed an armistice on 7 September. Germany invaded Italy and the forces were stretched on all fronts. Roosevelt, Stalin and Churchill met at Teheran between 28 November and 1 December 1943. Roosevelt presided and asked for strategic guidance. Stalin insisted on Overlord, Churchill argued against but lost. Churchill was disappointed with Roosevelt, who was very ill and as a result in no fit state to be representing the USA.

The sky early in 1943 was filled with April showers; the sound of sirens was almost sickening; for several weeks there had been a welcome lull, and then not an air raid, just sirens. The warden turned in his seat to face my colleague Len and I who were messengers in the ARP (Air Raid Precautions). The warden's centre was behind the college in the Butts. We both wore tin hats with a large white capital 'M' painted at the front. We were wearing blue, full-length overalls with double pockets and a whistle attached to a lanyard placed in the right upper pocket. Both Len and I had large torches of standard issue.

"As I am expecting the all-clear to be signalled at any time, please cover the buildings in Hertford Place," the warden instructed Len and I. It was the practice to ensure that there were no casualties, even if there was no air raid and the sirens sounded. We both moved off as the all-clear sounded. We decided to begin at the top of Hertford Place; the first building had been taken over by the Young Women's Christian Association (YWCA). It was a large house with a double driveway and a large walled garden. The front of the house was in Hertford Place and the back, enclosed by a wall, faced a road called the Poddy-Croft.

We approached the front of the establishment and tried the large door; to our surprise it opened. We entered cautiously, shining our torches along the hallway. We heard voices singing from a distance. It was not obvious where the singing originated but we moved stealthily along the hallway and came to a recess on our left, where a door of what seemed like a cellar was ajar. Shadows flickered in a sort of candlelight. It was a ghostly scene, as the heavenly voices increased in crescendo. We descended the steps and, deep in the cellar, we encountered a lady's choir, singing in perfect harmony. The ladies were singing in candlelight, wearing dressing gowns. The singing petered out as we came into view.

An Australian voice questioned who we were. We replied, "The warden of the local ARP centre has requested us to check that people in Hertford Place were alright, that the all-clear has sounded."

Miss Straker, the senior lady with a slight Australian accent, introduced herself and said, "If we make for the kitchen I am confident that you will take a cup of tea and a biscuit." We filed up the cellar steps, following the ladies of the YWCA along the passage from the hallway to the kitchen. The large kitchen was clean and spotless. There was a huge, well-scrubbed table in the middle of the room with a large porcelain sink at one wall.

A yellowish, covered-framed window over the sink looked out onto a passageway. There were cupboards mounted all around the walls from floor to ceiling, some with draws at waist height. The colour of the cupboards was pale green and beautifully matched the big airy kitchen decor and windows. All the ladies gathered round with cups of tea and some ate biscuits. Len and I were pampered.

Miss Straker opened the conversation. "You really are young to be members of the ARP. Do you live nearby?" I said, "Just a street off Hertford Place and Len here lives only two streets away."

"We are thinking of starting a youth club to meet once a week to plan rambles, dancing and socials. Also, we are anticipating being allocated a mobile vehicle to use to distribute sandwiches and hot drinks to the public. I am sure it could be of use to the youth club for transporting camping equipment. Please tell your friends in the area and come back for leaflets to give out," Miss Straker said in conversation.

This was the beginning of the Blue Triangle Youth Club, during wartime and beyond. A few old members still meet now (2004) and then. There are a few founding members still alive.

The Blue Triangle Youth Club was initiated. Roy joined, as did many of my pals from the surrounding area. I made many new friends, some of whom I know to this very day. There were other activities at the YWCA premises when the youth club was formed. One of those was the Fifty-Fifty Club. It puzzled me how they arrived at the name Fifty-Fifty; whether the origin of the club was regulated to have an equal number of females to males below the age of fifty or for some other reason I never knew. However, the Fifty-Fifty Club met on a different night to the youth club, but John Brown of the 'Fifty-Fifty Club' became the first leader of the boys, and Miss Johnson, a member of the YWCA, became the leader of the girls.

It was rumoured John was sweet on Miss Johnson, who other boys rudely described as 'boss-eyed' because she had quite a cast in one eye. When she looked you in the face it appeared as if one eye was looking elsewhere.

John Brown was studying for his solicitor's articles at the time and later became my Coventry solicitor until he retired. Other activities used the YWCA premises, one was the Girls Life Brigade (GLB).

I remember the club organising a trip to London one Sunday. Some of the girls from the GLB took part along with the other boys and I got to know them; as we did the girls of the youth club. John Brown was tall and had fair hair. It was said that he could not join the armed forces because of a gamy leg, whether this was true I never knew. I rather suspected that he had religious reasons for not joining up, as he was a member of the non-conformist church to the best of my recollection. He drove to the club once in a brand new blue, open-top Lanchester car. It could have belonged to one of his relatives as we thought he was quite well off.

A boy named Winkless from Earlsdon, whom we all regarded as slightly mad in the head, struck a match on the polished finish of the shining new car, causing damage. It was more in devilment than malicious damage, but we all rounded on him, as we knew it was wrong. Mad Winkless, as he was often called, had to own up on pain of death to John Brown that he was responsible. John, who was a perfect gentleman, tried hard to stomach the misdeed; but even this was too much for him. He tried hard to rub the mark off but made it worse in the process.

A camp was found at Old Milverton, near Leamington Spa, and the YWCA mobile van was used to transport equipment to the campsite. Both girls and boys camped quite often.

Occasionally we had dancing or social nights, usually on Saturdays, and we invited along the 'war workers' who were directed into war work from other parts of Britain. I usually had to be Master of Ceremonies for the night saying, "Please take your partners for foxtrot," or "a quickstep" or whatever, having placed the gramophone on.

During camp we gathered water and milk from the nearby farm; the campsite belonged to the farmer. The girls had a sizeable hut and the boys slept in an old barn. There were always two youth club leaders present. At night, we had a fire in the open air and around which we congregated to have a singsong. "Katie, beautiful Katie, when the moon shines over the cowshed she'll be waiting-waiting there for me, etc." was a great favourite. As was "There is an old mill by the stream, Nellie Dean, where we used to sit and dream, Nellie Dean etc."

The barn had open rafters and Mad Winkless would often climb them stark naked. Once, when a female leader opened the barn door, Mad Winkless was doing his usual climbing stark, egged on by the boys who were making an awful din.

Miss Lambert, a young leader of the YWCA, was requesting the boys "to please make less noise". She had the barn door partly open when she spotted Mad Winkless in his birthday suit. Without batting an eyelid, she held the door open for what seemed to us a long time; she then closed it saying, "Do be quiet!"

The 'Elson' toilets were accommodated in what appeared to be an old stable. The boys had the job of burying the waste matter at the close of each camp. I still have snap shots of these camping days at which we had many happy weekends, both winter and summer. During the winter we would toboggan in the fields on the slopes near the railway bridge. Most of us had to cycle there as the trains ran sparingly in those days. We boys played rounders and cricket when in season, and explored the riverbanks. The River Avon flowed nearby at the bottom of the field slopes. We actually found what might have been an old cave entrance on the riverbanks that sloped up to a field above. We imagined that Ancient Britons had inhabited the cave.

Often, the boys would be called up to enter the armed forces and they would receive a night to remember before reporting into one of the three services. Sometimes they would visit the club on their first leave. I recall some of the names of the founding group. There was Eric Kayser, Brian Pettipher, Maurice Poyser, Bob Watkins, and Frank Terry, and of course Roy, myself and others.

The founding girls were Doreen Cleverly, who married Roy, Joy Baker, Eileen Field, Mavis Perks, Rita Lancaster, Joyce Johnson and others. As people left, others would join and a year later there were many new names. I recall that we started a newspaper called 'GEN' that we published once a month.

An article I wrote caused the committee to resign. It was entitled "Disturbing News". It was critical of the serving committee, who at that time I accused of "looking through rose-coloured spectacles". It was great fun and I only ceased to be a member when I took a post as a civil servant in the North of England, but that story is for later in the book.

That Christmas, in 1943, looking back on my life I considered myself to have been immensely fortunate. First I had a wonderful mother who encouraged me at an early age to join the church choir. This had a profound influence on my early life and beyond. I enjoyed being in the choir and all aspects of church life. Whether it was because the choir took me away from a poor home environment, I will never know. My schooling had little effect on my life, other than my very early years because my defective eye remained untreated.

Looking after James Galsworthy I did quite naturally as a brother. Mrs. Edgar Jones taught me good behaviour and decorum that shaped my character. The naval training tested my fitness and ability for leadership. The higher technical education enhanced my capability to think for myself. Now the YWCA's Christian principles were complementary to the church upbringing. However, life with Vi and Ted was anything but refined; they made great sacrifices for Roy and me, taking us in during our greatest need. They were the very essence of honesty and good working people. So I was thankful for my life that Christmas 1943.

British and American forces landed at Anzio on 22 January 1944. Landing craft could not be released as a result of Overlord. The Germans had time to counter-attack. The Allied forces were nearly driven into the sea. On 11 May, the Allied armies launched an offensive and on 4 June Allied troops entered Rome. Two days after troops entered Rome, Allied armies landed in Northern France. The Soviet armies halted outside Warsaw, only renewed their offensive against the German army in January 1945.

D-Day arrived just at the right time. Germany, though hard pressed, was by no means finished. The German war production was still increasing. New weapons were being developed, some of which were in full-scale production. There were fast U-boats equipped with 'Schnorkel', jet aircraft, unmanned V1 aircraft (nicknamed Doodlebugs) and V2 rockets. The former fast U-boats and jet-aircraft never operated in great numbers. The V1s were aimed indiscriminately at London. Fast fighters could shoot them down and did so, helping to reduce casualties.

Field Marshall Rommel, who was in charge of defending the coast, knew that the 'Allied forces' must be defeated on the beaches. He had ordered the 'Atlantic Wall', using 500,000

# Automotive gas turbines

workers, and millions of tons of concrete and steel in order to construct tens of thousands structures. There were fortresses, gun batteries, machine gun nests and motor emplacements, as well as millions of mines and other obstacles on the beaches. In June 1944, he had 300,000 troops under his command.

Opposing Rommel and the other German commanders, was General Eisenhower, the Supreme Commander of the Allied Expeditionary Forces. General Montgomery was Commander-in-Chief Land forces. The Commander-in-Chief Naval Forces was Admiral Ramsey and the Commander-in-Chief Air Forces, Air Chief Marshal Leigh-Mallory; the last three of course were British. The British 3rd Infantry Division, code name 'SWORD', attacked with 28,845 troops covering an area from Ouistreham, moving east to Lion.

Towards evening, six gliders with 150 armed men from the 'Oxfordshire and Buckinghamshire light infantry' and 30 sappers, one of whom was my sergeant brother Albert, swooped on the twin bridges over the River Orne and a parallel canal, capturing the Pegasus Bridge that was vital to the invasion. The bridge was named later after the insignia of the unit involved. Albert went on to call for naval fire from a grid map from a hilltop. Approximately 4,000 British paratroops and 13,000 from the US landed in France on D-Day. The Canadian 3rd Infantry Division, code name 'JUNO', attacked with 21,400 troops an area from St-Aubia to La Riviere. The area covered by the D-Day landing in total was about 50 miles of coastal beaches.

Hitler thought the landings were a preliminary to the full invasion and mistakenly held back his better divisions. Valuable time was lost to Rommel, enabling the invading force to establish a foothold on the beaches. The British 50th Infantry Division, code name 'GOLD', attacked with 24,970 troops from the Le Hamel district to the Port-en-Bessin area. The US 1st and 29th Infantry Divisions, code name 'OMAHA', were 43,250 strong and were pinned down for part of the day between Colleville and east to Vierville. Finally the US 4th Infantry Division, code name 'UTAH', were 21,250 strong and fought between Pouppeville and Varreville. The total Allied forces in the D-Day landings were approximately 139,715 troops, excluding the naval force, air force and paratroops.

About 6,000 Allied troops went to France by glider. The most successful glider was the 'British Horsa' a plywood monoplane with an 88ft wingspan that could carry 30 men. The largest, the 'Hamilcar', could carry a seven-ton tank. The design of the flat-bottomed landing craft came from a New Orleans boat-builder, called Andrew Jackson Higgins. Some 9,000 boats of this type were used on D-Day. The ingenious design was based on a cigar box with a side that flipped open.

My recollection of the late evening before D-Day was of aircraft filling the sky, and all night long the noise of aircraft was apparent until dawn. Allied bombers opened up the assault late at night on 5 June 1944. As dawn came up on 6 June, the armada of ships that set sail from the south coast through the English Channel under the command of Admiral Sir Bartrum Ramsey, was the biggest ever known invasion force. Channels had to be swept through wide areas of enemy minefields. Paratroops were dropped on either side of UTAH and SWORD, widening the potential beachhead to 100 miles of the Normandy coast, from Deauville in the West to the Cherbourg peninsular in the East.

By D-Day plus 2, the Allied forces were just outside Caen; Bayeux had fallen to the British. Penicillin, discovered in the nick of time by Alexander Fleming, saved many lives in the D-Day invasion and since.

Partisan uprisings began throughout Southern France as the US spearhead was reported only a few miles from Cherbourg. A great tank battle was taking place with every type of Panzer in use, even a new experimental one. In Italy, the forces were pressing on to Florence, capturing more towns on the way. Most Italian people fought against the Germans, resenting the invasion of Italy. Teenagers took up arms against the enemy.

That summer, my brother Tom visited Violet and Ted's home. During the visit from the North, he stayed at the home of his wife's sister, Elsie and brother-in-law Joe, who lived at the bottom of Rudge Road. Tom mentioned it was rumoured that there was a German secret weapon in use against London. There were many random explosions and much resultant damage to buildings and dwellings occurred. He was referring to the V1 flying bombs and V2 rocket system. These missiles were unguided but directed towards London and, in the case of V2s the Germans gauged the amount of rocket fuel for a known distance. In other words, it was straight up and down towards London. It was quite indiscriminate as regards the exact target. Only a few officials in Britain knew that the German were using rockets. The London area had returned to a wartime basis but the worst aspect of the V2 was the unknown location in relation to where the next one would explode. Some Londoners joked that 'if your name was on it you couldn't do anything about it'.

D-Day lifted the spirits of the British people. Many had doubts of total victory but now unconditional surrender became a reality. By August 1944 a large percentage of the flying bombs were being destroyed.

The V2 rocket was a different matter. Over 1,000 rockets were fired and the number of fatalities reached nearly 3,000 people, but their cost-effectiveness was questioned by the German high-command as they were many times the cost of a bomber. The last rocket fell on British soil during March 1945.

Churchill met President Roosevelt at Quebec in September 1944 and declared that the British Navy was ready to join the war in the Pacific. There was some discussion on when Germany would be defeated; Churchill thought it would be in early 1945. Churchill went to Moscow, after his trip to Quebec in October 1944, and Stalin and Churchill agreed to share out the political control of Europe. Rumania 90% Russian; Greece 90% British; Hungary and Yugoslavia 50-50. There was a disagreement in respect of Poland. By October 1944 there was still hope of a compromise over Poland but agreement was never reached.

Despite a significant amount of study and homework, I became friendly with Joan Steer, who was a member of a group at the YWCA that prepared young ladies for a career in the armed services. I had met Joan on a trip to London organised by the Blue Triangle Youth Club. I was attracted right from the time we first met. Joan was the essence of friendliness and vitality. I soon stayed on after the meetings in the hope that I could see her.

When successful, I walked Joan home with another boy called Terry O'Neil, who lived near Joan. Joan's parents lived in Middleborough Road and were quite keen to meet me, possibly as a result of my seeing Joan home with Terry. Joan was just 15 at the time and was studying at the local art school. Joan's parents were very good to me and I ate their wartime rations. Joan was my first girl friend and I became head over heels in love with her. At the time, I felt my feelings were reciprocated. I learnt a great deal from Joan and I was completely besotted. In fact, my colleagues at college must have been fed up with my talking about her. I had many photographs of Joan. My sister Alice was quite concerned, as she must have sensed the very close relationship developing between Joan and myself. We quickly became inseparable; Joan shared my sense of fun and adventure. She had an outstanding voice and sang in an operatic group at the Catholic Church of All-Souls near Hearsal Common.

Joan's mother made many of her clothes. She had an immense skill as a dressmaker, making a number of Joan's dresses and costumes. At the time, good cloth was scarce and difficult to come by. I can recall two smart costumes in the best tweed. One tight fitting costume was an ochre tweed colour while the other was a kind of check, with red and white flecks. Joan looked a picture in both costumes; one had a grouse's claw on the lapel. I am confident that it was the yellow costume. It was the source of much good-natured bantering at the time. I pulled Joan's leg that it was a sign that she was one-sixth a Scot. Joan was a bonny girl. I loved her and the fact that to me she was just Joan.

# Automotive gas turbines

My studying work was taking a knock; unless I got down to it I would fail examinations that hitherto I had been so good at. In my experience, I observed good and bad teachers. The good ones explained the simple 'twists-and-turns', usually associated with jargon and catch words. They made you work from the beginning to the end of the class, focusing your attention on the subject matter in a sequential way and not being side-tracked. I also observed that being successful in examinations was as much about knowing the subject matter as much as it was an art, amounting almost to a science of revising or preparing for the examination; reading years and years of past examination papers and establishing the drift proved helpful.

As is the case where many examinations are to be taken, the subjects can be revised from the weakest to the strongest; spending more time on the weakest until you feel confident in relation to all the subjects to be examined.

This reminds me that one day I decided to play a joke on a friend with the connivance of his other classmates. Robert Woodfield lived at the bottom of Rudge Road. His parents owned the shop on the corner of Croft Road, just below where Alice my sister lived. The Woodfield's shop sold all things, as did so many corner shops in those days. Robert was about my age and studying an identical course as me. Although not in the same class as me he was taking the same subjects. We both had the same bad teacher for Heat Engines. This involved more homework and writing up those experiments conducted in the laboratory than any other subject. The teacher, called Chisholm, was nicknamed 'Chisel' or sometimes 'Chishole'. He was pleasant enough but he could not teach to save his life. He spent the whole time in the classroom or laboratory telling stories; some were interesting when you heard them once, but not when they were repeated over and over again.

One Christmas, in our experiments there was a requirement by the college for certain books and tables, one of which was a set of *'Calendar's Steam Tables'*. When Chisholm wanted to use steam tables he borrowed them from a student. There were two kinds of steam tables in temperature units: one in Fahrenheit and one in Centigrade. The numerical values were quite different from one another, depending on the temperature unit. Chisholm had made a great play at the beginning of term that all tables must be in Centigrade units. So we placed a set of Fahrenheit steam tables that had the same cover appearance on the small Christmas tree. It was presented to him. He was thrilled, spending the whole of the time in the class relating his usual tales. But in so doing he failed to spot that his Christmas gift was in different units from his insistence that all *'Calendar Steam Tables'* must be in Centigrade units. I doubt if he ever bothered to convert.

Between 1943 and the D-Day landings in France there were many strikes in Britain. Even the Communist shop stewards could not prevent them and did not try. Prosecutions by the Ministry of Labour were ineffective. The miners were restless and nothing would satisfy them, short of nationalization. In October, Churchill rejected this but did offer an increase in wages. The miners' federation of Britain changed its name to the 'National Union of Mineworkers' in order to claim the national wage.

The British gloom of early in 1945 cast a shadow over the meeting at Yalta where Roosevelt, Stalin and Churchill met in early February 1945. Churchill sounded the anti-Bolshevik alarm although he had to profess faith in Stalin's goodwill. Contrary to world reports, the conference at Yalta achieved nothing. Roosevelt was losing his marbles due to severe illness; Stalin was very dictatorial and Churchill, conscious of the 'red menace', was risking the loss of leadership of his party. In early March the Americans crossed the Rhine at Remagen and later that month Montgomery's armies crossed the Rhine further north and penetrated into the Ruhr. The Russians entered Berlin and resumed the attack on the Germans through Poland.

President Roosevelt died on 12 April 1945, having never witnessed victory. On 4 May 1945 the German forces surrendered unconditionally to Montgomery. Admiral Doenitz, who had been Hitler's nominated successor, attempted to end the war in the West while continuing to fight against the

Russians. His proposal was rejected. On 7 May Germany signed an instrument of unconditional surrender on all fronts. Churchill announced victory in the House of Commons on VE-day (Victory for Europe) on 8 May. Joan and I spent the time together; as it was a fine night we kissed and cuddled for hours. I suppose we were never closer. Someone once said that "Young love is the best love."

Crowds danced in the streets, church bells were rung, floodlights replaced the blackout and street parties were held up and down the country. On VE-day, Alice was one of the organizers of the party in Rudge Road; Violet arranged the street party in Hertford Place. Roy and I alternated between both parties. The adults, wrapped up in victory celebrations and the prospect of their returning husbands from the war, took little notice. It was a wonderful feeling that war was over to be concerned about anything bad.

The year 1945 was an historic one, apart from the death of President Roosevelt. My sister Hilda Roslyn had given birth to a boy in April 1945. Her husband Harold was away in India at the time as a member of the Army Military Police. We went to see the new baby; he was to be christened Clive, after 'Clive of India'.

Winston Churchill ended the wartime government by resigning. The most crucial result occurred in the history of the British Parliament. The General Election of 5 July saw the British people turn their back on their war leader, Churchill, and voted the Labour party into power. Labour won the election with 393 seats against 213 for the Conservative party and 31 for others. Demobilization of some of the conscripted forces began. Violet and Ted were offered the Swanswell Cottage public house to manage, while Beryl and my sister Norah moved with them. Roy and I would go to live at my sister Alice's house at the bottom of Rudge Road.

I bought a book from His Majesty's Stationery Office bookshop with part of my Christmas money. It would take me into the 'Department of Atomic Energy'. I explained to Ted, prior to the first atomic bomb being dropped, how the atomic bomb worked.

Despite my college work, I had found a girlfriend in Joan Steer. Her parents, I think, were in favour of the friendship and were kind to me. What more could happen in 1945? But it did! Rationing of foodstuffs during the war was not the only commodity for which ration books were issued to each person. Clothes and fuel were also rationed, though the latter were confined to essential professions such as doctors, nurses and essential workers whose journeys were vital. Most people thought, that with the end of hostilities, rationing would cease; however, after the Great War rationing continued until 1921. The average housewife was considered a marvel of management in the way families ate.

People were generally healthier than in peacetime. Basic foodstuffs were available on coupons, but tinned fruit and dried fruit were on points. Rations fluctuated but generally none of us went hungry. There was much bartering and exchanges of sweets, food and even clothing coupons. If you knew how and had the money you could buy anything on the black market. Ted, for example, never ate dairy foods, so he traded his butter and milk for other things more to his liking. Controls after 1948 tapered off and milk rationing was abolished altogether in 1950. Ted grew most vegetables and some fruit, so we were better off than most and had a fair amount to exchange for other things such as eggs or poultry. Christmas 1945 witnessed an anticlimax as thoughts remained with those fighting in the Orient.

H. D. Smythe authored the book I bought from the bookshop in Birmingham in December 1945. It was called "*A general account of the development of methods of using atomic energy for military purposes under the auspices of the United States government*". It is dated 1945 and I still have the brown-looking book; it is in a very tatty state. It has 13 chapters, including the 'introduction' and 'general summary'. It is well written and easy to understand. It concentrates on the separation of uranium to extract fissionable material. It discusses the critical mass required to create a chain reaction or nuclear explosion, and then debates the peacetime role for atomic energy.

I explained to Ted that the Allies were working on a bomb that was equal to many 1,000lb bombs. He asked to see the book and thumbing through said "It is well over my bloody head".

I tried to explain by using the analogy of a football match in search of the 235 Uranium materials capable of fission. I first had to explain what I thought was an atom; up to that time I believed an atom was the smallest particle of matter. Now I had to consider it as a small universe with a nucleus surrounded by a largely empty space in which electrons moved like planets round the sun. Each electron carries one negative charge and the number of electrons circulating round the nucleus is equal to the number of positive charges of the nucleus, so that the whole atom has a net charge of zero.

This was difficult to imagine for anyone brought up with the idea that an atom is solid. Nevertheless, I tried to explain to Ted that uranium has three atoms and as you cannot pick them out with a pin you first have to turn uranium into a gas. The fissionable atom is more mobile and this is the one you want to catch. So, going back to my football match, the more mobile ones leaving get out first and the infirm leave last. So you have to devise a way to catch the ones you need. However, I said to Ted, "You can get millions of them on a pinhead".

By this time Violet had walked in and later she had to hear Ted attempt to explain to his mates in the pub his version of my explanation. He was saying, "You can get millions of these little bleeders on a pinhead, and the bomb is no bigger than a thousand 'pounder and would destroy the whole of bloody Cov." Violet said to me later that I had caught Ted's imagination by spending money on a book that should be on the secret list. Some years later, when I applied for a post in the Department of Atomic Energy, the book was one of the reasons I secured a job to become a young civil servant.

Ted was the first man I knew who could swear in the middle of a word. He had several stock phrases. Two well-worn sayings of Ted and which Roy and I would repeat quite often, even before he could get it out, were "You will bloody do something you will" and the other one "Beat that bugger" (Naming the person). When referring to aluminium he would say "ali-bloody-minium".

Violet, my sister, knew exactly how to handle Ted. He would get slightly tidily but was always amusing with it. When he had had too much to drink, he would try to lift his leg to untie his shoelaces; repeatedly, he would fail so that his foot would hit the floor time and time again. Roy and I looked forward with great fun to this antic of Ted. He had 'green fingers' and grew most of our fruit and vegetables. Ted and Violet could turn their hands to virtually anything.

Ted gave good advice, but only when asked for: for example, on tending first aid, on parental guidance and on many other subjects too numerous to mention. Unfortunately, Ted died after months of illness fighting cancer. Later, when on a long airline flight and my notes of the trip were complete, I used to search my mind for the most remarkable character I had ever encountered in life. Ted was one of those that leapt into my mind. He was so remarkable that I started to write down what I imagined was his life. Whether the following is true I don't know, but Ted was brought up in Great Tew in the next cottage to Aunt Min.

"His father lived, I believe, in Oxfordshire and was the head gardener at a big house in Great Tew before going off to the Great War. When his father was killed in the war his mother, who was in service at the big house, was left pregnant. Domestics had to leave the service in those days when they were expecting but the owners, the Boultons, took pity on the pregnant servant girl."

"However, she was in turmoil and set off to walk to Banbury, looking for her husband's brother. Banbury was at least nine miles away from Great Tew. An old farmer with a

# Automotive gas turbines

horse and cart, many miles from Banbury, gave her a lift after first having difficulty getting her to accept. His farm was some way from her destination as the farmer pulled into a yard. The farmer got down from the cart and asked his wife if they could look after the girl, who was in a very distraught way. When the farmer returned to the cart to say she could stay, the girl had left. Ted's mother had wandered into a railway yard and was knocked down by a railway wagon. In the hospital at Banbury she recovered her injuries and when Ted was born he was taken back to the big house in Great Tew. The Boultons left the baby in the care of Ted's relatives, who lived in the cottage next to Aunt Min of Great Tew who had a daughter Betty."

Early in 1946, the Labour government under Clement Attlee, who was both Prime Minister and Minister of Defence, set the scene for the socialists. Their election manifesto extolled: 'Let us face the future'. The establishment of a socialist Commonwealth of Great Britain was their aim. Nationalization was top of the agenda. Coal, railways, airlines and health were the immediate targets. Labour adopted a 'cheap money' basic policy. Rationing remained in force and Harold Wilson, later to become Prime Minister, was the minister responsible for regulating clothing points.

In the early part of 1946 I was moved from one department to another, spending the last three months in the drawing office. I was responsible for the design of hydraulic pumps and rams. This was where I first met Mr. Vernon who had been chief engineer at the famous Rudge Company. He had many patents and had invented the detachable wheel for all cars and lorries. He was down in the world through illness and he walked with difficulty. He worked in the drawing office. Although senior people tried to turn me against him, I felt he deserved better treatment and respect.

To me he was a famous engineer. I learnt an immense amount just listening about his former engineering problems and the way he approached them. He used to stumble into the drawing office. He would assist himself using the palms of his hands on the two lines of drawing boards on each side of the office; but his hands sometimes left marks on the drawings, much to the annoyance of those whose drawings were marked. The members of the office used to ridicule him in his absence. At that time I was the youngest in the office apart, from the print girl, and several times I had to round on my colleagues for the belittlement of Mr. Vernon.

Frank Grime ran the office. He was a good draughtsman but sometimes he also would join in. One new draughtsman, Roger, was taken on. He had a stump where his hand must have been. Roger turned out to be a very good draughtsman and the most productive in the office. Roger lost his hand early in the war and as a result Mr. Vernon was never again belittled.

Roger was a most cheerful addition to the office and demonstrated that infirmity can be mastered. He had his own drawing board attachment, equipped with both vertical and horizontal rulers. At the point where the two rulers came together there was a four-inch wooden round disc, which he used to cup his stump. It also acted as a protractor so that rulers at right angles to one another could be turned together through angles. Later it became a standard drawing board attachment in place of tee-squares.

Joan laid on a celebration. Her father was coming home from the war. Before hostilities, he had been the manager of an Earlsdon store near the Astoria cinema. Then came his call-up papers, late for his age. I was shocked that the older ones were being called-up. He was not young but they married quite young. The provisions from the store were expertly wrapped in brown paper; it was an art that most stores of that kind taught in those days.

The older threshold reached demobilization in the early days of 1946, as there was no prospect of involvement in the Far East. The war with Japan would come to an abrupt but ignoble end after the atomic bombs dropped on Hiroshima and Nagasaki. Mr. Steer said to me that, "He did an honest job at the store, but would not return to his old job as manager when 'demobbed'".

Joan's parents seemed good people to me and I found Joan's grandparents likewise. Joan, in her youthful way, often used to correct both her parents and grandparents, much to my embarrassment.

Joan always treated me royally and fed with food that I knew well was rationed. This was decidedly against my wishes and caused me some embarrassment. I learnt many good things from Joan and we composed together a poem about the rebuilding of Coventry. It went something like the following:

Within the sound of Cathedral bells

A City new we render.

Undaunted by the fiery hell

Standing in all its splendour.

With brighter windows of modern shops

Long straight highways and flat rooftops,

The parks displayed in majestic array

Behold the Coventry of today.

To us God's banner has unfurled

The fury of the modern world.

Created, molded by his hand

Behold the prize of England.

Joan also arranged a birthday party, to which I was the only male invited amidst her school girl friends. I think at the time that she was attending the Stoke Park Secondary School, although I don't recall any of the girl's names. One evening, as we walked to the Operatic Society, Joan explained that our close relationship was due to be disturbed. It was her intention to go to London with the support of her parents to further her career. I was devastated and pleaded with her to change her mind.

That ended the association with my first real girlfriend. My sister, Lillian Sarah, thought I needed a break and she was due to visit the Porter family, near Brighton, so she invited me along. My closest friend at that time from the Blue Triangle Club was Phillip Newby, and so with my sister's agreement I suggested that he came along too. Mr. Porter senior chauffeured my sister, her young daughter Dianne, Phillip and I from Coventry. We were due to stay at his family's home in Brighton. Looking back on those years, young lovers seemed oblivious to the world around them. They become so much part of one another, as Joan and I were. We nevertheless did have thoughts for others. I loved Joan more than life itself. When we parted it took me a very long time to recover, but I had to grit my teeth again and continue with my life.

US President Truman decided atomic bombs should be used against Japan. Britain gave consent on 2 July 1946. The first atomic bomb was dropped on Hiroshima on 6 August. The second bomb followed that on 9 August on Nagasaki. There was terrible slaughter in both cities; for most of the living the worst was to come through their agonizing survival. On 14 August the Japanese government accepted terms of unconditional surrender. General MacArthur received the formal capitulation of all Japanese forces in Tokyo Bay on 2 September. Victory of Japan, VJ-Day, was officially celebrated on 2 September, marking the end of all hostilities.

The horrors of war in Europe were again related at the trial of Hitler's henchmen, recalling the German murder camps such as Dachau, Buchenwald and others.

Roy and I celebrated VJ-Day with the street parties that were renewed. People were used to arranging parties with the meagre rations available; it was quite surprising that sweets and chocolate cake, jellies and other good things were so plentiful at street parties. We had settled down with Alice and were both teaching her son and our nephew David to read and write. He was nearly five.

Bob Woodfield, as he preferred to be called, (his name was Robert) was sweet on Eric Kayser's sister. They were both surprised with their rapport. The Kaysers lived only a few doors away from my sister's house, next to the opening that was the access the back gardens. Eric now was training to become a sergeant pilot in Canada. Now that hostilities had ceased, we asked 'Dot' his sister what would happen, but she did not know as he volunteered for service and was not a conscript.

I called at Bob's place, entering through the back door in Rudge Road just below Alice's, to query a question we were set for homework. There was no sound at the house when I knocked at the back door. I wondered why, as I had previously spoken to Mr. and Mrs. Woodfield on their way out and asked "If Bob was in this evening?" They had answered in the affirmative saying, "You are very welcome to go to see him." After some time, Bob came to door looking quite flustered and said, "He was just entertaining 'Dot'." I excused myself and thought he was quite grateful. 'Dot' was quite a peculiar girl; to the best of my memory she was a blond but had a sort of odd form that I could only describe as a milkmaid shape, for her clothes seemed to fan out. Whether Bob's parents knew that he was entertaining 'Dot' I had my doubts. I was of the opinion that the older Woodfields were uncomfortable with the relationship; only Bob seemed to have rapport.

During the winter months we played for the Blue Triangle football team. Roy had organized a payment from each member of the team in order to purchase red jerseys. The girls sewed blue ribbon triangles onto the red jerseys. Equipped with white shorts we turned out with the appearance of a professional team. Our performance was not good, as I recall. We only won two matches during the whole of one season. We played other teams, but usually on their pitches as we were without a pitch that we could call our own. We were placed in the youth league; to the best of my recollection this encompassed the age group of 15 to 20 years of age. I recall also two of the football teams we played; one was Walsgrave-on Stowe, who always beat us, and the other was Tile Hill – we beat them once. Roy was a forward and I played at right back. When we beat Tile Hill by one goal to nil, I kicked the ball right up the field to Roy and he caught it on the volley and scored.

It started to snow in early November 1946 and continued through to May 1947. This caused most of the games to be postponed, as it was one of the worst winters in living memory. During the latter months of 1946, we used the camping area at Old Milverton to enjoy winter sports. The slopes down to the River Avon proved ideal for sleighing. At weekends, both girls and boys had a wonderful time enjoying the sport in the snow. A few snap-shots record how we spent our time in the winter of 1946.

The government raised the school leaving age to 15 in 1947, as a means of expanding working class opportunity. There were already signs that a greater adult education opportunity was being considered. All working people contributed to a national insurance fund that enabled them to receive benefits. The Health Service and state-provided education were freely available to all people. Hugh Dalton, then Chancellor of the Exchequer, feared a return to the Great Depressions of the 1930s. He kept interest rates low, to encourage investment, and he reduced taxes to speed up spending.

The increase in government spending on Nationalisation, together with the addition to the Chancellor's policies of expansion, resulted in a budget deficit. There was also a balance-of-payment deficit and signs of inflation; a change of policy was indicated. Sir Stafford Cripps replaced Hugh Dalton as Chancellor in the government reshuffle of late 1947. Also the USA channeled American resources to Europe, particularly Britain, through the Marshall Plan in 1947.

My new girlfriend was Sheila Oakley, who I often walked home through Spencer Park. Sheila lived in Earlsdon. Another young girl, Joan Spurgeon, had joined the Blue Triangle Club early in 1946; she also lived in Earlsdon and walked with us. I invited Joan to visit me in Rudge Road when I my sister Alice went out one evening; I was baby-sitting for David. Joan and I soon became good friends. Joan was only 15 but her deportment and good mannerisms decidedly were older than her years.

I was coming up to my 21st birthday at Christmas 1946. Knowing I was in the church choir since the age of 17, Joan bought me a prayer book for my 21$^{st}$ and a pipe with a tobacco pouch. She had saved up each week and this constituted quite a sacrifice as most of her earnings went to her parents for her upkeep. I well recall taking Joan home and leaving her at the top of her road. She lived in Huntington Road, Earlsdon, but she urged me to refrain from escorting her to the front door of home. Her father did not take kindly to her having a boyfriend.

We would meet again in 1985, as I will relate later.

Unbeknown to me, the late James Galsworthy who was tragically killed on the eve of his 14th birthday, was Joan's uncle. He was the boy placed next to me in the 'big school' and who I fed with liquorice wood in an attempt to bring the roses back into his cheeks. Joan was a distant relative of the famous writer 'Galsworthy'. She was tall and highly attractive. I must have been off my head not to take Joan seriously; instead I became engaged to Doreen Finch a YWCA trainee

As soon as Miss Maud, the then head of the YWCA, found that one of her trainees had become engaged to a member of the club she reacted immediately, explaining to Doreen Finch that she must leave at once. Her parents reacted adversely also; as their daughter's young life had been problematic. Doreen was the eldest of two children – both girls. Her father had a building company and her grandfather was once the Lord Mayor of Hull. Doreen had been found a post with the YWCA as a protective, but Christian move. She had been studying for an art degree at Manchester University and had tried to run away with a married American officer, who was returning to the US at the end of the war. They were upset that once more their daughter had seemly left the rails.

They never accepted me, so our engagement was doomed from the start. I was not sorry when the time came to end the relationship, as study once more became my first priority. I was to take an engineering degree course.

George Cooknell was an odd character. I first became acquainted with him during the war, as he lived opposite our house at the top of Rudge Road. His life's episodes were a fascination to me. He had sleeping sickness which caused him to shake all over. The drugs he took proved worse than his illness. I had the impression that at that time there was no cure. His life had been one of adventure. He taught me to navigate a windjammer on the sea by stars at night and the sun by day; it was decidedly zigzag in retrospect. He had a most interesting life story that began when he was thrown out of Bablake School, Coventry. He objected to the master who was teaching carpentry by throwing a 'jack plane' at him. He was expelled from secondary school. His parents, who owned their residence in Regent Street, were incensed with their problem son who had been in and out of trouble since early childhood. At the age of 15, he ran away to sea and became marooned on one of the Turks & Cacos group of islands.

The Turks & Cacos Islands form a small group of 28 coral islands south west of Java. George was searching for Spanish gold that was said to be buried there. The Spanish explorers were prolific adventurers of the wide world. Spanish vessels became well known for carrying gold round the world, so when they discovered a new island on their outward voyage the gold would be buried there and retrieved on the way home.

George stowed away on the first ship from Liverpool bound for Australia. Two days later, the crew discovered a starving teenage boy under the covers of a lifeboat. He was press-ganged into the task of stoking the ship's furnaces to enable payment for his food. He was not one to enjoy being down in the hold, especially stoking boilers, as he was a fresh air boy. So he applied to become a deckhand; this started his life as a seaman.

He enjoyed working hard, travelling the world and listening to old seamen relating tales of their various voyages. Some of these tales were of buried treasure, especially Spanish gold. It was while George's ship was in port in the Sudan that he thought he contracted sleeping sickness. His health deteriorated and he was forced to abandon his occupation. George's parents

# Automotive gas turbines

were well off and excused his troubled youth, setting him up with a caravan. For a time he had rented the house at the top of Rudge Road but now was living on his parent's land. He loaned me £40 to purchase Doreen's engagement ring, expected at the time. I returned the money over the course of several months, as my earnings were too low for a quick return. We agreed mutually to a prolonged repayment.

# Four
# Atomic energy

**In which I go for an interview and find myself in the exciting world of atomic energy.**

I HAD taken up work with a company at Longford, near Coventry as a stopgap measure before joining the Atomic Energy Establishment (AEA), which I was determined to do. I had applied for a grade in the Civil Service and was awaiting a response. I received an official letter asking me to attend for interview at Risley, near Warrington. A travel warrant was included and the letter stated that I would receive expenses for the trip, which would be paid on arrival. I decided to travel from Coventry to Warrington by train, via Crewe, and then by bus to Risley. A young, tall, dark-haired lad got on the train at Crewe and, in my excitement, I engaged him in conversation. Peter Whittaker was shy at first and answered in monosyllables, but slowly he developed a friendly rapport. He revealed that he also was due to be interviewed for a new post that day. Peter was about my age and, by sheer coincidence, he divulged that his interview was with the Ministry of Supply at Risley.

We were both going to the same establishment; my interview was at 11:30am and Peter's at 11:00. We both warmed to one another, exchanging notes in great excitement as we pulled into Warrington station. We had, I recall, ample time to seek out the bus to Risley, which stopped outside the headquarters of the Atomic Energy Establishment for the North of England. A sign stated that all visitors must report to the police lodge for passes to enter through an iron turnstile. We noticed, as far as we see, a high fence with barbed wire at the top on each side of the police lodge. Presumably the fencing encircled the establishment.

In front of the police lodge was a large board spelling out 'The Ministry of Supply' in large letters; under this appeared a warning sign "KEEP OUT". Peter and I had to sign an undertaking that anything we discussed must be subject to 'The Secrets Act', which was surprising as this was the first we heard mention of the Act. We assured them that we had no cameras and after giving our personal details we were issued with passes. I had to wait longer than Peter, as his interview was 30 minutes ahead of mine. A female receptionist, who asked whether we needed refreshment, showed us both into a side room. She returned with a tray on which were tea, milk, sugar and biscuits. The cups and plate were of bone china.

Peter and I agreed that we would take the same return train; Peter would leave the train at Crewe. We had a bravado time discussing every imaginable topic, without the likely questions, which could arise from the interviews. I suppose this was one way to keep alert, as we agreed later that we both had strong attacks of the 'butterflies'.

Three men faced me at the interview: Christopher Hinton, who was head of the establishment, a gentleman referred to as Jackson, who would become my boss, and another who I think was named Mr. Arms. Later, prior to his retirement, Mr. Hinton was knighted to become Sir Christopher.

I was young and anxious to please; I suddenly felt bright, alert and at ease. I was an avid reader and had read all about the atomic energy project in Chicago. I warmed to the group, particularly the central figure who started the questions. He asked me "Did I know the name of the first man to split the atom?" I replied: "I thought it was Rutherford at Cambridge in 1920." He smiled. Then he asked, "Did I know Einstein's energy equation." I said "Yes", and he asked me to explain. I said, "The energy in a piece of matter was equal to its weight times the velocity of light squared."

He corrected me by saying 'weight' instead of mass, as mass is preferred in the CGS system and is used in place of weight, which can be mistaken for work. Mr. Jackson then asked if I knew the phenomenon of radioactivity and which element was the most significant? I answered, "Yes, uranium".

"Is uranium material a metal?" "I think so." "How many isotopes does uranium have?" "Three, I think". "How do you think we can extract the isotope we want to keep, considering uranium is a metal, like brass?" I had to consider whether this was a catch question. I decided not and answered, "By turning uranium into a gas, such as UF6."

The look on their faces was one of astonishment. Mr. Hinton said "The mention of UF6 is classified and we refer to it as 'Hex' or 'six'. Where did you get to know that we were using UF6?" I replied, "I bought the 'Smyth' report from His Majesty's Stationery Office book shop for two shillings and six pence." "The book has been taken off the shelves and your copy should have been recalled," said Mr. Hinton.

Mr. Arms, who had not yet spoken, said "I understand that you served your apprenticeship with a company specialising in hydraulics. Regarding flow in a pipe, how is the velocity related to area of the pipe?" I thought about this question in relation to whether a liquid or a gas was flowing in the pipe and I answered, "The velocity of flow is inversely proportional to the area of the pipe." There were many other questions raised but these were the ones I remember.

The three gentlemen outlined the post, and salary for the grade on offer. I felt distinctly that I was hired, as the last thing Christopher Hinton said at the interview was "We look forward to welcoming you to the isotope team."

When I left the interview, Peter was waiting for me and we had a sandwich at the railway station. On the way back, Peter was rather quiet and I sensed his unease regarding his interview. At last he said he thought he would not get the post; but he did, as the Ministry was taking on many young engineers.

I managed to brighten his outlook in the journey back to Crewe. There was a loud banging and we both went into the corridor of the train to investigate. The noise was coming from the toilet and a voice said, "Where is this, mate. The door has stuck." We said, "We are travelling to Crewe." The raised voice shouted almost hysterically "I wanted to get out at Warrington."

Peter and I looked at one another. "Hold tight in there and I will get the conductor," shouted Peter.

Peter left me outside the toilet door while he went to look for the conductor. After some time, the banging started again, as if the poor man locked inside was hoping to break down the door. I called out, "My colleague has gone to fetch the conductor." After another delay the conductor appeared with Peter. "Can you lift the catch?" called the conductor. "It's stuck," said a voice in great agitation. "You will have to wait till we get to Crewe. They have the right tools to free the door," was the reply. There was more hysterical banging. "No, no. Get me out of here." The frenzied attack on the door brought smiles to most of the people who now gathered along the corridor. "Listen, mate, whoever you are. Be quiet and relax until we get to Crewe, then we will lever the door open," said the conductor to the door. The noise subsided as a little old lady proclaimed, "The poor man must be distraught caged up in there."

We soon reached Crewe and Peter and I stayed behind while all others alighted. I had to change trains to Coventry in about an hour's time. The station engineers arrived with enough tools to open a bank vault. Soon the door burst open and the smallest, most insignificant of men walked out with his head bowed. He left the carriage in silence and stumbled off along the platform. We all stopped momentarily and then broke into simultaneous laughter, accompanied by the station engineers and conductor. Peter stayed with me until I boarded the train to Coventry. He had been good company and I would see him again when both he and I were taken on as civil servants with the Ministry of Supply. Soon after, we were informed by letter that our interviews had been successful.

The time came for me to leave my sister's home; both Alice and Roy were quite upset. I left with few things and much sadness. On a cold winter's day I took the train from Coventry to Warrington, once more changing at Crewe. Peter Whittaker had travelled the day before, so I had no companionship on the journey from Crewe. The Ministry hostel at Risley, adjacent to the establishment but without a barbed wire fence, welcomed me on arrival. I was placed in the 'Hindhead' block in my own room equipped with a wardroom, a desk with a chair and a single bed. There was a large window looking out to other blocks and open fields. Peter Whittaker was also in 'Hindhead', but placed in the end room. Each block gave the appearance of a large barrack room, with a bath, toilets, showers, and lounges at both ends. There was a corridor running the whole length of the block with numbered rooms on either side. My room number was 15; other blocks separated by a large space, carried names of the Lake District.

There was a large dining room with a canteen and a shop that opened from 2:00 to 6:00pm during the week, and from 10:00am to 4:00pm on Saturdays. It was closed on Sundays. Two meals were supplied daily, breakfast and dinner. At weekends there were three meals on Saturday, and the equivalent on Sunday, except high tea was served in place of dinner at Sunday evening.

The excitement of my own room and the eating arrangements were wonderful compared with my previous existence. I felt that I never had it so good, using the phrase of the Tory Prime Minister of later years. My first day in the isotope department comprised paperwork and being shown round. A small group of raw recruits, two of whom was Peter and myself, were handed a copy of 'The Secrets Act' to sign, by an official-looking Higher Executive Officer (HEO). The HEO explained that if we were required to work on out-stations we could claim subsistence and he would pay any claims we might make. We were given pencils and notebooks and folders to keep all the paperwork. Mr. Jackson, the head of Isotopes, then addressed a number of the group and he explained that the other recruits were to work in different sections. He outlined what we would be doing for the next few weeks; that sounded like learning all about separation by gaseous diffusion. He introduced us to Mr. Dixon, would was to act as our mentor or teacher.

When there were no classes, we were expected to work at drawing boards laying out one cell of separation. Each cell comprised a membrane, centrifugal compressor, cooler, heat exchanger and pipe work, culminating in two open pipes, one of the impoverished cell and one of the enriched cell. Both led in different directions to the next adjacent identical cells, and so on and so on. The Smyth report had taught me much, but was no substitute for work in the department, initiating the system to separate uranium into its end product.

At this point, the British were way behind the USA, perhaps in parallel with Russia. The Americans dropped two atomic bombs with tragic results to end the Japanese war. The moral debate of these actions will probably go on for all time.

In the Roman Empire, when Caesar was confronted with the inventor of glass, he had the man put to death in order to stifle the innovation. For what purpose? It is a good illustration that destiny cannot be cheated and things will out in the end. Becqueral first discovered radioactivity and the atomic structure in 1896 and subsequently studied by Pierre and Marie Curie. Marie Curie lost her life for her early scientific curiosity.

However, the phenomenon of radioactivity has played leading roles in the discovery of the general laws of atomic structure and in the verification of the equivalence of mass and energy. Electrical power, derived from atomic energy, can do an immense amount of good for mankind. The biggest problem to be faced is what to do with atomic waste. The terrible hazard of radioactivity cannot be seen; it can only be measured and much of the waste will remain a serious hazard for hundreds of years to come. The worldly debate of military versus peaceful use of atomic energy is but a scratch on the surface at the moment and will go on for all eternity.

My early time in the department was a breath of spring; my difficulties were slight. I had to pawn my drawing instruments. Later, however, I recovered them. My pay was insufficient to cover what I paid for the hostel and my clothing needs, but being short of money in the excitement of the day was no

# Automotive gas turbines

hardship. Finally, my problem was resolved by my joining a team to operate at Springfield.

Jackson believed that the more we were exposed to experimental tasks, the quicker we would learn. So I was invited to travel to other establishments, including Sellafield, in the Lake District, and Springfield, near Preston. I eventually landed up in Springfield, although I did visit Sellafield a few times for the purpose of meetings on the pile. Some years later I gave a lecture on small gas turbines at the atomic plant at Sellafield. I made many good friends at Risley; they are too numerous to mention as I have lost touch. We were mostly young, keen and dedicated.

My time at Risley was most enjoyable. We were all treated with great courtesy and well respected. Most of the young recruits felt the same. "We were the cream of engineering", so Mr. Dixon emphasized and we all lapped it up.

I became friendly with the assistant manageress of the canteen; her name was Ida Borne-Jones. Her city of origin was Cardiff and she was just out of the 'Wrens'. She was tall and had an auburn complexion with a freckled face. When she was serving, which occasionally she did, I always felt pampered. She took an instant liking to me. We were allowed a guest visit at weekends, providing there was room at the hostel. I had invited Doreen Finch to stay as a guest some time previously. Doreen took up my invitation one weekend when Ida was on duty.

After Doreen's return to Hull, Ida was keen to find out all about my lady friend guest. It was an embarrassing moment, explaining to Ida that I dare not disillusion her. Ida became the manageress of the dining facility at Springfield atomic plant; whether she found out that I was moving and applied for the vacancy I never was the wiser. However, I put it down to destiny that our orbits were meant to stay that way. Of course, she was domiciled at Lytham Hall where our team and I resided.

It was fortunate I felt that I should break off my engagement with Doreen Finch as I had the most tortuous and

# Automotive gas turbines

horrible visit to her home in Hull. First, I had to take two bus trips and a train journey over two days to get from Risley in Lancashire to Hull in Yorkshire. On the way, I lost my bus ticket, which was for a return through to Pontefract where the train would take me on to Hull. I asked at the bus depot what could be done, having lost my ticket. After a long delay, a manager offered to ring through to Warrington to confirm whether they had issued a return ticket to Pontefract, despite the change of buses. Fortunately there was only one return ticket issued to Pontefract, so I was allowed to take the bus with a temporary return ticket. This caused a long delay, and as a result I missed the last train out of Pontefract to Hull. I had to hang around on Pontefract station all night until 6:30am and catch the milk train out of Pontefract.

When I finally arrived in Hull I rang Doreen, who seemed distant. Eventually I arrived at the 'Finch' residence, explaining what had happened. I received a cold response, to say the least; this persisted until I left. It was on this trip that I decided that Mrs. Finch and, to a lesser extent her husband who was a quiet little Yorkshire man, disliked the idea that I was suited to their daughter. So I came away from this awful weekend determined never to repeat it. It was soon after this that I terminated the engagement. Needless to say the return journey went well without any snags.

My friends and colleagues were numerous; we were all young and some of those joining were just out of the forces. Some had come from significant posts. One guy, named John Simpson, was from 'Bletchley Park', famous for the 'Enigma' story. He was an ex-Cambridge graduate. There was a good cross-section of talent, not all university-trained. Some of the ex-officer engineers, physicists and chemists went into the forces in place of going to university and were offered part-time study for a degree.

There happened to be an initiation custom when you went to Lytham Hall for the first time. An Irish lady, Miss Bentley, who had a gamy leg and walked with a walking stick, ran Lytham Hall. Miss Bentley was a good-tempered lady and kind to her lads and lassies, as she used to call them. But she ruled her staff with a rod of iron. I was placed in a large room with five others. Three were Scots and two others worked in my team with me.

I got on well with them all and we had many good trips to Lytham St. Annes and Blackpool, via Squires Gate. The initiation prank was a demonstration of the fire escape from the roof of Lytham Hall. It was positioned in such a way that a new member at the hall was caught in mid-air outside Dr. Wells' window. Dr. Wells was the oldest member of the Hall and he had a room to himself on the third floor. It seemed that he was always reading and he treasured his privacy. He was partly bald and, when agitated, had a habit of running his hand over his bald patch. The fire escape was a rope chair on a ratchet system that lowered you down at your own speed, provided you held a rope that disengaged the ratchet. The initiation process was to lower the poor unsuspecting victim to dangle in suspension outside Dr. Wells' window and, when Dr. Wells, as always rubbed his bald head, to lower the victim gently to the ground. This practice went on until Dr. Wells complained bitterly to Miss Bentley.

My days at the Hall were some of best I experienced while at working on atomic energy. Butter was still rationed and that was kept in small individual bowls with our names. These were replenished every week and we collected them as we filed through for our food. The food was good and we never went hungry. In our rooms we had a kind of comfortable camp bed and a wardrobe each. There was a large communal cupboard in which we kept our suitcases. The grounds were large and several of us used to practice golf, much to the annoyance of Dr. Wells who always complained that we were too close to his window.

There was uproar one night as ex-RAF drinkers returned; two were from my own room, namely Kerr and McDonald. They carried a marble statue from the rose garden; in total it weighed far more than their own weight. I think it was Diana-the-Huntress and they placed it into the bed of another Scot, named Bill Law, who was taking his girlfriend out for the night. The bed, containing the weight of Diana-the-Huntress, sagged to the floor.

In the dim light, the marble statue appeared as if some beautiful girl was asleep in the bed. It was a rule at the Hall that those coming in last could not turn the lights on, for fear of waking those sleeping. When Bill returned he was the worse for drink and we others were tittering with our heads covered, as if asleep. First, he tried to get into bed, not aware that Diana was a statue. There was an almighty commotion as he tipped the bed over and Diana-the-Huntress nearly went through the floorboards. There were several lounges downstairs; one was a writing room and on that particular night someone was writing in the room very late when Diana hit the floor above. Miss Bentley knocked on the door with her stick and said "Be quiet in there, kicking up such a terrible commotion, you will wake the whole place up." It was too much for Kerr and McDonald, who burst out laughing. Poor Bill had righted his bed and got in; he was asleep as soon as his head hit the pillow. I still have many photographs of my days at Lytham Hall.

There was a large billiard room, and many ladies' rooms. Ida shared a room with a little lady from the Ministry of Works, who operated on the site at Springfield. There were other ladies in the Hall. One middle-aged lady was called May; the young lads would tease her about her love life. She was consorting with another male, a Welshman named Jones, who was also staying at the Hall. May turned round on her tormentor's and said, "You imply I am on the shelf. Not a bit of it, I get Jones to take me off the shelf and dust me now and then, so put that in your pipe and smoke it." It was all good sport and brought a smile to life.

Springfield was a 'Uranium Refinement Plant' at the time. We were there to experiment with a small pilot separation system. It was accepted that we were misfits at Springfield, as the rest of the plant concentrated on the refinement of Uranium ore or pitch-blend; we knew we were on borrowed time. Indeed, although we did not know it at the time, there was great urgency to establish the feasibility of many elements of the system.

We were domiciled at Lytham Hall where a small number of civil servants also resided. Lytham Hall was requisitioned by one of the Ministries during World War 11. The splendour of Lytham hall, set in its own grounds, echoed the past glory of the Clifton family. Sir Cuthbert Clifton lived at Lytham Hall in the early fifteenth century. His son, Thomas Clifton, and two brothers were with Henry V at the Battle of Agincourt.

I recall the names of my young colleagues in the team seconded to Springfield. Walter Long was team leader; the other members were Bill Charnock, and a Scot, Bill Law. We were housed in building number 201.

The director of 'Isotopes' was called Mr. Jackson, He was a tall, slightly balding man with bushy eyebrows. He was a man of great bearing and stature; he was always dressed in an immaculate dark blue suit. Jackson would put the fear of God into us by calling at the top of his voice. He always used our surnames. Later, as the years rolled by and the pilot plant was operating, I recall the first verse of a poem in honour of the director of isotopes. It went like this:

A woman sat on a fluorine cell

Sucking a bottle of CFL

The cork flew out and she gave up hope

And she gave birth to an Isotope

And they called the bastard Jackson

They called the bastard Jackson"

CFL is one of the electrolytes used in the electrolysis process to make fluorine. It is slurry made up from hydrogen fluoride, known as HF, and a binder. It is highly toxic and dangerous. Fluorine is the top member of the halogen family; iodine, bromine, and chlorine being the other family elements. Fluorine and HF are among the most highly reactive substances known. Since Uranium is not a gas, some gaseous compounds of

# Automotive gas turbines

Uranium must be used for separation by gaseous diffusion. The two main isotopes of Uranium are 235, the lighter one with fission potential able to release great amounts of energy; and 238, the heavy one or throwaway, as we used to call it.

I must explain the simple basic principles of separation by gaseous diffusion. A mixture of two gases of different atomic weight can be partly separated by allowing some of it to diffuse through a porous barrier into an evacuated space. Because of the higher average speed, the atoms of the lighter gas diffuse through the barrier faster than the gas, which has passed through the barrier and is enriched in the lighter constituent. The residual gas is impoverished in the lighter constituent.

A typical gaseous diffusion plant consists of many stages, with the enriched gas passed one way and the impoverished gas the other way. We used to explain it simply by describing a football match. At the end of the match it is reasonable to suppose that the young and more mobile spectators would leave the match more quickly than the old and the infirm. However, if it is imagined that the faster ones are passed to another match and the slower ones to a different match and so on for a series of matches, a new situation arises.

At the conclusion of the matches, at one end there are more of the faster spectators, while at the other end there are more of the slower ones. The process gas, uranium hexafluoride, contains two main isotopes, U-235 and U-238, and only one isotope of fluorine. Actually, uranium has three isotopes; U-234 isotope is not deemed significant because of its minute amount; however it was soon demonstrated that U-235 was the isotope susceptible to fission by thermal neutrons.

A significant part of the enormous cost of atomic energy was the manufacture of concentrated fissionable material for use in the atomic bomb. Several hundred stages were estimated, simply to double the enrichment of U-235 from 0.7%, which is approximately the percentage of the U-235 isotope in raw Uranium. The scale of the problem could be appreciated knowing the percentage of the U-238 isotope, which is estimated at 99.3%. Finally the exact number of stages depends on how much unseparated U-235 could economically be allowed to go to waste, as well as other factors such as hold-up and start-up problems.

Returning to the sequence of events, the department decided to second me to out-stations, as we called it, to gain experience of the manufacture of fluorine. I was asked to report to a secret ICI plant in Runcorn called 'TAR'. I appeared there with a black dispatch case with a HM crown embossed. The chief chemist of the plant, Ray Jones, said, "The first thing is for you to obtain some digs, as I am told you will be here for two to three months." I was taken aback as I had not been informed of the duration of my 'out-stations' or even what was going on in the 'TAR' plant.

I found digs on the first day with the help of the chief chemist. Mrs. Ball, my new landlady, lived in a Victorian house with many rooms; it was not far from the 'TAR' plant. I was briefed not to let on what I was doing, or what was going on at Runcorn plant. Before I became a civil servant I was obliged to sign the official Secrets Act. Mrs. Ball, who was in her 60s, treated me as a long lost son, which she never had. It soon became obvious that she lived alone in this great house. One daughter visited her during weekends when she could get away from her final year at Manchester University.

Ray Jones showed me round the small, secret compound that he operated with another senior manager who I assumed to be his assistant. The process workers, who were limited to just a few, were not told of the true nature the operation.

It was suggested that the experimental manufacture of chlorine for commercial use was in operation and the secrecy was to avoid competitors obtaining a lead over ICI. A small laboratory was housed within the compound containing the usual jars of chemicals and test equipment. My attention was drawn to a large jar of acetone and slim glass burettes with glass taps at each end. These were graduated in 0 to 100 divisions over a 12-inch length. "We test the purity of the fluorine in these," Ray told me.

This was the first mention of fluorine. We then moved to the test cell inside of which an iron tank measuring about 5 feet long and 3 feet wide with the same depth. The thick iron lid appeared to be hoisted above the open tank by an overhead crane system. Many pipes of all sizes had been detached from the lid. There was evidence of electrical bus-bars attached to one side of the tank.

"This is a 1000 amp fluorine cell," explained Ray Jones. The far side of the cell contained a small inspection window of very thick glass; on the other side of this in a narrow corridor was a big electrical make-and-break switch. A sample pipe was disconnected from the hoisted lid passing through the wall into the narrow passageway. Another similar pipe, but lower down, was also disconnected.

When we moved into the narrow passageway, Ray explained that the other end of the sample pipe would be where we took a sample of fluorine to test its purity in the lab. Both sample pipes, the feed and return to the cell, had small steel taps from which a small length of rubber tubing extended.

Nick explained that the glass burettes were placed with the glass taps open into the rubber tubing. Then the steel taps on the sample pipe and its return were opened together. We were standing well back, because over 50% of samples went up in flames as the fluorine reacted with the rubber tubing. The sample pipe and its return taps had to be closed swiftly if the sample flashed. If the gathering of a sample of fluorine was successful, the inside of the rubber tubing was fluorinated with a protective coating. This always struck me as a dangerous 'hit and miss' operation.

This process of manufacturing fluorine was by electrolysis. A 1000-ampere electrical current was passed through a solution of electrolyte, resulting in the migration of ions to the electrodes. During electrolysis, the ions react at the electrodes, either by receiving or giving up electrons. In this case, hydrogen

# Automotive gas turbines

passed to the positive electrode, called the anode, and fluorine to the negative electrode, called the cathode. Fluorine reacts with hydrogen with some considerable force.

Therefore, a baffle plate was put in place before bolting down the lid, prior to operation. The electrolyte, in solid blocks at room temperature, was melted into slurry. The amount of gas produced is proportional to the current consumption. Each gas flows into a pipe from its electrode by means of a suction pump, which is frequently replaced. The fluorine pipe, made from the nickel-based material Inconel, was replaced every 10 hours of operation of the cell.

I enjoyed being at Runcorn, not only because of the work but the people there were so friendly and welcoming. I spent several weeks there making many friends, including Mrs. Ball's daughter, who came home at weekends when she could. There was an old transporter bridge across the River Mersey from Runcorn to Widnes. We took this at our peril, as it was a music hall joke.

My work there came to an abrupt end when I received an accidental whiff of fluorine. The cell had been in operation for many hours and all was well. A faint smell, almost like cabbage, became stronger so Ray Jones and I went to investigate. We were not allowed to enter the cell when it was in operation unless wearing a mask, to which an airline was attached. We opened the solid iron doors to the cell armed with a white absorbent paper, which we soaked in potassium iodide, a colourless liquid. If there were any slight traces of leakage the paper would flash brown on coming into contact with fluorine.

Detecting no trace of fluorine we decided to take off our masks. These were cumbersome, as the masks were tied at the waist with the air entering at the top. This was at a slightly higher in pressure than atmospheric and leaked outwards. We stood at the iron doors and looked into the cell with our masks off.

During less than a second we observed the sample pipe turning white hot for a three-inch length near to the lid. The pipe comprised quite thick tubing in Inconel material and the part, which was white hot, vanished altogether leaving the burnt out end hanging. The cell was immediately shut down. The explanation at the time was that one of our PVC gloves had a trace of grease.

As we were searching for leaks we must have touched the point where there was a small trace of fluorine. Such is the high-risk of a chemical reaction. It was reported at the time that the cows in a certain farm miles away in Cheshire had loose teeth. We continued until later that day when I became short of breath. I was seen by a doctor and spent a night in hospital; I was discharged the next day with loose teeth.

My gums bled for years later and I was told to carry a note. "To administer oxygen if found breathless." I never saw Ray Jones again as I was called back to Springfield, ironically to lecture process workers on the hazards of fluorine. Dealing with any highly toxic chemical was highlighted, in my case, by three phases. The first phase is one of shear fear and dread of handling such materials. The second phase is over-confidence that the procedure is child's play. The third, however, if one lives to the third phase, is treating the material with just the right amount of respect to avoid accidents.

My work on atomic energy was nearly at an end and the establishment sent me to ICI's plant at Kings Norton to study the manufacture of the membrane or porous barrier that is essential for gaseous diffusion. I was intrigued that the thin sheets of nickel were graded with a hydrogen probe. One gaseous diffusion cell consisted of a two-stage centrifugal compressor, a heat exchanger to reduce the compressor heat, and a membrane for diffusion suitably piped to pass U-235 in one direction after diffusion and U-238 the other way. All of these were working at a lower pressure than atmospheric so that nothing leaked out. However, all earth points to the exterior had nitrogen purges, so that the inert gas would go inwards.

## Five

# Rover and gas turbines

**In which I join the Rover Company and find myself in the world of gas turbines.**

AS I was to travel near to my place of origin, so my brother invited me to the Rover Company to see the secret work going on there on small gas turbines. I arrived with my Ministry of Supply dispatch case and discovered they spoke a different language; I was used to the metric system while the Rover people working in gas turbines at that time used the inch system.

I well remember one of the team, Charles Spencer 'Spen' King, who was much my age, called me a 'bloody scientist' because I referred to one atmosphere as 760mm. I took an instant disliking to King who later, however, became one of my best friends. I must have impressed them with my knowledge of high-speed bearings and seals as later they offered me a post as technical assistant.

The main separation plant was to be built at Capenhurst near Chester. The Atomic Energy Department offered me a post to become a plant manager at Capenhurst. I did consider this offer but turned it down due to my desire to remain within research and development. The department, knowing I was wavering, asked me if I would like to work on the new cyclotron being developed at Birmingham University. This was intended for research on another form of separation by alternating electric fields while following an outward spiral or circular path in a magnetic field.

Unfortunately, it had taken years to bring the cyclotron at Birmingham into operation, so I decided to take the post with the Rover Company, thinking it would be a walkover due to the lack of toxic and radioactive materials. All the components I had been used to, namely, compressors, heat exchangers operating in the laminar flow regime, high-speed bearings and diffusers involved the same theory, although they were different in detail and without the hazard of toxic and radioactive materials.

After the war, the feeling generated by the new Labour government was one of euphoria: anything could be done. Massive schemes were started, to name but two: the 'Ground Nut Scheme' and the 'Brabazon Aircraft Liner' with eight Rolls-Royce turboprop engines per aircraft. This rubbed off on to the British people. In 1950, Labour's massive majority was reduced to just nine and the following year the Conservatives won the general election. This brought Winston Churchill back as Prime Minister for the first time since World War 11.

The Korean War began in 1950 and ended in 1953. International trade for Britain then started to pick up. This revived the feeling that anything could be achieved, particularly amongst the young. One example was the enormous finance and scientific resource going into nuclear technology. Thus, in 1952, Britain detonated its first atomic bomb and three bomber aircraft were developed, namely the Victor, Valiant and the Vulcan, all were capable of carrying atomic bombs.

Meanwhile, two other children were born to my brothers' wives. One of these, my nephew Stephen, was born to Olga and Tom in 1947, and my niece Julie to Doreen and Roy in 1949. I had been working in the North of England with the Ministry of Supply at the time of these births.

I started work at the Rover Company in May 1952 with a department called 'Water Pumps'. The department's name was a 'cover' for gas turbine development. I was employed as a technical assistant, responsible for high-speed bearings and fuel system developments. This was an odd mixture. However, my background in hydraulics and the experiences I owed to atomic energy in high-speed bearings had helped me get the post at Rover.

## Automotive gas turbines

I reported to Frank 'Tinker' Bell who led gas turbine development; but I soon learned to my cost that the objective of replacing the piston engine with a gas turbine in a car would prove to be no walk-over. When I joined Rover, Frank was quite friendly towards me, despite Spencer King's attitude that was seen as arrogant and unfriendly. Spen had earned the nickname 'the Monarch'. Spencer King's behaviour in those days was decidedly bossy, but he seemed to be absent quite a lot. He appeared sometimes to run the department alongside Frank Bell.

Frank Bell's country of origin was New Zealand. He had come to Britain to work for Rolls-Royce where he met a young man, Spencer King, who was serving his apprenticeship. Frank was of average height and had thick lips and a large nose. I suspected that he suffered with some form of breathing trouble. I hardly ever saw him dressed in anything but an old brown tweedy shorts-jacket. During bad weather, he wore an old overcoat of the same material and an equally old trilby hat.

It was some time later that I learned that Spencer King was a nephew of the 'Wilks' Brothers. S. B. Wilks was chairmen when I joined Rover, and his brother, Maurice Wilks, was chief engineer but he later became managing director. Rover Company was run on paternal lines; everyone who joined eventually became part of a family. As such, you were referred to as the 'new boy', even years after joining. Rover buses operated between Coventry and the Rover works in Solihull.

It was when I returned to Coventry that I met Sybil, the sister of Geffrey Dunford, who married my niece, Beryl. Sybil Dunford lived apart from her family, having just left the Women's Auxiliary Air Force, or the WAAFS, as it was called.

Initially we found a home with my sister at Rudge Road until we were married at St Thomas's church, about a year before I joined Rover. After a short honeymoon at Dawlish, in Devon, we moved into Clarendon Street.

I often caught the bus from Earlsdon to Rover. Coventry was the city where the Rover Company first began developing its business from that of J. K. Starley. The company was then called the Rover Bicycle Co. Ltd. Starley first made early bicycles in Starley Road in the 1800s; Starley Road ran parallel and near to Rudge Road.

I was referred to as the new boy for as long as I used the bus. The Rover Company had worked with Frank Whittle during the war. The Ministry of Aircraft Production (MAP) had cause to switch gas turbine developments to Rolls-Royce. There was the typical basic difference of opinion between Frank Whittle and the 'Wilks Brothers'. The entrepreneur, Frank Whittle, naturally wanted to keep improving the design by 'rubbing-out-in-hardware' whereas Maurice and Spencer Wilks, who rightly were used to freezing the design for manufacture, disagreed with Whittle. When the gas turbine work was switched to Rolls-Royce, many of Rover's gas turbine engineers were transferred also. Rolls-Royce, in exchange, switched their work on the tank transporter engine, the Meteorite, to the Rover Company.

One of my earliest recollections was that of 'clocking-in' at one end of Rover and walking through the factory, passing the experimental department, the trim shop and the small track where the first Rover 75 cars were assembled, and then through the various roadways to the 'Water Pump' department.

The work on small gas turbines, when I joined, was confidential, even to the name of 'gas turbines'. Later this was rightly changed to 'Project Department'. The work of the department was varied. For example, being developed alongside the small gas turbines was an early torpedo engine, called the 'swash-plate-engine'. This worked on a combination of hydrogen peroxide and fuel. There was also the 'Neutron Chopper' for use on atomic energy research.

Jet 1, the world's first gas turbine car, was fitted with the T8 gas turbine engine. It was being developed when I joined. In fact, it was one of the reasons I was given the post because of my supposed knowledge of high-speed bearings. I soon learned that what I had known as high-speed was decidedly not so, as rotational speeds in small gas turbine in revolutions per minute were much higher.

The T8 engine's maximum speed was 40,000rev/min, whereas I had worked in the Atomic Energy Department with speeds up to 10.000rev/min.

I operated from the technical office that was separated from the drawing office by a room without windows, which we called the 'black hole'. At one end of the drawing office was 'Tinker' Bell's office; it was used also by Spen King. This was cordoned off by green partitioning with glass in the upper panels. Leading from the technical office, in the same brick building, was a space for rigs.

A Merlin 500bhp motor driving a Merlin supercharger compressor supplying air for turbine rigs, installed later, was housed in the same building. Three test beds were used for testing the T8 engines.

In a separate brick building, just opposite the test bed exhaust to the atmosphere, the fitting shop was located. It also housed areas for test and rigs.

Ron Hill and I became good friends. Ron was in the drawing office and lived in Coventry. I bought a 1932 Morris Minor from my brother and Ron agreed to ride with me to work and back on 'L' plates. I knew how to drive cars, as I used to drive an old Austin seven in the fields during my apprenticeship in wartime, but never on the roads. My test in Coventry was due to start near Greyfriars Green; it is no longer there. I remember, during the test, backing into a drive near Coventry station and the driveway sloped upwards. As I reversed, the clutch filled the car with fumes and I tried to pass it off by saying to the examiner, 'It's an old car'. To which the understanding examiner replied, 'None of us gets any younger'. The examiner gave me a full license first time and that weekend Sybil and I motored all the way to Lytham, staying at the Clifton Arms.

Also working in the drawing office, I recall, was Harry Knowles, the turbine designer who drew out large-scale turbine blade forms using numerous blade sections taken between the root of the blade to the tip. This was one way of laying out for manufacture. Johnny Garrett in the technical office carried out the design. Johnny was young and about my age. He had red hair and was shy, and he had a slight stutter. Fred Pickles, responsible for heat exchanger calculations, also worked in the technical office; he was aided by Phil Phillips. Fred's brother, Jack Pickles, who worked on the swash-plate engine for the Royal Navy together with a middle-aged designer, Mr. Oliver, operated from the drawing office. Oliver was an immensely good designer and used to working out his calculations alongside his designs.

I had great respect for Oliver and I used to walk with him from the top offices to the 'Water Pumps' gas turbine operations. Oliver was short and walked in small steps, almost like a penguin. He was quite rotund and I never remember him wearing anything other than an old, stained brown suit with a matching waistcoat. Like the other outstanding designers I ever met, he landed up in Hatton where Mr. Vernon the inventor of the detachable wheel for vehicles had finally died.

Charles Hudson was foreman over the testers and fitters; he bore responsibility for engine build and discipline of the weekly staff. He was quite a character; I warmed to Charles. He was one of the old foremen. All the men had total respect for him, as he was skilled in all the operations that he could expect from his men. He ruled with a rod of iron but was absolutely fair. He drove a Land Rover and used to give me a lift to Coventry when he and I had to work over. He had to pick up another foreman, Bernard Davis, from top experimental, as we used to call it, as Charles transported Bernard to and from Coventry.

I was given a section with fitters covering the topics of combustion, fuel systems, and high-speed bearings. We started with two rigs, one of which was a drive for pumps that formed part of the fuel system, together with flow meters and pipes. The other was a bearing rig driven by a small axial turbine from the air mains. It had a cathode ray tube for measuring high speeds using the waveforms appearing on the screen. Charles allocated me three fitters: a man named Arthur Griffith, who

# Automotive gas turbines

was in the army; and another, Jack Belfit, who was very bright and therefore argumentative. The other fitter, Larry Hicks, was somewhat brighter even than Arthur.

Unlike Jack, Larry would not argue with Charles. One day, as I passed Charles's office and could see in through the glass in the upper panels, Charles was wagging his finger at Jack Belfit, who looked angry. When they emerged from the office I put it to Charles, "What was the difficulty with Jack?"

Charles spat out, "I caught Belfit in the bog and I don't mind him washing his hands once, but I am not standing for washing his hands twice. He is an argumentative bastard, protesting that he is doing it is on the grounds of hygiene." From that day on, Charles had a down on poor Jack; it did not make my life easier.

Jet 1 (above and p.439) was little more than Rover P4 saloon with the top chopped off; it dated from the late 1940s or early 1950s. All Rover cars had a 'P' before the number; the P stood for 'prototype'. Jet 1 had a small gas turbine, the T8, installed in the rear. The car is now on show in the Science Museum in Kensington, London. It can be taken out and driven, once toffee papers are removed from the exhaust and intakes in the course preparing the car to run. The fuel system is very rudimentary. When the accelerator is pressed, a tap closes in the spill-line of the atomizer, allowing the pump to supply more fuel to the engine.

The tap is so set that idling is possible when the car is started by a normal starter motor using an electrical cut-out, once a certain speed is reached. Lucas tried to replace this system with an automatic control system but when I tried it out in Jet 1 with Les Goddard, the Lucas man, the car ran away. Fortunately I managed to switch off in time to save the car from crashing.

The T8 engine had a Ricardo combustion chamber of a very simple type. As I was also responsible for combustion, I visited Whitchurch, near Aylesbury, to discuss with the powers-that-be whether there was an alternative injection system. I came back with several 'Lubbock' sprayers without spill-lines. The Lubbock sprayer has a central spindle that closed off the sprayer when no flow and pressure is applied. The fuel pressure works against a piston, opposed by a spring, so as the fuel pressure increases so fuel flow to the engine is also increased.

About that time, discussions centered on a small gas turbine design to obtain experience of engines in service, albeit in industrial markets. The Ministry of Supply was willing to discuss a contract with Rover Company provided their representatives were involved in the requirement. A gentleman named Reggie Sholtell, who at the time was director general of the appropriate department, was involved with George Roach, the former chief engineer of Sunbeam Talbot. Roach could have been foreign. He was outspoken in relation to design.

Spen has often told a story about George when he was working at Sunbeam Talbot as chief engineer. Like most chief engineers, George wanted things done 'yesterday', so to speak. He had designed a new piston engine and, thinking that the foundry that was casting the cylinder block was dragging its feet, he rang them. They reported that they were just about to arrange the cores for the cylinders. George was in a rage. "I don't want to hear about the cores, I want the bloody blocks at once." So the foundry delivered solid blocks of cast iron next day.

There was some discussion in relation to a 60bhp gas turbine engine. I recall heated arguments with the Ministry men about which turbine rotor to use; whether this was to be an axial of Rover's design background, or a radial type, like those used in turbo-blowers of the day. The design was to be called a '1S60'. The engine had one main shaft and was nominally rated at 60bhp. It was designed eventually with an axial turbine rotor.

Another Rover saloon was designed to have a hollow chassis that could handle the exhaust gases from the engine that was installed under the front bonnet. The T8 engine was running in Jet 1 with its heat exchanger assemblies removed. These were mounted in a radial form around the main casing of the engine. With cross-flow heat exchanger assemblies the air leakage must have been too great and as a result the power was too low. Without heat exchanger, the exhaust temperature was simply too hot, so the second gas turbine car attempt became known as a 'White Elephant', one step in the development of gas turbine cars.

We never knew the cause of Frank Bell's decision to leave Rover, but we rather suspected that it was due to the 'White Elephant' design attempt to install a T8 in the front of the P4 saloon car. I was extremely sad to see Frank leave. I had total respect for him. Now the man who first suggested I join the team was going. On Tinker's demise, a photograph of the whole team was taken (see photograph, p.442). Frank Bell's epitaph might well be 'That he did excellent work in launching Jet 1, and previous small gas turbine developments, but ended his days at Rover on the cross of exhausting to atmosphere'.

Returning to the 1S60 gas turbine engine, it was conceived with a so-called 'baffle-combustion system'. This was the Lucas name for it. It was designed at Burnley in Lancashire and was a 'high-intensity-combustion-system'. It was too clever to work. Maybe if we had persevered with its development it might have improved, as the Lucas designer intended. His name was Harry Cheetham.

There was another man, whose name escapes me. This young team reported to J. S. Clarke. He was a bombastic man who became chief engineer of the entire Lucas group after Dr. Watson's retirement; but at that time he was head of Lucas Burnley. There were two doctor Clarkes at Burnley: one was J.S.C. and the other Bert Clarke, who was younger than J. S. Clarke.

The fuel system used on the new small gas turbine, first called 'Neptune' but was later dropped for the 1S60, comprised a mixture of the design of two companies. It was called the 'Lucas-Plessey fuel system'. It was designed around the spill burner.

I had developed a respect for Spencer King and I think he respected me. Spen travelled with me to the various companies and took immense initiatives. He was also supportive in all of my developments. Spen and I became good friends and had many ideas working together.

We were both disillusioned with the baffle combustion system and decided to design our own system. A new boss appeared in the form of Stewart Hambling. Hambling took the place of 'Tinker' Bell. He was quite clever and invariably was seen with a cigarette hanging from his lips and one eye closed. He was short with dark hair and would appeal to the ladies in a handsome way. He was authoritative but softly spoken, and we warmed in his presence. At first, he was like 'a breath of fresh air'. As time went by we came to recognise him as a likeable rogue, but he contributed much to the department with his enthusiasm and encouragement.

I was taking two evening classes at technical college to help with our financial situation. They paid £5 per class and that was a great help as we were newly married. Sybil's father had found new lodgings for us in Meriden Street as we had moved from Clarendon Street in Earlsdon where we started our married life.

One evening, I was taking class at technical college and Mrs. Hambling rang Sybil to say that she had heard from her husband that I was in an accident but was not seriously hurt. If she could do anything she would be pleased to do so. Sybil was worried out of her mind and took the message seriously. She was greatly relieved when I returned home unhurt. I said that 'Ham', (as we called him) was a rogue and was probably at Molly Barker's place and had given his trusting wife an excuse for not coming home. Molly Barker was one of our tracers and we all knew well that 'Ham' was having it off with Molly. When I entered work next day, I said nothing, as I thought they could sweat it out. But nothing was ever said.

Douglas Llewllyn said Ham's special vehicle, an experimental car of usual design, smelt of the six 'Ps': Passion, piddle, paint, perspiration, periods and petrol. Ham always used to carry extra fuel in the back of the car and which he stole from the stores. Ham was always short of money. Numerous members of the department were tapped for cash, from the lowliest to the highest. I was approached twice and was forced into saying I had no cash on me.

I learnt years later that Ham worked for Alvis and was regarded there as a 'con-man'. He was soon asked to leave. We all thought Spen had a lot to do with it, as his uncle was then managing director.

Before Ham left, he asked each technical assistant to write a paper on his main subject. He was to deliver it in lecture form to the director-general of the Ministry during an evening session that coincided with the visit.

I wrote a paper on my main subject: combustion. I had acquired a new recruit, Joe Poole, to work on fuel systems. He joined at the same time as a lad working with Johnny Garrett. Mark and Joe Poole were fresh out of University. Mr. Shaw, the Rover apprentice supervisor, had a tendency to select bright graduate apprentices. Shaw was a retired naval officer, who walked with a stick due to an old war wound. He spoke in a cultured voice consistent with giving commands. Mark was due to join my team, following his round of the technical office. The paper he prepared was on turbine design. The poor lad was obviously nervous and as a result I wanted to help him. He was asked whether the velocity triangles of turbine blade design were vector quantities and he had to get Johnny Garrett to give the answer. Despite the fact that his knowledge of chemistry was scant for a time I put Mark to work on combustion.

The development of the Plessey-Lucas fuel system proved troublesome. We had many sets built for the early 1S60 engines. I racked my brains and eventually decided that there were too many things that could go wrong. The fuel pumping part consisted of two Plessey fuel pumps with compensating end blocks to minimise end leakage. These were located in one casing. One fuel pump was called the circulating pump and the other the metering pump. In theory, the system should have worked but the accumulation of numerous detail faults caused the problem, amounting to a fundamental reason for it being abandoned. The spill burner set the scene with the flow of the two pumps being delivered to the spill atomiser. The theory was that the circulating pump flow would be about equal to that of the spill flow and only the metering pump flow would be delivered to the engine. But in practice it failed to work.

In 1952, Rover Gas Turbines Limited was founded – the same year that King Farouk of Egypt was deposed by the military. Eventually Abdel Nasser emerged as leader. Churchill was reluctant to vacate Suez as a sign of British weakness.

The newly formed Rover Gas Turbines Limited was proud of the 1S60 and we took part in the British Industries Fair at Castle Bromwich. About the time of the Festival of Britain in the early 1950s the first marine application of the T8 gas turbine took place on the Thames. Two T8 engines, each developing 200hp, were installed in M.Y. Torquill. The Admiralty bought several T8 gas turbine engines to operate at West Drayton on endurance tests at their experimental establishment, AEL.

Mr. Rigby was in charge of the establishment and often rang me up to seek advice in relation to their endurance testing.

# Automotive gas turbines

Charles Hudson tells a story of when Torquill was out on a river where Dr. J. S. Clarke, the head of Lucas Burnley was on holiday. A man in shorts started to wave from the bank of the river and Charles recognized him as J. S. Clarke. Charles said to Tony Poole, who was aboard as a fitter, "Look at that silly so-and-so waving his heart out". Charles pretended not to recognize him and gave him the two fingers with conviction. Dr. Clarke rang up M. C. Wilks and complained. M.C. joked with Charles that he must show more respect to the 'head boy' of Burnley.

Spen and I visited Burnley some time afterwards in relation to the baffle combustion system which, on acceleration, exhibited a phenomenon of rumbling noises. I had previously examined several running engines and associated this with a rotating flame front. On the return trip, we both came to the same conclusion that we should design our own fuel and combustion systems. We went miles out of our way discussing ideas. We brainstormed an invention for the fuel system at least. The combustion issue was another matter that would take some time to resolve.

I was impressed with Solar's (of the USA) combustion design for their small engine; it was similar to our own. I discussed this with a man from Solar and he reported that it worked well. So I designed our first Rover combustion system based on a two-dimensional flow rig that was transparent to view the flow pattern. First sightings indicated that the vortex forming mechanism must be strengthened.

I increased the volume of the primary zone design, making it roughly spherical with eight jets to form the vortex. I argued with myself that as long as I covered all surfaces with air there would be no carbon-forming process, due to cracking of the fuel without oxygen. Combustion is rapid, high-temperature oxidation, which means the addition of oxygen and the removal of hydrogen in a chain reaction. Although there are rules that cover the theory of combustion, it is a complex subject, involving chemistry, fluid mechanics, physics and thermodynamics.

The special problem facing the gas turbine designer is how to slow the reaction down to the limiting speed of complete combustion of the fuel. All hydrocarbon fuels impose a limiting flame speed of generally a few feet per second, so that a toroidal vortex or smoke ring is vital in order to slow the speed to that at which the air mixes with the fuel. This is a very simplistic outline of gas turbine combustion. In the absence of oxygen fuel will form carbon.

So this is a basic difficulty in the development of any system: how to get rid of carbon. There is a loss associated with combustion as the air passes through holes in the primary, secondary and, of course, the third zone called the tertiary or dilution zone. Probably the biggest loss is associated with forming the toroid or Taurus vortex. There is a cold loss and a burning loss; in general, these add up to 4-6% of the compressor outlet pressure.

It used to be said that combustion is a 'black art' with 99% covered by design over the board, and the remainder spent in years of development to perfect and eliminate the carbon. Only about a third of a gas turbine's air is utilized for complete combustion, and approximately two thirds of the air from the compressor is used to reduce the temperature of the gas from the primary zone to a level acceptable to the turbine's materials.

The first run of my design of combustion system was completed in mild steel, as engine running was short-time testing, simply as a trial only of its function. The engine test was so encouraging that I had three flame tubes made in Nimonic 75 and a stainless steel combustion outer casing. I devised an adjustable convex ring round the atomizer so that it could be moved in and out of the flame tube. From this it was possible to gauge the setting that was clean of carbon. The combustion for the 1S60 was working at last. It was a revelation what a novice in combustion design could achieve.

I had been given the task of designing combustion systems as I was the one in the gas turbine department who knew chemistry. Together with the new fuel system, avoiding the spill burner and the Lucas-Plessey fuel system, we were well on the way to success with the 1S60. However, years of testing and development would lie ahead.

Bearing tests continued with a form of plain bearing called the 'thermal wedge', invented by Dr. Bertie Fogg who was at that time head of MIRA, the Motor Industry Research Association. I was not impressed with the results.

I was surprised, based on tests we performed, that the heat of friction generated in relation to speed increased so rapidly. The T8 locating bearing fully loaded at maximum speed absorbed over two horsepower and the equivalent heat had to be removed by the oil.

The quality of bearings, particularly rolling element bearings, left something to be desired during and after the war. An ex-Rolls-Royce bearing expert by the name of Alf Towle, emphasized this in a technical article in 1945.

His graphic illustration was that of a nail driven through a steel ball from a rolling element bearing of the time. I contacted Mr. Towle and arranged a meeting at Derby.

From the meeting we were able to devise a list of general requirements compatible with his and my tests on high-speed bearings. The general requirements are in relation to deep groove ball and roller bearings at the time. These are as follows:

1) Tracks, steel balls and rollers need to be designed for a given life and material selection made accordingly.

2) Top quality is essential for gas turbine engine bearing systems with respect to hardness, finish and selected clearance.

3) Flexibility in mounting is a vital feature; too rigid a location must be avoided to prevent the risk of failure.

4) Adequate amount of oil must be used to maintain the temperature of the bearing at an acceptable level. Only a small quantity of oil is necessary for lubrication.

5) The oil outlet must be free of any restriction and a means of extracting the heat of the oil is essential.

6) The oil selection is vital; generally, oil must have excellent load-bearing and heat-carrying capacity.

These are some of the general pointers established from my meeting with Alf Towle, based on the T8 and 1S60 developments of that time.

The Conservative government found the aftermath of Labour's six years in power a daunting task to reverse. The wartime legacy in relation to the economic position of the country set the scene for any British government for many years to come. The new Labour government of 1945, however, had obligations that were defined in the party's election manifesto. Promises to expand the welfare state and nationalize certain key industries, as well as maintain full employment, imposed crippling economic strain on the government's purse strings.

The first law of economics, as I understand it, is to find people productive work. If Britain was unable to produce enough goods and services that its people consumed as a nation it was living beyond its means. Yet the excess goods and services had to be paid for somehow, either in dollars or Gold which would weaken the pound. Demand beyond the capacity of the economy to meet it would result in inflation or a balance of payment deficit. Careful management according to 'John Maynard Keynes' of all the economic factors involved would enable the Conservative government to juggle and just about get out of the mess left by the former government – in theory.

In practice, however, I doubted whether the government could juggle all the factors at once, as it could not predict the future. In any case, all the factors were anything but revealed at once; sometimes, for example, tax returns might remain not collectable for years to come, and so on.

Politics were also not far from my work with the Rover Company. I needed a rest from my concerns and worry over a fuel system that was, in my view, basically unsound, namely the Lucas-Plessey system that was dictated by the spill atomizer. Fuel systems were the only device I knew that could have a build-up of detailed faulty aspects amounting to a fundamental error.

So I decided that I must get a break. Sybil and I decided to visit Dawlish, staying for one week at the same place where we stayed for our honeymoon. There is a snap shot somewhere of Sybil and I on the lawn of the hotel during this second visit. I returned refreshed from the one-week break away from my fuel system difficulties, only to learn that Hambling was due to leave and start a small gas turbine establishment at Standard-Triumph in Coventry. It crossed my mind that he would try to poach staff from Rover and indeed one or two people from the Rover team did join him.

A letter arrived from Peter Wilks, who was the first production manager of the newly formed Rover Gas Turbines Ltd. I was surprised that Peter, who was a nephew of the Wilks Brothers and therefore related to Spencer King, signed it. Everyone liked Peter, as he was friendly with all. He was almost six feet in height with dark hair and quite handsome.

The letter explained simply that, as a key member of the Rover gas turbine department, I had been placed on the monthly staff at a new salary of £52 per month. It was a most handsome raise from my weekly wage. Furthermore, the Rover management would be glad to locate a house for me at Coventry or nearer the place of work.

I did not need assistance to find a new abode as Sybil and I found a new house being built in Cambridge Avenue, Solihull. I had for some time been negotiating with Pearl Assurance, through a manager named Mr. Sharp from the Hertford Street branch in Coventry, to purchase a new house through an insurance system whereby there is a fixed interest of 4½ percent throughout the twenty-five year term. The idea is one of an endowment with tax relief, so the sum advanced is covered with an insurance policy for the same amount. Both the insurance policy and the endowment agreement or mortgage attracted tax relief. However, in different Chancellors' budget announcements, taxes were subject to change. I complemented myself on a good deal and when we were due to move to Alderbrook Road in 1967 the tax had remained unaltered.

Drabble Brothers, who lived in the next road, were building all the properties in Cambridge Avenue, as well as other roads off the avenue. There were two Drabble Brothers, one was short and of normal size and the other was taller, rotund and inclined to be fat. The arrangements took from January to August when the house was complete. We moved in late August. The first task was to furnish the house, as we had little furniture having lived in furnished rooms. Mr. and Mrs. Bates, who owned the rooms we occupied, were sad to see us depart. They had two daughters and Mrs. Bates often said they were, "happy as pigs in muck".

Fred Bates had a motorcycle and sidecar. How on earth Mrs. Bates ever rode in it was a mystery, as she was a big woman. She was a good-natured soul and Sybil and I had many laughs with her, and in private. We joked when we left in the morning, as she cleaned the step, which she always seemed to be doing. She was on her knees and it was not difficult to escape seeing her stout legs adorned with stockings. It looked as if the stockings were tied just above the knees, with the ends hanging down. Her enormous fat legs were bare from the top down to where the stockings seem to be tied.

She was seen often in her dressing gown, shuffling from the kitchen to the front room where she and Fred slept. We often joked that it must have been a miracle of dismantling before Mrs. Bates retired. Not that she had a detachable wooden leg or a retractable glass eye, but with her teeth out she gave the appearance in her dressing gown that Fred must have long given up.

The Bates' daughters were named Ruth and Jill. Ruth, a nurse, was to be married to Ron Lord, who worked locally as a metallurgist. We enjoyed their company and went to see them in their new home near Redhill, when they moved from Coventry. We were going on holiday together that year to Lymington, but we changed our minds when we saw the place that we had booked. We moved instead to Swanage. We had a jolly time and visited the local theatre to see an actress in a musical attempting to play Doris Day in a well-known Western with Howard Keel. We were in 'the Gods' and the actress's attempt was pitiful. We could not stop laughing. I think we laughed through the whole performance.

Fred Bates had a hard life, but I do not recall his occupation or whether he had retired. They were the 'salt of the earth' and we enjoyed staying there and were indeed sorry to leave. Sybil worked at Alvis as a secretary to the sales director during our time at Meriden Street.

February of that year saw Harry Cox join the team at Rover. Harry was in one of my classes at the technical college. I taught him about gas turbine engines on our journey to and from Rover, whenever I gave him a lift in my Morris Minor. I garaged the car in a lock-up in Minister Road, opposite Meriden Street and across the Holyhead Road from where I used to pick and drop him. He had been assigned to the technical office as a development engineer, but his first task was to scale up parts of the 1S60 engine for a new engine, named 'Aurora', destined for the third gas turbine car to be called T3 (illustration, p.439).

I well recall Spencer King rounding on Harry because the radii of Harry's 'scale-up' design appeared out of proportion. A fierce argument ensued, with Spen shouting at the top of his voice, "Imagine a small giraffe and a big giraffe, you would bloody well get the curvature of their hooves right and in proportion." To which Harry replied, "I used your scaling factor and I am confident that the radii are right despite your small and big giraffes." Spen walked away from Harry's drawing board arrogantly, believing he had made his point.

That night, when I gave Harry a lift, he said that Spen had been round again and had apologized. He must have checked that Harry was correct.

Harry Cox had served his apprenticeship at Alvis and also knew Hambling when he worked there. Harry's view was that Hambling was an excellent engineer but would borrow money from anyone foolish enough to agree. He got to know Hambling quite well and he said, "'Ham' always settled his debts eventually".

At the time, Alvis also employed Alec Issigonis before he joined the Nuffield Organization. Hambling and Alec Issigonis were working to place Alvis into prominence in the car business with a new design. Alvis's management decided to shelve the new car design when Issigonis left to join Nuffield in Oxford.

Fred Hulse, who had been at Rover for some time, began to work on development. He was ex-Armstrong Siddeley.

Rover Gas Turbines Ltd was beginning to take shape and now had a managing director, Mr. G. Searle. Peter Wilks was production director, and George Cowan was head of sales. Derek Aslin, was head of service, and others joined too. Also a new building, called the Coseley Building, was in the course of erection. Later, Leyland Gas Turbines would occupy it.

Britain had adopted what was known as a 'stop-go' economy, mainly as a result of the balancing act that governments from 1950 onwards played as they sought to protect the pound. The enormous military burden overseas at the same time as maintaining employment caused the Labour Chancellor of the Exchequer, Hugh Gaitskell, in 1950 to take-up a 'stop-go' policy in the first place. He invented the policy when seeking ways to pay for the Korean War.

Britain was already in full employment and the result of the wartime demand was to divert production from civilian to military goods; that, in turn, proved to be a drain on exports. Britain's defence spending as a percentage of the gross domestic product or 'GDP' increased from 5.8% in 1950 to 8.7% in 1952. The gross domestic product is the value of goods and services produced by the nation (GNP) less the amount invested overseas; the remainder is the product left within the nation. Such was the strain of fighting a war overseas.

At Rover, we had a successful trial of the new three-piston pump with the leaf spring governor. We were able to adopt a 'Simplex Danfoss atomizer' and an accumulator for starting. I had to invent a bleed valve attached to the atomizer to get rid of the air from the system. The trial was performed with the new combustion chamber I had designed to replace the Lucas 'high-intensity' design. We nicknamed the Rover combustion system 'Penny's Pepper Pot Can'.

The 1S60 was conceived with a bearing system based on the salient points from the meeting that I had previously with Alf Towel. We reached 10,000 hours of endurance testing on all development engines. I had previously established what became known as the 'ten times rule'. The rule implied that adequate endurance testing must be accomplished before releasing an engine into service; it was based on the life declaration. In order to safeguard taking engines out of service for faults or defects, a 'rule-of-endurance' is vital because in the early 1950s small gas turbine engines could not be designed over-the-board. My rule stated that a development must have at least three engines. It went on to state that at least 10 times the declared life endurance hours testing must be completed. Within those tests at least three engines must perform faultlessly, completing the declared life according to the duty cycle of the engine.

All of this implied that we still had to endurance run the 1S60 for a further 90,000 hours, carrying out three type-tests faultlessly, according to the duty and life in service.

It was soon obvious that sales of the 1S60 engine needed design for installation in order to make it suitable for the many different markets. The basic engine development was only one main aspect; the other was the prime purpose for which the engine was to be used. For example, a prime mover could be used for a multiplicity of applications. One of these was for fire-fighting equipment, usually in conjunction with an engine-driven water pump. An instructional set for colleges could be installed with a device for power measurement and suitable instrumentation. Fixed or mobile sets for emergency electrical power supply needed an electrical generator, and so on.

The 1S60 engine was no challenge for the equivalent piston engine in relation to fuel consumption, but it had many other advantages. The 1S60 had a reasonable power to weight ratio. It could be easily started in cold weather without any assistance. It was a continuous system and once started needed no further ignition and it had exceptionally low oil consumption. These were all the advantages that one would associate with gas turbine engines or jet engine of the day over the heavier piston engine. It was more expensive than engines produced in volume. The most natural use was as an auxiliary power unit fitted an aircraft, but careful installation was vital. One Rolls-Royce man in one of my lectures commented "Installing an auxiliary small gas turbine engine into an aircraft was like getting a grand piano into the attic."

A new department in Rover Gas Turbine Limited was created; it was known as Installation. Douglas Llewllyn had performed excellent work in production on the 1S60 and was in the running for the head of the new installation department. Instead, the management of Rover Gas Turbines decided to recruit for the post of head of installation, and another Armstrong Siddeley ex-engineer won the post; he was Len Harvey. Len set about staffing his department and the first man he recruited was Fred Jones, who became the second-in-command. Then there was a tracer 'Big Pam', as she was known; these were just a few of the first members of the team.

The theory of the small gas turbine's expansion into fields other than cars was justified by gaining experience of the technology in all respects. Whether this pursuit was viable or not I doubt, even if it was really quantified totally. But my estimation was decidedly profitable, considering that this resulted in Ministry contracts with the gas turbine department of Rover Company. In later years the chairman invited me to the Rover Annual General Meeting to answer any awkward questions on gas turbine development expenditure.

The heat exchanger was a vital component for the gas turbine car and in 1953 that was added to my responsibilities. I thought about the heat exchanger cycle and I came to the conclusion that there were two main issues: thermal efficiency or fuel consumption, and cost. I formulated that for a gas turbine car to sell it had to be at least as good as its rivals. It could sell as a novelty in limited quantities but M. C. Wilks was aiming for a volume quantity production that was at least equal to the ranges of other Rover cars.

I should explain the reason why a heat exchanger is so vital, yet can be troublesome. The heat of the exhaust that would normally be thrown away into the atmosphere is used to heat the compressed air entering the combustion chamber; in this way less fuel is required to be burned to heat the gases. This is but a simple explanation, but in practice there are many problems to be overcome. There are two types of heat exchanger: one is fixed, like a radiator, but using gases instead of water and called a recuperator; the other rotates and is called a regenerator.

The early work at Rover concentrated on the static recuperator type but, when a new ceramic material was introduced by Corning Glass Company in the US, Rover changed to the development of the regenerator. First, Rover developed the secondary surface contra-flow type but changed due to the cost of brazing and the extra weight. Then the primary surface contra-flow was adopted; that could be argon arc welded, a process of less cost and better leak-tight joints.

My lectures were enhanced by the invention of a five-sided figure, based on a series of graphs done by Phil Phillips. Along the 'Y' axis was thermal efficiency, or the inverse of fuel consumption; along the 'X' axis was pressure ratio with each curve representing thermal ratio or the heat exchanger effectiveness. I pointed out that our target was the centre of the five-sided figure.

The Ministry of Supply sought an Auxiliary Power Unit (APU) for the Vulcan bomber, which at the time was being developed at A. V. Roe in Manchester. Preliminary discussions were in hand. The Vulcan bomber's electrical system was of the AC type and the four Bristol Siddeley (later Rolls-Royce) Olympus jet engines supplied the power for the aircraft. The flying control power was supplied through engine-driven

## Automotive gas turbines

electrical generators. So, in the event of an all-engine failure, which was unthinkable, the aircraft had no flying control capability. The aircraft was designed with a 'ram air turbine' or 'RAT', but this depended on forward speed and, in any case, was hardly powerful enough and it was dependent on the attitude of the Vulcan. This explains why an APU was desirable.

If the 1S60 could be installed in the wing it could generate 30kW of power at altitudes of up to 35,000ft, and then 15kW up to 60,000ft, depending on 'ram'. It would be an immense undertaking to develop a ground-level engine to start and function at altitude, let alone 60,000ft, to say nothing of the installation in the aircraft.

When the Ministry's secret cardinal points specification was issued, we just laughed. One main requirement was that the engine could be flown at 60,000ft for several hours and, in the event of an emergency, the engine was expected to be generating electricity within 5s.

Thus started an immense challenge that became not only an outstanding engineering achievement, but its development was spread over many years. The main responsibility of the 1S60 for the Vulcan lay with the new installation department but that of the basic engine was on my shoulders.

Meanwhile, the end of rationing was in sight, but was prolonged to 1954. Events however in the Persian Gulf had damaged Britain's supply of oil in the wake of the nationalisation of the Anglo-Iranian Oil Company. Following the advent of the Eisenhower administration in 1953, the Americans agreed with the British view and used the Central Intelligence Agency (CIA) to restore the Shah. The foreign secretary, Sir Anthony Eden, brought Iran into the Baghdad pact.

I was introduced to a man named Geoffrey Middleton and Spen said he would be working on ignition and other electrical systems. He had been responsible for the electrical system for the Marauder car, makers of which had a small outfit at Kenilworth. I soon learned that there were four people involved, Peter Wilks, Spencer King, a man named George Mackie and Geoffrey Middleton. The small company had been dependent on Rover to manufacture a handful of cars that were sold and today could be at a high premium. The firm was closed down for reasons that were not known and the members were to be found work within Rover. George Mackie was to take over special Land Rover designs and excelled in turning out various vehicles applied to special duties, such as fire-fighting; there was even one fitted with a breathing device to move under water. It became clear where Spencer, and to some extent Peter, had spent time when absent from the department.

At first I did not warm to Geoffrey, who had 'a-chip-on-his-shoulder' that I believe dated back to his service in the RAF during the war. In time I did became friendly, as we had to work closely together. In fact, I went home with him once and met his wife who kept a hairdressing saloon at the Gosford Street end of Gulson Road.

Following the success of the combustion system for the 1S60 I was approached by Spen to address the 'Aurora' engine for T3, in relation to the combustion and fuel system. The 'Aurora' engine was later designated the 2S100 (see p443), in line with the 1S60 first called the 'Neptune'. The 2S100 engine had a contra-flow, secondary-surface stainless steel recuperator. The air mass flow was higher than for the 1S60 engine and as a result the flame tube design I had formulated was bigger than the 'pepper-pot' design but similar in concept. At this time, we had made great strides with the 1S60 Rover fuel system, the development of which was based on Lucas-Plessey design but made suitable for the requirement of a gas turbine car.

Charles Hudson was quite ill and we were worried about him. He seemed cheerful enough and resigned to his fate. I have a great respect for him and I felt a good friend was slipping away. Charles has been diagnosed with cancer and although treatment could begin imminently he commented that he had been smoking all his life and would not stop now.

Meanwhile, the situation in the country looked grave with another recession looming and there was trouble brewing in the Middle East, especially Egypt.

The time came for Sybil and me to move into our new, detached house in Cambridge Avenue with its long, uncultivated garden. The neighbours, on one side in a semi-detached house, were Mr. and Mrs. Winch. They were an older couple with no children. On the other side, in a detached house, were the 'Grays'. By a strange coincidence, one neighbour was Millicent Gray and the other was Millicent Winch. We were all in the same situation; many couples were of our age but without children. The blind were leading the blind in making homes livable.

If one couple rotovated the garden, then generally most did. And if one couple varnished the floorboards, then we all did, with the exception of those with hard wood floors. The craze was to have a concrete solid fuel store outside on a concrete plinth with a sliding opening at the front. The sliding opening of the store I installed stopped working, due to the weight of the fuel. So we had to fill the coal hod by lifting the top.

I recall that I installed a hard wood surround in our living room, letting the carpet fill the centre, where the hard wood finished. Another craze of the time was to have 'Baxi' fireplaces, in which the air was drawn from under the floor; this had a deep ash box that could be taken out and emptied. Those were good days. We were young and ready to raise families. Furniture downstairs was sparse, though one room upstairs was equipped handsomely with a Wrighton bedroom suite bought in Coventry.

Slowly, we came to know the people in the street. There was Allan and Pamela Cross, and across the road lived the Tichiners. Oddly, both their first names were also Allan and Pamela. The Hood's lived nearly opposite to us and Sybil and I referred to their house as the Rhubarb House, because it was painted in red, yellow and green. Living next to the Crosses was a Mrs. Bishop; she was a little and rather rotund lady. Across the road, in a detached house, were the Beresfords and, just below them were the Warners. At the top of the road, at the corner house with Heathcote Avenue, Tom and Janet Lindsay moved in after the people who first bought the house had moved away

When we moved to Cambridge Avenue we had a Morris Minor Series Two. We bought the Morris Minor second-hand from Parkside Garage in Coventry, using as down-payment money left to Sybil from her great aunt Ginny.

We had visited her once in Holywell, and stayed with the Cooks who had responsibility for taking care of Sybil's great aunt for a price. On this occasion I crossed a wide yard to the outside lavatory and a little girl followed me. She was only a small tot. I entered one of a row of lavatories built against a wall and was just seated when a long stick was thrust through the gap of the ill-fitting slatted door. It was resting on the floor when a rapid banging movement from side to side started up. As the stick-gathered speed I was forced to lift my legs to avoid the stick, I was astonished that the tot had the strength.

A voice said, "I know you're in there." I had to keep lifting my legs in rhythm to the movement of the stick in order to prevent my shoes and ankles from being slashed. I eventually gave up and replied, "Please go away. I will give you some sweets if you do." There was a slight pause and I heard a shout that I assumed was the tot's mother calling her. When I opened the wooden slatted door the little girl had moved away, leaving her stick shoved through the gap at the bottom of the door.

When the Morris Minor started to burn oil, I ventured to take off the head and 'de-coke' the top of the pistons, which was the practice in those days. Afterwards, I gave the car a run and it seemed that it was burning as much oil as before I did the de-carbonisation. We eventually sold the car to my niece Beryl, ex-wife of Sybil's brother. For some years afterwards we suffered the wrath of Sybil's brother's wife who dithered over whether she should buy the car or not.

There was trouble brewing in the Middle East and in April 1955 Anthony Eden became Prime Minister. Eden became convinced that Nasser was playing the Soviets and Americans off against one another. Nasser, he thought, was pro-Soviet. Eden became obsessed and nervy about the Middle East issue. The crisis came to a head in July 1956 when Nasser reacted to the American withdrawal of aid for Egypt's Aswan Dam project. Nasser nationalised the Suez Canal to secure its revenues.

The Frenchman who built the canal founded a company that was registered in Paris, but in law both the company and the canal were Egyptian. The company held a lease from the rulers of Egypt that expired in November 1968. Both the British and French Governments owned most of the shares in the company. Eisenhower was hostile to Nasser but thought the canal was not the issue to take him on. The French proposed that Israel should invade Egypt, enabling Britain and France to intervene and protect the canal. When the campaign was launched on 29 October it turned into a disaster. Britain was in violation of its obligations under the United Nations Charter, as a result of consorting with the French to attack Egypt. America and Russia opposed this humiliating disaster. The Suez Crisis stimulated immense debates at the time, the biggest hardship of which being fuel rationing, but it was soon over.

Eden never recovered from the setback, even though he was seen long before as doing excellent work with the old League of Nations, not as a warmonger but a man of peace. The general feeling about Eden was one of sadness, as most people in Britain thought at the time that for far too long he had been a good number two to a strong man, namely Churchill.

# Six
# The T3 gas turbine car

**In which Rover's T3 gas turbine car is unveiled to the public and this leads to other things.**

THE success achieved with T3, shown at London Motor Show in October 1956, was quite amazing. It was announced to the public as offering fuel consumption better than 14 miles per gallon at 60 miles per hour (photograph, p.440).

The Rover T3 car (photograph, p.441) was such that it enabled all advantages of a turbine car to be designed into one vehicle. The low-mass glass-fibre bodied car with the absence of radiator, cooling equipment, clutch and traditional transmission had disc brakes all round and a De-Dion rear suspension. The disc brakes were in-board to reduce unsprung weight. This was not a legal feature at the time.

T3 had a rear-mounted engine coupled directly to the rear wheels (illustration p.441). As a four-wheel drive car also, there was a power take-off connected to a prop shaft providing power to the front end. It was possible to take advantage of the rear-mounted engine by having a low bonnet line; this, together with a good windscreen and rear window provided safe visibility. The car was able to take a hairpin bend and stay glued to the road at high speed. The engine was separated by a firewall from the passenger compartment. It was my favourite gas turbine car and provided an immense amount of data during route analysis.

One of the difficulties that came to light during the many miles covered on the road was the secondary surface heat exchanger that exhibited a rise in gas-side pressure drop – an indication that it was becoming blocked with a fine carbon deposit. When we dismantled the engine, the compressor and diffuser were clean, although the car had been operated through

## Automotive gas turbines

foggy atmospheres. Later this was attributed to the linings in the rear intakes that were integral with the glass-fibre body. A soft, polyvinyl-foam, about two inches thick, lined the intake and the rear wings acting as 'momentum-separator', trapping the dirt and filth in the foam. The foam of course could be taken out and washed. The heat exchanger was opened up and cleaned in 'Ardrox' to remove the deposit on the gas side.

The inter-rotor shrouding, joining the outer shroud of the compressor turbine and the power turbine, showed signs of distortion. I suggested to Spen that the shroud should be separated into two and designed to mitigate leakage with a sliding joint. He adopted this suggestion and it overcame the problem of distorting the turbine outer shrouds.

For a time, the car was operated without a heat exchanger, increasing the jet noise. It was interesting how adults were oblivious to the noise of the jet yet small boys would recognize it immediately and seek out what was different about it. We used to think it was due to the fact that the younger people have a greater audible range, coupled with the fact that they are far more observant.

Once, when we drove T3 through Oxford, a don on a sit-up-and-beg-bicycle with a basket in front on the handlebars, leaned on the rear of T3 at the traffic lights and was completely oblivious that the car was anything unusual. Only when we pulled away and accelerated did he turn his head with the increasing jet noise.

The T3 was transported to the Motor Industry Research Association (MIRA) facilities, near Nuneaton, to be driven over the 'Pave'. Styled on Belgium blocks, the pave could wreck any suspension system. We used a transporter so as not to draw attention to the car before it had been announced to the public.

My team's hand in the development of T3 was quite extensive. The combustion was satisfactory, apart from the micro-soot deposition building up on the gas side of the heat exchanger. This triggered a unique development in which we were able to grade combustion with an instrument that measured soot sucked through glass paper. This was related to the rate at which a certain matrix would increase its blockage over time. Also, we made a decision to move to a primary surface heat exchanger. The fuel system, which was our own design, functioned perfectly. I had suggested a major modification that must have improved the 2S100 engine.

In parallel, my attention was drawn to how I could meet the Vulcan Ministry requirement in the development of the 1S60 combustion system. I had already visited Lucas Burnley's high-altitude plant with Warwick Bloor to discuss altitude tests on the 1S60 up to 60,000ft. We had to book a slot in the use of the chamber for our first high-altitude test. The first test at 60,000ft proved a disaster. The engine failed to accelerate. The engine performed only when we reduced the altitude to a much lower level. We deduced, when the engine was dismantled following the attempt in the altitude chamber, that burning was taking place in the exhaust. The rest of the engine was in a very bad state. One crude fitter put his finger on it, "The engine looks quite shagged".

Lucas was very proud of its high-altitude plant and had every right to be so. It was based on the injector principle that required large quantities of pressurised air. Two large piston compressors, which came from Germany after the war, supplied the air. The Lucas people boasted that the compressors belonged to Herman Goring, as he was head of the German Luftwaffe. We took it that they were referring to the German air force generally. A great deal of energy was required to cool the chamber to the right condition equivalent to high altitude.

I can never forget that we were working to 'ICAN' conditions. This was a standard of tables defining atmospheric pressure and temperature corresponding to altitudes. When the tropopause is reached, about 35,000ft, air temperature remains constant from thereon up to 60,000ft and somewhat beyond. The tropopause is the interface between the troposphere and the stratosphere. I can remember that the temperature was -40, which was convenient as Centigrade and Fahrenheit are almost the same at -40.

I racked my brains and read almost every article I could in relation to combustion re-light at high altitude. One article in particular, by Dr. Watson, the Lucas chief engineer, written for the James Clayton lecture to the Institution of Mechanical Engineers, caught my interest. Together with a National Gas Turbine Establishment (NGTE) report on re-light trials in Canberra bomber aircraft at high altitude, a clue was forming in my mind.

I could not believe my eyes. A little elderly gentleman, wearing a gabardine raincoat entered the test area and introduced himself as Dr. Watson from Lucas. I must have blushed, as he said to put me at my ease, "I thought I would come to see what you are doing. I lost my way walking from the top offices."

I stammered that I had read his James Clayton lecture and that I was working on a problem in the high altitude plant at Burnley. I put to him the difficulty of the 1S60's failure to accelerate having been cold-soaked at -40 for one hour. He was quite interested in the problem and suggested the engine may need an increase in the heat release to warm up the sheet metal in order to supplement the combustion at altitude. He emphasised that air was very rare up there. Then I twigged it!

"If I added oxygen could that solve the problem?" I asked.

"If you add oxygen, it will increase the heat release and assist with ignition. But as far as solving the problem, you must try it," he replied.

I thanked him and showed him to the gas turbine offices where he could meet Spen King and others. I could not wait to get to Burnley again to see if adding oxygen would work.

It proved a long wait until we were able to reserve the altitude chamber again. I arranged to be seconded to the plant at Lucas Burnley. The Lucas staff were very helpful and I stayed at one of the 'local pubs', arranged by Bill Gregory and his wife.

Bill Gregory lived in the hope that Rover once more would return to Lucas for the design of the combustion system. He had worked on the design and development of the original baffle flame tube.

Bill was not very tall and had an almost an eastern-looking face that was slightly yellow. He reminded me of the Japanese in the likeness of his appearance, although I believe he was quite British. I never discovered whether this was due to having served in Asia during the war. I intended to ask his wife but never got round to doing so. I was the regular visitor to Bill's home during the time I was at Burnley. They were darts champions and I played with them at their local pub.

The temperature and pressure were drawn down to the equivalent of 60,000ft over the course of two days. The 1S60 engine under test was already installed. A cylinder of commercial oxygen was piped to the primary zone of the engine's combustion system using a valve that, when opened, would admit oxygen, albeit very cold. At a suitable signal, Warwick started the engine and I turned the oxygen valve to the 'on' position. A loud cheer went up as the engine accelerated away and reached full speed in 30s. Great activity was then turned to taking all the readings. This was our first successful start at 60,000ft, the engine having been cold-soaked at -40 for an hour.

When we subsequently dismantled the engine, all the sheet metal work from the combustion chamber through to the turbine nozzle appeared to be oxidised with a white deposition. The flame tube was starting to show slight signs of distortion, despite having been fabricated in 'Nimonic 75'.

I was driven to experiment with the amount of oxygen required at the point of injection. Also, I had to locate ways of coating the flame tube against heat. Eventually I discovered that a ceramic dip, effectively a green coating, could be easily applied and this prolonged the life of the flame tube under the rigours of accelerating the engine to top speed at 60,000ft with the engine frozen at -40.

# Automotive gas turbines

Finally, oxygen was injected around the atomiser. This was supplied by means of a 2-litre cylinder pressurised so that a small orifice could meter the oxygen for several starts.

Bill Gregory at Lucas Burnley introduced me to a colleague known as 'Carbon Clegg'. Bill explained that Carbon Clegg was having difficulty in coming to terms with the doctor's verdict that strangely there was nothing wrong with his health. This was playing on Carbon Clegg's mind and 'the-powers-that-be' got to hear about this. Consequently, Clegg was looking round for a new post. Knowing that I could be in the market for a combustion man, Bill implied that Clegg would be a good man to develop combustion.

I have mentioned that the subject of combustion is a 'black-art'. At Rover, I was saddled with it because I was the only one who had a bit of chemistry. I thought that 95% was in design and the last 5% took years and years to develop into an acceptable form. I hired Carbon Clegg with all his faults. I was known throughout the engineering industry as a collector of strays and odd balls.

Carbon Clegg was beginning to settle when he surprised me by returning to his old obsession. He always seemed to wear the same old shabby light blue suit. His hair was a fine golden colour, cut near to his head like a baby's. He had a round face and wore spectacles. His mannerism was decidedly taciturn. I gave him the Aurora car engine combustion system to develop. In fit of temper one day he dropped the flame tube. It distorted slightly taking up the shape of a banana. He didn't have time to straighten the flame tube before the next engine test. It performed excellently. Unfortunately we were not able to reproduce in manufacture the same distorted shape. Such is combustion. After several months of good work, Carbon Clegg became homesick for his home at Burnley and left Rover.

I thought at the time that one of the main reasons for the Labour party's defeat in 1955 was years of constant bickering. Attlee resigned soon after defeat, but the struggle between Bevan and Gaitskell continued. Bevan was known for ranting and raving against those who opposed him. Gaitskell, on the other hand, was a man of intellectual leaning. Matters were resolved with the victory in the leadership election.

Gaitskell won easily; Bevan then made peace with the new leader. Gaitskell resented the 'far-left's' view that theirs were the only true socialist policies. He wondered whether Labour could ever win another election. Gaitskell encouraged Tony Crosland, the MP for Coventry, to make a case for changing Labour's policies in his 1956 book, *'The Future of Socialism'*. Crosland put forward a view that the nature and structure of capitalism had changed and could no longer be portrayed as socialist myths made out.

Sybil and I were expecting our first child in the autumn of 1957. Many wives in Cambridge Avenue in that year had become pregnant. The Morris Minor, still with its bad oil consumption had to go, as we needed more money. When Sybil became pregnant she was due to take up a new appointment as secretary to Mr. Mell of the Solihull Borough Council. I think he was director of education at the time. We announced to relatives our intention of selling the Morris Minor car, and both my sister-in-law, Shirley, and my niece, Beryl, were interested. I was obliged to move quickly as I knew I could purchase a second-hand post-office van that was on offer by the Solihull Post Office.

Both my niece and my sister-in-law married my wife's brother, Geoffrey Dunford. Beryl and Geoffrey became divorced and then Geoffrey had since married Shirley. As you could well imagine, there was no love lost between Shirley and Beryl. My niece bought the car and later we received a strange letter from Geoffrey, breaking off all friendly relations with us. This was their loss, we thought, as we had more than enough to do to prepare for the new arrival.

Meanwhile, at Rover I was going from strength to strength; sales for the 1S60 flourished for many varied applications. A new engine was being designed for the fourth gas turbine car and the basic engine design for the Vulcan APU was attaining the Ministry requirements and developing well.

I compiled a draft book on the performance calculations for the instructional set that was selling well worldwide. The book was intended for sale in its final form. Little did I know that the book would result in much correspondence and comment for years to come. Some was critical, but mostly it was constructive.

My good friend Charles Hudson had died and Harold Smith had taken over. He was a charge-hand under Charles but he was not such a great a disciplinarian. Charles Hudson was respected throughout Rover for his forthright manner and his outstanding ability. Few people held any grudges against him.

The chairman of Rover Gas Turbines Limited was George Farmer. He was a very able man; it was even alleged George could make more money on the share market than Rover could in vehicle profits. I never was able to prove this one way or the other, but George Farmer was an excellent accountant, as I learned later, and he was always a good friend. When I attended my first board meeting he took me to one side and gave me some good advice.

"Remember, Noel," I can hear him saying now. "Anyone can say anything about anything, what you have to do is seek the truth".

I bought the old post office van and it motored perfectly, although I felt as if the mileage was excessive. The two front seat backs were supported by an iron tube that I was able to wedge to the sides of the van. The van was quite respectable with two windows on each side of the body.

The time came when I had to take Sybil to Solihull Hospital. It was morning and I had been briefed on where I should take Sybil on my journey to Rover, which was located in Lode Lane. Sybil was admitted to 'Netherwood-ward' and an enema had to be administered. I rang up several times during the day, only to be told by the nursing sister that Sybil continued to be in labour.

At around 5pm, I rang to be told that I had a daughter, born on the first of October 1957, at some time in the late afternoon. I enquired whether I could visit on my way from Rover and did so. Sybil understandably was quite sleepy and I was shown where I could peep through a glass screen at a dainty little mite. A large stiff white card was wrapped round her little wrist. I was amazed how red faced and small she looked.

"She was only born just over an hour ago," said the nurse, almost reading my mind. I asked about visiting times and whether I could visit on my way to work. The nurse became dubious, so I confined my visits to the normal visiting times. To my best recollection, I came to see Sybil the following afternoon and evening. I do not recall how long Sybil remained in hospital but I well recall driving my old van and taking Sybil and the new baby home. I was very careful with the tiny tot and I treated her as if made of china.

We were not prepared for the effects of post pregnancy, and Sybil gave up breast-feeding after a good start. She had difficulty getting rid of the milk. I am not using the right terminology, even if I ever knew it. Nevertheless we had many neighbours who were in a similar situation and we listened to sound advice, ignoring 'old-wives-tales'.

That Christmas proved to be a simple affair. Sybil was hardly over the worse time but spring was just round the corner, I hoped. The new baby girl thrived on the bottle and she began to increase weight slowly. We used to sterilise the baby's feeding bottle in a solution of Milton diluted with water. I became quite adept at flushing out the bottle with boiling hot water and pricking the teat so the baby did not suck in vain. We used to take it in turns to get up in the middle of the night, so that the burden of loss of sleep was not too onerous. For the first three months the baby had to be fed every four hours, and then it could be fed every six hours until baby slept throughout the night. Children have been known to send their parents into a fright by developing an excessively high temperature and then rapidly going back to normal.

The Duke of Edinburgh visited Rover Motor Company in 1958 and asked to visit the gas turbine department. He was accompanied by George Farmer, Spencer and Maurice Wilks, and Geoffrey Searle, who at the time was managing director of Rover Gas Turbines Limited. The fire-fighting water pump set was demonstrated, as were many other rigs and equipment. The airborne auxiliary power plant for the Vulcan was shown without its side panel. The Duke had the same devilment that has got him into many difficult scrapes throughout his life.

Someone asked him, over the boardroom lunch table, what he thought of the present Rover 75 car. He replied quickly, "I would sack your stylist if I were you". Maurice Wilks, who was seated across the table and had had a good hand in the style of the car, looked very sheepish. There were other things said, 'about little men with trilby-hats peeping over the dash board while driving' and that the car was designed for six-foot men. George Farmer remarked that the car was well designed for pipe smokers as the ashtray was large enough to rest a pipe.

The Vulcan airborne auxiliary power plant, in its installed form, looked every bit a Christmas tree, unlike the basic engine. On several occasions I visited Focke-Wulf in Bremen Haven and on one visit the chief designer there said the 1S60 looked like a flying desk.

We were negotiating use of the 1S60 as an APU on the new Transval commercial airliner that was called 'Wasser Flugsbau'. Little did he know that packaging would change the shape completely. A basic engine is one thing, but an installed engine is another. For example, the Vulcan APU was fitted with a small oxygen bottle as well as an axial compressor to act as a fan for cooling. There were also two cartridges to detonate directly on to the turbine to operate a rapid start, as well as firefighting wires throughout the package. It was literally a stainless steel box with intakes and exhaust suitably located to suit the aircraft. In other words, there are many systems that need to be added to any basic engine for a specific application or purpose.

A. V. Roe, the Vulcan's manufacturer, said that they were approaching the Ministry to extend the use of the Rover 1S60 APU on the ground. If they were successful in obtaining Ministry agreement, they would need the best of both worlds: large quantities of air on the ground, with no change to the electrical performance in the air. We were against this extension because of fouling the engine on the ground, to the extent that when there was an emergency in the air there would be a risk to performance despite the log of running hours.

However, one of the unwritten rules of engineering is to accept a challenge; for by so doing an engineer can achieve the requirement, remembering that nearly anything can be done, given sufficient money and time. We had to ensure that starting the main turbojet engines on the ground *and* compass swinging could be accomplished without risk to performance. We introduced routine compressor cleaning and a two-position-diffuser.

In one setting of the compressor diffuser, and at a slightly higher speed, we developed the unit to deliver the electrical performance plus. In the other setting of the compressor diffuser, we achieved the right amount of compressed air at normal speed. The trials were successful and the Ministry approved the modification.

In 1958, the Labour party, it seemed, was having difficulty portraying a united vision of socialism for the masses. And so the party lacked credibility with the electorate to win the next election. Harold Macmillan led the Tories to victory in 1959 with an increased majority, but after his 'never-had-it-so-good' speech the party was beginning to fall apart.

I met Bob Weir, a Scot who was Director General of my branch of the Ministry known as 'Eng-RD-4'. Although Bob was much older than I, we got on well together. His visiting officer was 'Buster Brown'. I knew him well. Buster was a pragmatic man who liked to think in simple terms. He and I were great friends. One of his sayings was, "show me your scrap", whereupon I had to display all the components that we had in quarantine stores.

The Air Ministry had a ruling, that if parts were not A1 or subject to the modification system, they were to be placed in a special store. By that means, sub-standard components that the Ministry had paid for were not used for front line aircraft. Only components of the latest frozen and approved design standard could be built into an engine.

This system grew up despite Frank Whittle, who wanted to keep altering designs without any organisation to address the changes. Frank Whittle was often accused of rubbing out in hardware with no system.

There were three modification standards: A, B and C. A was the most serious and to call up an 'A' meant all aircraft must be grounded at once until the failure was rectified and a solution fitted to engines. A 'B' modification was less serious and must be applied at the next opportunity. A 'C' modification was downgraded to 'as and when' parts were available. It was the chief engineer's responsibility to call the grade of modification, and the heads of engineering, design, manufacturing and procurement to approve the solution. There was a special full-time section attached to the design department dealing with the Ministry's modification system.

I had two assistant chief engineers: both were quite different. Harry Cox was one and the other was Mark Barnard. Harry, I knew well. He was a straightforward engineer, scrupulously honest and outspoken. He was educated in Coventry at King Henry, the secondary school, and later at technical college. Like any famous Coventry engineer, he had 'swarf in his veins'.

Mark, I had selected as a graduate apprentice. He was bright, but much more complex and had some surprising gaps in his knowledge of engineering. Chemistry was one of them. That was always a mystery to me as I had met his parents, who taught at a public school near Derby and were very good people. He obtained a good degree from Cambridge and his early contribution was excellent.

Fred Hulse became chief development engineer. He had worked at Armstrong Siddeley and consequently became a friend of Len Harvey, head of the installation section. Fred had a Rolls-Royce vehicle. I think it was a Silver Cloud that had been converted into a 'shooting-brake'. Spen used to tease him that he travelled in the middle of the road at 10 miles per hour, because the fuel consumption was so bad. Fred was a refined gentleman and an excellent engineer. He proved a great help in the development of the Vulcan APU.

One specification requirement was for the Vulcan to fly inverted for one minute and with, of course, the 1S60 engine running. The engine had a wet sump so it was incumbent on maintaining the oil during inverted flight. Fred invented an anti-gravitational device that blanked off the oil breather and consequently the oil. The device was of simple design, just a circular piece of heavy gauge metal that blanked off the pipe when the engine became inverted.

Fred mysteriously disappeared one day and never returned to the department. One rumour suggested that Fred's fuel bill remained unpaid. Salaried staff were allowed to buy available material from Rover, as long as the material was subject to billing and paid for from salary. Fred always said to me that when he was taken on he had an agreement that fuel would be supplied. Unfortunately that was not in writing. However, Fred Hulse was a completely honest person and had no reason to be untruthful. He was a member of the council where he lived, just outside Knowle at Chadwick End. He had a lot to lose. The difficulty was that Harold Smith, the gas turbine development foreman who had taken Charles Hudson's post, had also been using fuel for his own car. Harold, who paid retrospectively, dealt with this, but the principle must have been uppermost in Fred's mind, I assumed.

Sometime after this event, I was promoted to chief project engineer over the entire gas turbine department. Coincidently,

Spen had been moved across to design a new vehicle that later became the famous Range Rover. This position of mine was in addition to the post I held in Rover Gas Turbines Limited that eventually led to my joining the board as the company's technical director.

Len Harvey, head of the installation section, was in an argument with Mr. Searle and they both came to a mutual agreement that Len should leave. Len went to Bristol Siddeley, before it became Rolls-Royce. I never discovered what happened to Fred Hulse, but the chief metallurgist of Rover, Sidney Heslop, was related to Fred by marriage, and he felt it wrong for Fred to leave Rover. Douglas Llewllyn became the new head of installation and Bob Myring, his assistant, became the new production head.

Bob and I used to alternate between running cars to Rover, each getting a lift with the other as he lived nearby in Longmore Road. He became production manager at Reliant, the makers of the three-wheel Robin vehicle.

A 40kW 220V DC generator set, powered by the Rover 1S60, was installed in the Brave class high-speed patrol boats that were sold to various Navies throughout the world. There were two Rover engines for each vessel.

In 1959, the 2S140 engine was undergoing endurance tests as it was planned to power the fourth turbine car, the T4 (photograph, p.444). The development of the 2S140 engine is dealt with mostly in the technical paper I gave to the Society of Automotive Engineers in January 1963.

Douglas Llewllyn was largely responsible for the adaptation of the 1S60 gas turbine engine to the Rover TP60, first flown in the Currie Wot biplane (photograph, p.445). Vivian Bellamy, chief flying instructor of the Hampshire Aeroplane Club, was the first test pilot of the most uncomplicated and trouble-free airframe imaginable. Starting was entirely automatic, with virtually no vibration and hardly any noise. The first flights were on a special 25 hour category, granted by the Air Registration Board. Later, both the TP60 and the TP90 were certified in my name, as I was the technical director of Rover Gas Turbines Limited.

The 90bhp version of the 1S60 had the same weight and scantlings but the air mass flow had been increased; so too was the pressure ratio. The turbine rotor had slightly longer blades and the engine delivered about 123bhp at sea level static on a 15C° day.

Two aspects of the Currie Wot stay in my mind. The first was the uttermost simplicity of the controls: there was just a key to switch on, and a throttle lever. The other aspect was that the propeller feathered when coming in to land and taxi. The Birmingham airfield control tower called down and said, "Excuse me sir you have broken your crankshaft".

The gas turbine engines were also installed in the Auster and the RAF Chipmunk. The RAF wanted to replace Gypsy Major piston engines with the TP90; the piston engines had experienced broken crankshafts. The Rover board decided against becoming purveyor of turboprop engines because of the potential liability costs.

Nevertheless, the engineering effort that Douglas Llewllyn and his team demonstrated with the TP60 and TP90 installations was second-to-none and has never been equalled.

It was an immense disappointment to Rover Gas Turbines, Limited, particularly those who had spent their time and engineering effort, not to be able to make an entry into the propulsion world with the 1S60 and 1S90.

In 1958, life in Cambridge Avenue flourished as newly born children appeared. On 1 October 1958 Deborah was just 12 months old. Spen had taken photographs of the baby after leaving hospital when she was about two weeks old. We both looked forward to our second child when Sybil was confirmed pregnant, just after Deborah was one. Deborah was a pretty baby, so petite and small, unlike the girl down the road who was burly like her father and unlike her mother who was quite trim. Deborah was walking long before she was one; she would not keep her bonnet on her head for long. Sybil placed Deborah in

# Automotive gas turbines

the pram at the front of the house so that she could get on with the daily chores. One day, Deborah was crying out and Sybil came out running, just in time to send off a cat that had just jumped on the pram.

Looking back to those days they were some of the happiest of our lives. During Sybil's second pregnancy, she often visited the children's nursery and the doctor's surgery. In those days, mothers had to walk pushing their prams and thought nothing of it. One day, when returning from the doctor's, she was pushing the pram and obviously pregnant. Several builders were working on the roof of a house at the top of the road where we lived; when Sybil passed they started singing, 'One night of love'. Sybil hurried past, smiling as she did so.

People became more outspoken in the 1960s; but the writing had already appeared on wall in the late 1950s. In 1958, the Campaign for Nuclear Disarmament (CND) was founded; it was a serious movement with an equally serious cause. What surprised me was the depth and feeling of the following, as if the stirring of ordinary housewives aroused the very essence of middle-class England.

The 1960s became characterised by a 'cultural revolution'; the release of social control was one thing, but it set the scene for loose sex, self-expression and many other changes in our behavioral patterns. Every aspect of change came about from the BBC to films and flower power, the Beetles, and so on. It was if some god had taken an indelible pencil and ringed the world, particularly the UK, with a new freedom of thought. 'I'm All Right Jack' was typical of the social satire of the time. '

Stanley was an earnest young man, gormless but he had been brought up a gentleman. He was instructed by the university appointment board to be more adventurous and to adopt an air of confidence. His great aunt, played by Margaret Rutherford, addressed him as 'Spencer'. Stanley failed to obtain scores of suitable posts. Finally his uncle, played by Dennis Price, managed to fit him into the factory, for which he is the owner. Margaret Rutherford remarked, "I expect you just to supervise, dear." Stanley landed up as a forklift driver, however. Peter Sellers was the communist shop steward and he spoke in a polysyllabic manner. He deplored the class society in which he grew up. Peter Sellers took Stanley home to tea and suggested that he came to lodge there. Stanley fell for Peter's daughter, Cynthia, a spindle grinder at the firm. He takes the lodgings, much to the disgust of Major Hitchcock, the personnel manager, played by Terry Thomas. The whole play was a satire on the class system. You must have seen it at some time!

Mrs. Ashton was a midwife preparing for Sybil to have the second child at home. Sybil blossomed in 1959 and the craze for crisps with vinegar heightened. The sale of Smiths' crisps, with a small dark blue paper containing salt, was still around but without the wide variety that exists today. I had to avoid eating the salt when we were in the dark of the cinema. Of course, I carried around a small bottle of vinegar. All weekend, Sybil had the warning signs that the baby was due at the latter part of June. So I decided to stay behind during the week and deem it as vacation. On the morning of 30 June, pre-birth signs were evident and I called Mrs. Ashton to ensure she was available. The midwife asked many questions and assured me that all was satisfactory, and that she would call mid-afternoon. If however I needed her before then, I could contact her by telephone. I think it was on Tuesday that Deborah, the sweetest little girl, was quite able to walk with me round doing the household chores.

Sybil was resting and she called down that I should call the midwife, who at that moment fortuitously was at the door ringing the bell. The midwife asked me to boil a large saucepan of water, which I did. Deborah and I waited anxiously downstairs until the midwife came hurrying down the stairs for the water that by that time had boiled. I offered to carry the saucepan upstairs but the offer was refused; instead, Mrs. Ashton was asking for me to boil more water. It seemed like an eternity, with Deborah standing on one of the kitchen stools with my arm round her, before we both heard a little cry. The midwife called down, "You have a son". I whispered to

Deborah, "You have a baby brother". And she repeated, "Baby brother". Deborah was only 21 months at the time.

When Mrs. Ashton came down stairs she presented me with the afterbirth, all wrapped in paper toweling to burn. She said, "I could go and see my newly born son". I went to see Sybil, after I had taken the waste to the bottom of the garden to burn later. I washed my hands. Mrs. Ashton remained to look after Deborah while I went upstairs. There was no need of the second saucepan of water as the birth was straightforward. I saw that Sybil was both relieved and pleased that it was all over. Nestling with his mother was the little face of a sleepy son. That night I nursed our baby son to sleep; he fitted into my shoulder perfectly. My first reaction was that he was every inch a boy, and longer than our petite daughter, Deborah, when just a few hours' old.

News soon reached Rover that we had been blessed with a second child – a boy. For, on the following evening, a group of rather rowdy members of my team arrived to wet the baby's head. They landed up several doors down the road, watering the Beresford's lawn. Jonathan David, as our eldest son was to be christened later, was born on June 30 1959.

Back at Rover, the gas turbine engine for T4 was making good headway. The long, power turbine shaft, operating at 40,000rev/min top speed, was quite a success. It was supported by two bearings: at the turbine end a deep-groove ball bearing and at the pinion end a roller bearing. Because of its length, there were three damping sections along the shaft, with three rollers in each section. The outer part was contained in the light alloy extrusion.

This brings to mind a Rover research experiment on a thin prop-shaft that would take the torque, but was solid, in contrast to the conventional hollow prop-shaft. The solid prop-shaft, of course, would whip, having a lower first critical speed; but it could easily be damped by the edge of a piece of cardboard and would operate at speed in perfect symmetry as long as the cardboard was held lightly on the running shaft.

# Automotive gas turbines

The T4 had its own crown wheel and pinion with disc brakes built in on either side. Initially, it was designed with one forward and one reverse ratio to form a novel transmission. When reverse was engaged, the compressor discharge was deployed to pressurise a large piston that was sealed by an equally large rolling 'O' ring. This system was abandoned, because of the difficulty of making the 'O' ring roll every time reverse was engaged. The transmission casing was cast in Electron magnesium for lightness. Later, this was changed, due to cold creep.

This was the first design with an overhung gas generator – a back-to-back inward radial turbine with a radial compressor. We had a lot to learn in relation to this rotating system, especially with regard to the amount of overhang. The heat exchanger was arranged in two banks on either side of the engine. One bank had 50 heat exchanger elements, measuring about 10in long, 5in wide and 0.3in thick. We adopted the contra-flow primary surface recuperator rather than the secondary surface on the grounds of weight and durability; it was a welded construction.

The whole engine package was neat, with two intakes on each side of drum-type filters installed into the two front wings of the car, behind the headlamps. The exhaust was split into two, also ending in a bifurcated duct. A tube within a tube ran the whole length of T4, culminating in a fishtail outlet at the back of the car to diffuse exhaust gases to a tolerable velocity.

Rover decided to introduce a car lease scheme for monthly staff. We would be notified of the rules and our entitlement in due course. When I read the note, circulated to the heads of all departments, I made a mental note to teach Sybil to drive.

I became known throughout Rover as one manager who was particularly exacting on secretarial work. Spen's secretary, for as long as he was working in the gas turbine area, had been doing my secretarial work, but once he left to take up office in another part of Rover, his secretary moved with him. I think her name was Enid Coverdale. Janet Star came to be my secretary. She was married and, with her husband, worked a smallholding in the outer part of Shirley. Janet was an excellent secretary, quite on a par with Enid. Regretfully, Peter Wilks poached Janet when he was promoted to executive director (cars); this followed the retirement of Robert Boyle who had held that post. Shortly afterwards, Janet Star left to help her husband, and Rover lost a first-class secretary. I kept in touch with her husband and Janet long after she left.

Janet Star had devised a code for salary increases that was known only to us two. The French war was raging in Vietnam at the time so our code was 'WHY VIETNAM'. The letter 'W' was one and 'H' was two and so on. By using these letters, Janet and I were able to decipher salary increases by referring to the code. We could do this without the bright gas turbine engineers, who could read anything upside down, finding out when the oral assessments took place.

Jet 1, T3 and T4 enabled valuable data to be accumulated but the difficulties in trials with T4 remained. The problems centred on the life of the heat exchanger, mainly due to thermal fatigue. Route analysis on T3 and T4 indicated that the percentage of time spent over 60% of engine load was small, but the number of accelerations was higher than expected. This had a detrimental effect on the heat exchanger, due to the high metal temperature of the leading welded edge of the elements; this caused thermal cracking. The best result indicated that the onset of tiny cracks was anything from 3,000 cycles to 10,000 cycles, depending on the manner in which the gas turbine cars were driven. In the case of bad cracking, compressed air could leak into the exhaust, causing a radical change of overall efficiency.

Although the production of gas turbine cars in the form of T4 had been planned in volume quantities of 2,000 a year, the top management had a hard decision to face.

At that time, one or two glass companies in the US were researching a new material called glass ceramic. It was rumoured that with this development, the new material's properties would allow stability up to 1,000°C, without the brittle behaviour of glass and with next-to-zero expansion at temperature. If this was true, it would remove the objection that Rover had to the use of steel regenerators, as adopted by Chrysler in its gas turbine car developments.

As I have mentioned, there were two different basic types of heat exchanger used for gas turbines engines, recuperative and regenerative. They both had the same function, namely to transfer heat from exhaust gases to the incoming air after compression and before combustion, thus leading to a smaller amount of fuel being required to create the desired turbine inlet temperature (TIT) than if the engine had no heat exchanger. The principle behind the rotating regenerator was to leave a hot matrix, in disc or drum form, to heat the incoming air while the exhaust gas was being passed through one section of the regenerator as it rotated.

The exhaust gas was sealed from the higher-pressure incoming air. Both air and gas streams flowed in a contra direction, so that one face of the disc was the 'cold' side and the other the 'hot' side. There were three engineering design tasks for the regenerator, as opposed to the fixed, more simply-constructed recuperator. The first required a slowly rotating disc or drum to be supported. The next demanded sealing the high-pressure air from the exhaust gas and, thirdly, causing the whole device to rotate in a high temperature environment. There were also secondary difficulties, such as carry-over and seal losses. These losses leaked previously compressed air into the exhaust stream, in turn to be wasted into the atmosphere.

Evidently, Chrysler had overcome these problems in steel, even though the expansion of the steel made sealing the gases an immense task, as the steel regenerator could bow and dish significantly.

"So near, and yet so far", was how Maurice Wilks summed it up. I felt quite sorry that his dream of producing gas turbine cars was delayed once again because of the heat exchanger.

# Automotive gas turbines

Although I was young at the time, and still at college, I found the story of Frank Whittle fascinating. Frank Whittle did not have the knowledge of gas turbine design nor the enormous development problems involved when he invented and patented the jet engine; unlike his counterpart Herr Doctor Hans von Ohain at Heinkel, who suffered at the hands of the German Gestapo. Herr Ernst Heinkel saved von Ohain from their clutches. The German patent was refused because of the Whittle patent. The turbojet patent of Frenchman Maxime 'Guillaume' should have caused rejection to Whittle's patent, as the Frenchman's invention was much earlier.

Back tracking to 1940, Power Jets Limited was apprehensive about the manufacturing capabilities of British Thompson Houston (BTH), which was responsible for the first design. Volume production of the W2 series engines was uppermost in Power Jet's minds and as a result contacted Rover; not that the Power Jet's design was ready for production. At that time, Rover's limited knowledge in relation to production of turbine engines was in every part equal to the immaturity of the W2, an engine design that was fluid to say the least. Power Jet's kept changing a design that was never frozen for production. There was always a better design solution in the pipeline. Later, this was called 'rubbing out in hardware'.

It is possible to imagine the friction between the two principal characters. On one hand, Maurice Wilks, who in wartime was chief engineer of Rover with responsibility for volume manufacture a design that constantly changed. On the other hand, Frank Whittle was inventing new aspects of design, and as a result of development was throwing up modification after modification.

My personal impression of Maurice Wilks was that he was a perfect gentleman. Frank, on the other hand, was an entrepreneur who could tell a good story. At one of his dinners at the RAF Club, where I was present, he related the tale of Stanley Hooker of Rolls-Royce adding pre-swirl vanes to the inlet of a Power Jet's design and getting a medal for doing so. A year later "he received another medal for taking it out". Frank Whittle, of course, was against pre-swirl in this case.

Rover Gas Turbines Limited was not the only company facing an uncertain future. On the political front, the country under the Tory government, was shaky. It was as if Harold Macmillan had 'knocked-off', to use a union expression of the time. Tory chancellors could not decide whether the economy should expand or contract. Macmillan, as Prime Minister, had just about recovered from the 'Profumo Affair' when it was announced that the P.M. needed emergency surgery. Selwyn Lloyd, as Chancellor, gave the impression that leadership at the top was spare; he was to be changed for Reginald Maudling, in a later reshuffle by Macmillan. During the time Macmillan was in hospital, R. A. Butler assumed the leadership of the country. 'Rab' Butler had been number two for years and was becoming another 'Eden', following his years under Winston Churchill.

On the domestic scene, I leased a Rover P4. It was called a Rover 100, as distinct from the Rover 75 because of its more powerful engine. Sybil at that time had recovered from the last birth of our son and was thinking again of taking driving lessons.

The 'liberating 1960s' emerged and most people who could afford them bought such items as refrigerators, washing machines, vacuum cleaners and television sets. Most of these items were a boon to the female population, albeit a strain on the household purse strings. Of course, the 1950s set the scene and most of these items appeared before the 1960s; it was the rate of progress of technology that brought these items more quickly within the reach of many.

The quality of life increased for most people. Only the very rich suffered a setback, due to inheritance tax; but a few, wishing to avoid the inevitable, passed their estates to the National Trust or English Heritage. Life in the late 1950s and early 1960s was good. We lived in a street where most families were of the same age and faced the same tasks of rearing children.

The Rover T4 was not the only gas turbine car in the world at that time. Chrysler in Detroit was also involved in a gas turbine car programme under the leadership of George Huebner. Consequently, Rover decided to interrupt endurance testing of the T4 on English roads and, in early 1962, sent the car for exhibition at the New York Motor Show. I took part at London airport to see the T4 loaded onto the Seaboard Canadian CL-44 cargo plane for its flight to New York. Jet 1 also was transported across the Atlantic and exhibited at the New York Motor Show.

In June 1962, the T4 was shown in France and driven around the course at Le Mans, prior to the 24-hour race. Little did we know that it was the forerunner of things to come.

That September, Sybil was pregnant with our third child and we enrolled Deborah into the infant's class at the local church school, Saint Alphege in Solihull. The headmistress accepted Deborah's enrollment and that autumn she was found a place at the start of the new term for the infant's class. I recall her first day at school; she looked so small, petite and pretty in a dark blue uniform. Sybil had made a little school bag for her, not knowing whether she could stay for lunch. However, the day for new intakes was short. It was just before Deborah's fifth birthday.

## Seven
# Racing gas turbines

In which Rover had the opportunity to join forces with the Owen Organisation to race the Rover-BRM at Le Mans.

I WAS busy at the office when the instruction came down that I would be preparing a lecture that I would to deliver at Cobo Hall in Detroit to the Society of Automotive Engineers (SAE). Len Raymond, President of SAE, had visited Rover in the summer of 1962 and issued Maurice Wilks with the invitation to talk about the case history of small gas turbines. Maurice Wilks handed me the task of preparing the paper and giving the lecture in Detroit. I can hear him now saying, "Be friends with all" and you will succeed.

Bill Martin-Hurst, who recently had joined Rover from 'Teddington Bellows', a company in South Wales, approached me to ask my views on racing against the clock with a gas turbine car. I explained that we were developing a ceramic regenerator in alliance with Corning Glass and any idea of racing had to be with a non-heat exchanger engine, the exhaust temperature of which would be very high. There was another difficulty, namely that of noise as the heat exchanger acted as a good silencer and attenuated the exhaust noise.

Bill then owned up. He had had preliminary talks with the Owen Organisation, which supported Formula 1. The company was prepared to form an alliance with Rover to modify an existing car for the purpose of racing a gas turbine vehicle at Le Mans. Application to enter a gas turbine car in the Le Mans event was accepted in February 1963.

Bill had joined the Rover Company as an executive director. He was made a director on the main board as deputy managing director. His wife was one of the Hilman sisters. Two of the sisters were married to S. B. and Maurice Wilks. So Bill was related by marriage to S. B. and M. C. Wilks; in fact, he was their brother-law.

In January 1963, I gave the lecture on *The Rover Case History of Small Gas Turbines* to the SAE in the Cobo Hall, Detroit. The paper was acclaimed a great success and I was inundated with invitations to visit the big three motor companies in Detroit, namely General Motors, Ford Motor Company and Chrysler. I made many good friends.

It was one of my first meetings with Sam Williams and his boss at Chrysler, George Huebner. I was invited to George's home and met his wife, Trudy, who was a Regent at a University near Detroit. I left Chicago in a snowstorm and arrived at Phoenix in glorious sunshine and where I was visiting AiResearch. From there I visited Paul Pitt of Solar in San Diego; I then retraced my steps to Corning Glass. I stayed with Rushmore Marnier after a delay due to snowstorms.

When I returned to Rover, a 1962 BRM Formula 1 chassis was suitably modified for the installation of the 2S150 non-heat exchanger engine. The 2S150 engine was an adaptation of the 2S140 installed in the T4. Modifications to the 2S140 engine were essential as the 2S150 engine at that time had no heat exchanger and the combustions system was different for an engine with no heat exchanger.

A telephone call came to my office late in the afternoon told me that Deborah had been knocked down by a sports car in Cambridge Avenue; she was being taken to Solihull hospital. All manner of things went through my mind as I rushed to the hospital; fortunately it was not far from the Rover plant. On the journey to the hospital, I died a thousand deaths, metaphorically speaking. I arrived and Deborah was just being wheeled from the ambulance and Sybil was with her. I cannot recall where Sybil had left Jonathan. Deborah was in quite a state; she had a fractured wrist and was unconscious for several days. The consultant, whose name was something like Mr. Keena, was quite dismal about Deborah's condition. I think I something must have left my life.

## Automotive gas turbines

I sat by her bedside for day after day; all thoughts of Le Mans seemed to have evaporated. After several days, I will never forget the joy when Deborah stirred and, with her mouth on one side, she murmured, in a loud voice, "Daddy". A loud cheer went up from all over the ward. When Deborah was at last discharged from hospital, her little mouth was all lopsided and she had lost the use of one side of her body. At the time, Sybil was pregnant with our third child. We decided to take an early holiday, as all thought of Rover had gone from my head. The only thing that mattered to me was to work with Deborah to improve her life.

In early May 1963, we travelled to Bognor Regis to a Butlin's Camp. Sybil was able to rest up while I strolled with the two children. Deborah found walking quite difficult. At meal times we ate at communal tables. The noise when we were eating was deafening and I well remember Jonathan holding his little hands to his ears.

That year proved to be a difficult one. Roland was born on the day of the Le Mans race on 19 June 19 1963. The race was actual a race against time, as there was no formula that allowed gas turbine cars to race against piston-engined cars. The Rover-BRM was designated the number '00'. It was not a contestant in the race, but it had been entered and accepted as a starter. There was a special prize offered by the L'Automobile-Club de L'Ouest for the first gas turbine car to compete in the race at an average speed of not less than 150km/h. The Rover Company, in alliance with the Owen Organisation, was written up as making motoring history when '00' easily won the special prize, but raced with exceptional reliability.

This immense success established many records of the time. The car, if it had been competing, finished in seventh place. Nearly two-thirds (65%) of cars taking part did not finish the race.

It marked an alliance between the Owen Organization and Rover Company. On Saturday 20 July 1963, when Roland Paul Penny, my youngest son was just one month old, we staged an

# Automotive gas turbines

acknowledgement of Rover's achievement at the Silverstone Race track. Jet 1, T3, T4 and the Rover-BRM completed three laps of the Silverstone circuit (picture p.451). Graham Hill and Richie Ginther, the team mates who drove '00' at Le Mans, were there to drive the Rover-BRM and the T3. Tony Worster drove T4 and Spen King piloted Jet 1. I had the task of ensuring that the gas turbine cars all finished at the same moment.

Had any of the cars crossed the line first the event would have been deemed a race and payment of duty resulted. Peter Spear was the coordinator from the Owen Organisation; I was constantly acting as peacemaker whenever he visited Rover's gas turbine department. Mark Barnard and Peter Candy were important in any Le Mans project. Peter was responsible for the car side while Mark was my assistant. Harry Cox had taken up a post with Wilmot-Breedon in Birmingham on fixed-type heat exchangers. Both Mark and Peter had a habit of rubbing Peter Spear up the wrong way; that was not difficult to accomplish. I suspected that it was six of one and half a dozen of the other, so to speak, and I should bang their heads together.

Peter Candy, I suspected, had a chip on his shoulder and was very ambitious. For some reason he had a down on Tony Worster. I often wondered what Spen would do in the circumstances.

Peter Spear complained that the Owen Organisation were not getting very much out of the alliance and, to placate him, I agreed to join with him to draft a paper for the SAE about Rover-BRM. I soon learned that Peter was finicky about the slightest detail. We were both keen to write the paper in our own time and questions arose about a suitable meeting place.

Peter Spear insisted that we find a venue half way between his office at Darlaston and my office at Rover. So we consulted a map and drew a line between Darlaston and Lode Lane in Solihull. We bisected the distance and the result came out at Bearwood in Birmingham. I went there with Peter Waters and we found a public house on the corner in Bearwood called 'The

Bear'. We explained to the landlord that we wanted to hire a room. Fortunately, a big room upstairs vacant for most of the week was available. There was no fee discussed; just an arrangement that if we rang up with a day's notice the room would be left in readiness. That is how we drafted the first of two papers the subject of a lecture to SAE in Detroit in January 1964 entitled *The Development of the Rover-BRM*, Paper 795B.

There was no way we could race at Le Mans again until there was a method of classifying piston engines against gas turbines. So Peter Spear and I racked our brains to think of a suitable formula. I came up with the idea that the area of the first turbine nozzle could be measured in a straightforward manner but at the time Ford in the US had a high pressure ratio engine, designated the 704. Most small gas turbines had low pressure ratios for the purposes of the heat exchanger, but the Ford engine was a complex engine with reheat and heat exchange.

The basic consideration suggested that the higher the pressure ratio the less a heat exchanger is justified, but there is no way of attaining good part-load fuel consumption without adopting a heat exchange cycle. So we had to devise some way of including a selection of current engines and to include pressure ratio. We prepared a paper at Peter Spear's insistence. I had the task of travelling to Paris almost once a week to convince the Le Mans committee that the formula invented by Peter and I should be agreed. The committee consisted of a group of elderly men; I guessed their average age of well over 70. One day, when I visited the chorus went up at the meeting 'Wee'. I ran out of the meeting having thanked them and later the World Racing Authorities accepted the formula.

It was just about this time that Rotax, part of the Lucas empire, contacted us to design a gas turbine starter. There was an aircraft called the P1154; it was the forerunner of the Harrier. Eric Earnshaw, then chairman of Rotax, joined George Farmer, chairman of Rover, on fishing trips to Ross-on-Wye. They discussed the prospect on one of their weekends together and the next time Eric visited Rover I had a call from the chairman's office to say that they would like to visit me in my office.

The outcome showed that the 2S150 profile required a redesign to meet the specification for a gas turbine starter. I proposed an annular combustion system and a lower weight fuel system; the transmission would be Rotax's responsibility. Lucas Burnley would take responsibility for changes to an annular combustion system and Lucas in Shaftmoor Lane would be responsible for the fuel system.

In the brochure presented to the Ministry of Technology, marked 'Secret', I was referred to as chief engineer of Rover Gas Turbines as well as of Rotax Gas Turbines. Regretfully, the P1154 was cancelled but the annular combustion system was developed.

Attention turned to the P1127; that proved to be the Harrier, as we know it today. The requirement for that was just half the power of the P1154, so the 2S150 was scaled down as a gas turbine starter. The 2S75 was designed in record time. Rolls-Royce was responsible for the main engine; that became known as the 'Pegasus'. The 'Jump Jet' Harrier could land virtually anywhere, even on a road. The gas turbine starter was connected to the main engine through a clutch so that the gas turbine starter could be operated with a 12V battery and in turn the main engine could be started.

We were entered for the 1964 Le Mans race and we devised a system that put the engine through the race without moving from the test bed. It was a simulated 24-hour Le Mans test that started from cold; the duty-cycle was then repeated up to each pit stop. There was a dial indicating where the car was on the circuit of the Le Mans racetrack. For example, when we were accelerating into the Mulsanne Straight and slowing down for the Mulsanne corner, the dial indicated as such. Graham Hill visited to view the set up during the short development period up to the 1963 race. He commented that it was all very well for development purposes. "What he would do during the race is to keep his foot flat down on the throttle and use the brake."

He was anxious to win but we explained that he would be up against more powerful cars in the 3-litre class and above, for example the Ferraris.

The main rotating parts of the 2S150 engines had performed faultlessly for over 100,000 hours. We were using the twin Corning glass-ceramic 18in cores made up of a multiplicity of triangular passages that acted like a honeycomb. The engine was designated the 2S150R. We had three engines that were subjected to the full 24 hours race on the test bed.

When we dismantled the first engine we noticed fine cracks at the base of the radial turbine blades at the exit. We placed the turbine in quarantine stores pending further metallurgical examination and we carried out testing on two engines; one of these was on the 24-hour Le Mans test bed and the other on a normal test stand. We stripped the engine set to be installed first on the Le Mans simulated test bed and found more fine cracks at the base of the radial turbine blades. The tests were complete on the other engine and we disassembled it only again to find more cracks at the base of the turbine blades.

We were becoming concerned; nothing like this had been experienced in all the similar engines run over tens of thousands of hour's endurance. We received the metallurgical report on the first case of this kind. It suggested either that the cracks were due to stress relief or further examination was warranted. The turbine would have to be scrapped and cut up to rule out failure due to vibrations. We suggested that as there appeared to be two other cases of this cracking, they could continue to pin point the cause, if possible. The practice sessions at Le Mans loomed ahead and the party set out for the circuit with one only one engine installed in the car.

After completing the practice satisfactorily, the transporter set out on the road back across France, and the boat across the English Channel to England and the Midlands. Regretfully, the normal transporter driver, one of the experienced fitters/drivers, relinquished the driving to a graduate apprentice. The roads in France at that time were heavily crowded and the apprentice lost control of the transporter when a heavy vehicle was travelling towards him. Unfortunately, the gas turbine racing car was damaged and the newspapers got hold of the story. The headlines in the most of the English papers spelt out "Le Mans car damaged on the road in France, race likely to be cancelled." The Owen Organisation carried out an investigation and, unbeknown to the media, considered the car could be made ready in time to race. There was criticism of Peter Candy, who was in charge of the party at practice and transport.

Before practice, Peter Candy argued that the transporter, which was Rover property, should be under his charge, and not the responsibility of W. Wilkinson, the appointed Owen team manager. This was agreed, despite Peter Spear's objections to the contrary. This further emphasised the rift between the two groups. I agreed that in this case, the Owen Organisation's criticism was justified.

We continued to be baffled by what seemed now to be an epidemic of cracks. We had seen nothing like it before in relation to radial turbines. It was the most difficult time in all my experience of small gas turbine engines. The time came to make a decision on whether we should race or not. In my heart of hearts I thought the risk was too great. I was in a state of trepidation to face the board and express the view that, in my professional capacity, "we should not race."

Bill Martin-Hurst expressed a view that he would like to visit the gas turbine shops to see how well we were progressing. It was early in May 1964. This was my opportunity to unload the burden of the 'epidemic of turbine cracks'.

I gathered all the evidence I could muster, including the batch numbers of the inward radial flow cast turbines. I observed for the first time that it was a new batch that had been rushed through 'Springfield', the experimental machine shop. Subsequently, it was spun to just above maximum speed and viewed under ultraviolet light to determine any signs of cracking, using a suitable coating of 'Ardrox'. All the turbine rotors now under investigation had been approved as passed for use in engines.

I met Bill Martin-Hurst during his visit with A. B. Smith to the gas turbine department in early May 1964. A. B. Smith was on the board of Rover Gas Turbines Limited and chief buyer on the main board of Rover Company. A. B. Smith was a good friend, as was Bill Martin-Hurst. A. B. Smith was now assistant managing director and periodically joined the meetings with the Owen Organisation. When we got over the 'time of day dialogue' I pitched in that we had encountered serious problems. At the culmination of the last six Le Mans tests, all three engines had been found with fine cracks at the base of the blades of the inward radial turbine rotors. I showed both gentlemen the rotors involved, with the exception of two rotors that had been cut up by the metallurgy laboratory. I then dropped the bombshell that, in my opinion, until we understood the reason for the cracking we should call off the race; there was insufficient time to track down the cause.

The implication of my remarks did not at first sink in; both men were thinkers and asked numerous questions. Fortunately, they accepted the answers. They asked me what I thought was the reason for the cracks, and then it came to me. I said that my best guess was 'built in stresses' that were manifest only when the turbine rotor became heated. The next question implied that previous engine running would surely have revealed the same thermal effects if the precision casting process had remained unaltered? I agreed that there must have been a change in the casting process. The last question queried whether the race should be at this late stage abandoned.

I said, "The risk is too great and until we track the cause of the cracking we should cancel the race in my opinion." "Other considerations had to be taken into account," said Bill. "The risk may not end the race in failure".

At this conjuncture, A.B. Smith commented "Peter Spear would have kittens, to say nothing of Sir Alfred Owen's view". Bill Martin–Hurst said, "The final decision was mine, on the grounds of engineering, but he would put the case to Sir Alfred". Following that there would be another meeting between the key people at Owen and Rover.

The 1964 race was subsequently called off and some of the media blamed this on the crash of the transporter on the way back from the Le Mans practice in April. This was despite the statement put out by the alliance that "Both parties had to call off the race due to Rover engineers being unsatisfied that not enough time remained to ensure that the engine has had adequate development time".

Bill Martin-Hurst was by this time Rover managing director. He placed on record Rover's appreciation of the outstanding efforts of Corning Glass in the USA. Corning Glass was responsible for the manufacture of the ceramic cores of the heat exchanger. "Development work has continued day and night, and at weekends on both sides of the Atlantic. Weekly telephone meetings have taken place over the loudspeakers at each end to enable contact and progress of solving problems in running endurance tests".

"In the acknowledgement of the hard work and untiring co-operation of Corning Glass and its staff, the Rover Company and the Owen Organisation express their thanks for the partnership between all other companies in the UK for their support in sponsoring the project".

We continued with the investigation into why turbine rotor cracking had occurred using a new batch of cast turbines. Just as we were drawing no tangible reasons, news came from Deritend, the precision casting firm that supplied the new batch: a change had been made to the way the moulds were cooled. Normally, the hot moulds were allowed to lose temperature naturally by placing them outside in the open air; when cold, the ceramic mould shells would be removed. But, because of the urgency, an airline had been used to cool them rapidly. The super alloy used for the turbine rotors had different expansion characteristics from the ceramic moulds and, as a result, built-in stresses were locked-in at the weakest point.

## Automotive gas turbines

These built-in stresses were relieved when the turbine rotor had hot gases passing through the blades during normal running, resulting in fine cracks. Rotors from other engines were then built into Le Mans engines; these rotors had exhibited no blade cracking earlier and operated without fault. However, it was too late to reinstate the race.

The time came for the annual dinner of the Society of Motor Manufacturers & Traders (SMMT). That year, Lord Kings-Norton was present. He was on the board of Deritend. Unbeknown to Bill Martin-Hurst and Sir Alfred Owen, they were sat at the same table as Lord Kings-Norton. The two men discussed the background of the Rover-BRM car and its withdrawal from the 1964 Le Mans race.

"Some fool changed the casting process at Deritend," said Bill to Sir Alfred, loud enough to be overheard. "If that is so, then I will take issue with the management of Deritend. Please convey the details," remarked Lord Kings-Norton joining in the discussion. Sir Alfred introduced Bill to Lord Kings-Norton and it soon came to light that the latter was on the board of Deritend.

I was instructed to pass a complete review of the details of the turbine cracking to the office of the managing director of Rover. I included letters from Deritend from their general manager, Neville Jones, who was a good friend of mine. Later, those responsible at Deritend were retired.

Looking back on the 1960s, it had been an eventful time and we were only coming to the end of the fourth year. I now had three wonderful children. There was Roland, the baby, who was just a few months old. Deborah, my first child, was making excellent progress and for whom I invented a right and left hand diary so she was nearly ambidextrous. Jonathan was my first son; he was a handsome lad and would soon be starting school at St Alphege. It was a good thing as Deborah was due to return following her accident.

# Automotive gas turbines

The signs were already there, as young as they were. All were very different but we knew how they would fare. Our children would be successful in many different ways.

Looking at the political scene in the country during the early 1960s, John Profumo, the Minister for War in the Tory government had been having an affair with Christine Keeler, who in turn had been sleeping with a Russian diplomat. The jokes about the affair were rife at the time. Another call girl Mandy Rice-Davis of the time coined what would become a phrase that caught on all over the world. Mandy was a confederate of Christine Keeler's and was involved in the scandal that rocked the world. In the court case under cross-examination she said about one of her clients, "He would say that wouldn't he".

One of the jokes current at the time referred to Miss Christine Keeler going to the doctor's surgery with a wooden splinter in her backside. The doctor, after inspecting the matter, uttered, "That's not a splinter my dear, that is half the cabinet".

Macmillan, who was Prime Minister, had a continuous battle with the media and now it was their turn to seek vengeance. His hand was forced when he required emergency surgery on the eve of the party conference. Rab Butler was called in to deputise, but he could not remain leader because he suffered, as Eden had, in being number two for too long. Rab Butler gave the impression of being a weak man but this was not true. He was an intellectual with a good mind and could be decisive on most issues. The only snag was that he lacked Macmillan's superb oratory. Once, Macmillan was asked how he became so successful in making speeches. He replied, "The secret is pause, poise and purpose."

An announcement was made that the Rover-BRM car would race in the 1965 Le Mans race. As for the new formula, this would not be in force until after January 1966. The Automobile de l'Ouest ruled that the gas turbine car must be classified in the 1.6-litre to 2.0-litre class, this despite the fact that the formula would actually place the car in the 1.3- to 1.6-litre class. This set-back was only to be accepted as a misgiving at the time.

Graham Hill and a new BRM Driver, Jackie Stewart, were chosen to drive the car with Wilkie Wilkinson as team manager, as he was in the 1963 race. I would be at the practice sessions, barring any mishaps. The plan for the actual race was that I should share a caravan with the two drivers.

After we resolved the turbine cracking difficulty with precision casters Deritend, the pre-race tests became a formality. So we decided to deliver the paper on the formula in January 1965 to the International Automotive Engineering Congress in Detroit. The paper, entitled *The Classification of Gas Turbine and Piston Engines for Competition Purposes*, was Number 996B.

In January 1965 I visited Corning Glass and met the Houghton family. Rushmore Mariner was vice president of Corning Glass and had married one of the Houghton family. He became a good friend although I felt at the time that his marriage was headed for the rocks.

It was within his domain that the glass-ceramic core of the heat exchanger was being developed. The type of core we were adopting for the Le Mans car was a disc of 18in in diameter and 3in thick. The engine was designed to accommodate two discs. The discs, or cores, had a multiplicity of triangular passages of the order of 0.038in base and 0.020in high; they were of a standard size and were just like a honeycomb. The honeycomb was cemented to a rim around the outer diameter for a radial depth of about 0.75in.

There were many types of Corning cores; they varied in what was termed 'throughway-ratio'. That is, the triangular passages were bounded by thin walls of approximately 0.005in general thickness. The honeycomb passages varied in size, some were more open than others. A flow test was used to establish the throughway ratio. Cores with a higher throughway ratio as a rule increased power by a small percentage at the expense of efficiency; this equated to higher fuel consumption and vice-versa.

Generally speaking, there were three standards of cores with varying throughway ratios, so we had to estimate which standard we would adopt for the race. The race engine came down to the selection of two engines; whichever was used the other would become the spare.

My life was busy, to say the least. One of my many regrets of the time was not being there to spend more time with my young family. When you are involved in a busy life you tend, selfishly, to become oblivious to anything that is not in the swim of a dedicated aim. A dedicated mother, for example, has been there in all circumstances when rearing children to shape their characters and teach them right from wrong, despite a busy father. Sybil was an excellent mother; she was always there when the children needed her.

Evidence of this came when the children were young. Occasionally they would call from their beds for their busy mother. I used to joke and call back to them. All of them, without exception, would cry out, "It is not you, I want mother."

Our social circle was beginning to widen. This was partly through Deborah's friends, although she had been at school only a short time and she had been away for months as a result of her accident. We grew to know the Cross and Lindsay families well; they lived in the same avenue. Alan Cross was a quantity surveyor and he had difficulty with one subject of his professional examination. The subject was Strength of Materials and the Theory of Structures. I used to spend night after night, going over and over old examination papers. When the time came for Alan to sit the examination he passed with flying colours. This was more from studying examination practice than anything I taught him.

Tom Lindsay in contrast was an accountant. He was a good one and eventually he became managing director of Smith & Nephew, but at the time was financial controller of Churchill's.

I managed to learn a lot from him about accountancy. Janet and Tom Lindsay were Scots and the family was about the same age as ours. The Warners, further down the avenue, had a daughter of Deborah's age, and so did the Drabbles who lived in the next road. Pat and Brenda Langmead also had children of a similar age to our own. All their children were invited to Deborah's birthday parties. Deborah was invited to theirs.

Pat Langmead was a schoolteacher at a comprehensive school nearby. He taught mathematics; his other subjects escape me. He was my chess partner – when I had time. We would alternate between his house and mine to play chess. Sometimes he would take a time when considering his next move so I kept a book at my side and would read. One evening, he became quite upset that I was reading; I must have been near to checkmate. Brenda Langmead had taught English and wrote novels under a 'nom de plume'. I met them years later and spoke to them at the Berkswell parish hall one Sunday afternoon when we went there for tea and cakes. Pat looked unwell and Brenda said, "He has been ill". I exchanged notes with them and I was reminded that Pat was working for the Liberal party. Brenda was still writing and it was good to see them as a family once more.

Sybil was taking driving lessons and I helped by allowing her to drive the Rover 2000 with 'L' plates, of course. The driving instructor was a large man, and when he got in the front passenger seat the car took on a list, as if the car's springs would not accommodate his weight. We had many laughs over the fat driving instructor. He redeemed his medals when later Sybil passed first time, as did our three children when they reached driving age.

The death of Aneurin Bevan and Hugh Gaitskell in the early 1960s was a time for the rise of new leaders of the Labour party. Harold Wilson had been an Oxford economist, as was Gaitskell. The two men were quite different in all respects; Wilson was a man steeped in the left, Gaitskell on the other hand was at the centre of the party; he was quietly spoken, almost refined. Wilson, in those early days, reminded one of a 'wide boy'. Wilson issued a challenge for Gaitskell's leadership and won one-third of the votes. George Brown was also a candidate but lost because of his funny behaviour. Wilson could plaster over the cracks when he was called to do so and his stature as a great politician grew ever stronger with his Labour colleagues. While Gaitskell appeared to be snobbish, Wilson could appeal to the workingman; and he had the intellectual background of the former Oxford Don. Of this the honest working man in the street was proud. They simply loved the story of 'the boy who rose from the back streets of Yorkshire and kept his accent'.

Sir Alec Douglas-Home was preferred as the leader of the Conservatives by Macmillan. He was the Earl of Home before he dropped his peerage to lead the Tories. His reign was a disaster. Although he was a presentable gentleman he was ill-equipped to lead a political party. His premiership lasted only from 1963 to 1964, during which time most ordinary people in Britain thought that he had no appeal. He looked like a 'skull' when appealing to the nation, or making a speech. Two up-and-coming potential leaders in the Conservative party were Enoch Powell and Iain Macleod, but at that time they were not to make their mark. Harold Wilson of the Labour party understandably became Prime Minister from 1964, a post he held until 1970.

Whenever I travelled to the US during the period Roland was in his baby phase, he would be waiting in his high chair for me on my return. The beam of his smile was all I needed to welcome me home. It was sheer pleasure. He was a wonderful baby. He always amazed us that he could vault out his cot long before he could walk. We were in ecstasy that he could pick up the smallest of crumbs, hardly visible to us, between his small fingers.

Roland had a small toy slide that stood about six feet high in his bedroom. He was able to climb up the slope only to slide down again. He loved to repeat the operation over and over again, never climbing the ladder. When we thought about it, the reason that he never took the ladder was simply that he could not even walk. Sybil had card pictures of everyday items. For example she had 'A' with a coloured picture of an apple, and 'B' with a picture of a boy, and so on. She went through the alphabet. He soon was able to master all of them, saying in his own language the letter and its name. Deborah sometimes had to translate what he was saying; she had an uncanny way of knowing what he said. Paul Dunford, Sybil's half-brother, painted one of the walls of Deborah's bedroom in colour. When Roland was just a few months old we sat him in the middle of the room and he gazed at the coloured artistry and would have stayed there for what seemed hours on end if we had let him. His diction was difficult and we enrolled him into speech training in his early years. He soon made an excellent recovery.

Roland would be four or five when we discovered that he had the same eye defect as me. He was shooting in the garden, near to the backdoor, with his brother Jonathan, who had an air gun. We noticed that he was having difficulty hitting the target, while Jonathan was on the target every time. We soon had his eyes tested and later he could see through my spectacles, sometime quite clearly. The left eye had a very strong thick lens. Later a lady in the street behind us bordering on our back garden, complained that we were shooting across her garden. I knew that was untrue as I had seen another boy adjacent to the garden shooting into it. I sent the lady away from our front door with 'a-flea-in-her-ear' so to speak. I saw the lady sometime after the incident and she walked passed me, looking quite sheepish; she must have discovered the error in accusing Jonathan and Roland.

The week before the Le Mans race in 1965, Rushmore Mariner from Corning Glass visited Rover. John Lanning, his assistant, was already with us. I travelled to Paris on the Wednesday prior to the start of the race on Saturday. The intention was to meet up with Rushmore on Thursday, but he was delayed and we missed him. Victor (if I ever knew it I have long forgotten his surname), who was Martin-Hurst's 'dog's-body', hired a Peugeot car in Paris and drove me to Le Lude

# Automotive gas turbines

where we had hired a large spare garage adjoining a village inn to work on the car and engines. The car had left the works on 11 June and arrived at Le Lude on 13 June.

The car underwent scrutineering on 15 June; the race engine and the spare engine having been sealed at the works. I left on 16 June for Paris and arrived in Le Lude on the 17 June. The roads in France at the time were heavily crowded so we were near to having mishaps. I doubted whether Victor was used to driving on crowded roads in France, but we arrived at Le Lude in one piece. On the way, he told me about how he worked for MH (Martin-Hurst) at Teddington after leaving the Navy, where he was a sub-lieutenant. He protested that MH believed that all Frenchman were supermen and anything to do with Rolls-Royce in France was a super-super. Franco-Britannic, who had the franchise for both Rolls-Royce and Rover in Paris certainly were super-super.

I met Rushmore in Le Lude. He was with a lady. She was young with dark hair, immensely smart and beautifully dressed. I discovered I had met her at Rushmore's swimming pool in the US; then she was with a group and wearing a swimming attire. Her name was Marie and she was the daughter of the local countess; she was of French noble birth. She lived nearby at Dijon.

Peter Candy drew me to one side and reported that the two practice days had been performed with minor corrections to the car to suit Graham and Jackie. However, the race engine had experienced a rise in exhaust temperature that caused concern. I queried how high the reading was; had it been a sudden rise? Had it now settled? And, if the rise was gradual, did it occur when running the engine or did the rise occur when the engine was warming up on start-up? He answered that it had occurred during the recent practice session; it was showing 650°C and gradually reached 690°C. I asked who was driving and he believed it was Jackie. I also explained that the exhaust temperature varied with the ambient temperature; that could have been high at the time. He agreed that the temperature was hot, in the upper 20s.

# Automotive gas turbines

Despite the excitement of being there, my misgivings heightened. My instincts suggested the engine was changing due an unknown interior variation due to any one of a number of reasons. A gas turbine engine is a continuous airflow device unlike an intermittent reciprocating engine. The thermodynamic efficiencies of compression, expansion and pressure losses obey the same laws. Changes to the cycle conditions, component efficiencies or losses, even ambient temperature variations could show a change in exhaust temperature. Exhaust temperatures remaining substantially steady but altering with ambient temperature only are acceptable as long as temperatures remain within the maximum. It all depends on the reason for the variations in the cycle.

The type of engine powering the Rover-BRM was subject to inspection by those details that could easily be viewed. Short of dismantling the complete engine there is little that could be done to ascertain the cause of rising exhaust temperatures. The combustion chamber cover with the flame tube could be removed. An inspection of the compressor inlet could be accomplished but with the engine installed in the car, apart from minor adjustments, it was virtually impossible to give any accurate diagnosis.

I had a disturbed spell before I could sleep that night. My feelings were one of excitement and concern. I suspected that one or two members of our Rover team were causing ill feeling with the Owen team. This was amounting to an undermining of the whole project to satisfy their games of power and self-importance. I thought it had been brought about by the reactions and lack of understanding of the supposed weaknesses of the opposite team.

I had seen this wrong attitude vented against poor Tony Worster previously. I deplored their treatment of 'Wilkie' Wilkinson, the appointed BRM team manager. Also, their behaviour towards Peter Spear, that was questionable to say the least, although unlike 'Wilkie' he was well able to hold his own against them. My instincts told me they were capable of making unilateral decisions without referring to anyone more expert or senior, so I was determined to keep an eye open for any stepping out of line during the race.

The day of the race arrived all too quickly. I received a lift from Rushmore Mariner of Corning Glass to the Le Mans racetrack from the inn at Le Lude. The Rover-BRM arrived about the same time on the low-loader. I stayed with the car and saw it unloaded from the trailer. The day was sunny, without a cloud in the sky; it was going to be a scorcher.

When I lectured on the subject, I used to say that the cooler the atmospheric temperature the more power the engine gave and, of course, vice-versa. So I used to say that we designed small gas turbines for Arabs and sold them to Eskimos, smiling. Of course, the ambient temperature also effects the performance of other cars in the race but some of them were quite powerful, especially those in the over 4-litre class, namely the engines from Ford, Ferraris and others.

As the time approached 3.59pm on Saturday, 19 June 1965, I thought of my youngest son's birthday I don't think my mind was on the start of the race at all. Somehow, the race started. Graham Hill was one of the last to move off so as to avoid the mainstream of vehicles. The first lap was uneventful and Graham signalled when he passed the pits. Our pit was next to that of Porsche and we had previously discussed their prospects with their German manager, who said, "Porsche that year had an all-French-crew". I could go into the pit area but not over the line; only two fitters/engineers were allowed into the area where the car would arrive. Peter Candy and John Harbidge, both of whom alternated with Bert Hole and George Perry, performed this task.

Graham came in for the first stop in the race but on time for a change. The exhaust temperature had increased substantially and although the ambient temperature was quite high there was something amiss. It was decided to lower the speed of the engine and thus the power. So the maximum speed was reduced to just over 61,000rev/min from about 63,400rev/min and Jackie Stewart took the car out with the power reduced. That pit stop was nearly 12 minutes and was referred to as 'unscheduled'. There were two further pit stops, 2 and 3, when the speed was reduced by an amount to limit the temperature. Pit stop 4 was only less than one minute to rectify the headlights – the plug for the electrical system to the headlamps had been dislodged earlier. The pit stops were now running at no more than two minutes as dawn came up and, coming in for the eighth pit stop, Jackie reported that the exhaust temperature had risen to 750°C yet had fallen back to 690°C following two bangs.

Both drivers reported vibrations through the seat of their pants but somehow the car passed the finishing line in tenth position at 4.00pm Sunday 20 June 1965. It was the first British car to finish out of the 55 that started in the race. However, only 14 cars finished. What an achievement!

This performance was acclaimed all over the world and secured many awards. However, only those with the knowledge of the innermost details understood that it had been a miracle for the car to complete the race. We did not know that we had the best fuel consumption of any of the cars that finished. Nor did we know what was causing the high exhaust temperature on the day the car completed the race, or for many days for that matter. On Sunday night we were invited by the Countess to be the guests at Chateau-la-Valliere and we stayed the night.

Next morning, while I overslept, the Countess and Rushmore rode on horseback. There was no prospect of food as I roamed through this magnificent Chateau. I was becoming hungry, as I had not eaten for what seemed a very long time. In fact, the last meal I had was in the Le Mans driver's restaurant with Graham, during his break from the race on Sunday morning while Jackie was driving.

When the Countess and Rushmore returned I decided to go to the French equivalent of the post office to send a message to Sybil and the children. Rushmore offered me a lift and introduced me to the Count, whom I recall was in his long shirtsleeves; these were hoisted by armlets and were a source of great fascination. Before I left, I agreed that Rushmore would drop me at the Hotel du Maine in Le Lude where I was staying. Victor had left when Rushmore dropped me at the hotel and only Peter Spear and his wife, who I knew only slightly, were just checking out. Peter agreed that they would travel back to Paris and he offered to give me transport.

Rushmore and I, on our way to the French post office, did a short tour around. We went past Chateau-du-Loir on the way back to Le Lude. From the Chateau-la-Valliere we motored to Neuille-Pont-Pierre, then north past du-Loir and on a minor road to Le Lude. I hadn't seen much of the French countryside since I had been in France; albeit it was a brief glimpse. I was attracted to the countryside around Le Mans; it was everything I expected, even watching the cloth-capped burly individual cycling nonchalantly with a French loaf tied to his saddlebag. The loaf was sticking out, gathering flies from all over the neighborhood.

In July, following the Le Mans race in June, I was invited to Corning Glass. I was there to address the top men in the US company on various aspects of developing small gas turbine engines. The general idea was to relate how we were progressing with the development of the Corning Glass ceramic cores, and the problems associated with the regenerator, including sealing, support and the problems associated with developing regenerative heat exchangers.

At the last minute it was decided that Sybil could travel with me, and Rover would pay the airfare, with Corning Glass paying the balance of expenses. We ran round to find babysitters and the solution we found was that my sister-in-law, Doreen, would move in to our Cambridge Avenue house while my brother Roy would manage the home and visit with his three girls overnight.

Sybil needed clothes so we bought a neat stylist suit from 'NOELS' in Broadgate Coventry. The suit was of yellow-greenish tweed. Shoes, handbag, gloves and lots of underwear were to follow. Sybil, who had regained her figure after having three children, was like something out of *Vogue*, the magazine of fashion. When we walked off the aircraft at New York airport the warmth hit us; it must have been over 80°F in the shade. We caught a Mohawk Airlines plane to Birmingham, at which point we needed change to travel to a place called Ithaca, where Rushmore was picking us up.

Returning to the race aftermath, the results of the dismantled race engine examination became known as we left for the US. My feelings were quite mixed. What we found was that a tip of one compressor blade, called 'a splitter', had fractured in a 'vee-form' from the outer diameter. The fractured piece had damaged another splitter as well as the compressor diffuser. The debris had lodged in the airside of one of the Corning ceramic regenerators. The Corning core had partly saved the day and the engine went on running because of the overhung bearing system; in other words, the flexible suspension. The flexible bearing system enabled the mass of the compressor assembly to find its true mass centre. All Jackie Stewart knew was a different feeling through his backside and a rise in exhaust temperature.

Tests indicated that during dynamic balancing of the assembly, too much material had been removed from the balancing land on the back face of the compressor rotor. The disc mode must have changed and the natural frequency of the splitter also. It could have been that there was a thinning of the material sufficient to generate a crack, which developed further with the engine running. The vibration theory was inconclusive from tests carried out, but this was considered the most probable cause as no other was identified.

My spirits were raised when I learned the Rover-BRM had returned the best fuel consumption of the race, despite the drop in power. The global publicity was immense and, given that a batch of Rover-BRMs had been made, they would have sold many times over. It was not to be; the architect of small gas turbine development, Maurice Wilks, had already died. He would have been pleased that the car did so well under the circumstances, but he would have been even more delighted if the engine had not been so plagued with problems.

# Automotive gas turbines

I had already dispatched a set of regenerator seals – for both the high and low pressure sides – for my talks with the American heads of the gas turbine department at Corning Glass. Regrettably, the seals never reached their destination. It was a great puzzle to me how an airline could find no trace of the package. I checked, but when I heard that Rover had allocated Indian Airlines to transport them to the US, I gave up thinking anything could result.

All went well at Corning Glass; it was good to see so many friends from the US. We stayed at Rushmore Mariner's ranch house, near to where Jamie Houghton had his home. Rushmore had sailed across the Atlantic when his was a younger man at Yale University. He told how they used to wear long fur coats in the Northern winters, and one day trappers in a field took a shot at him in mistake for a bear. Fortunately, they missed but he was knocked down to the ground by the blast. Rushmore was a great horseman and asked whether he could bring his hunting outfit to England to partake in the Stow Hunt. He was a good friend of the Pillingtons; one brother, Edward Pillington, lived at Stow-on-the-Wold.

Rushmore was about 5 feet 10 inches tall, but he had a stoop that made him appear less so. His head was slightly inclined to one side and I suspected he was deaf in one ear. He was quite slender and his hair was greying. He was quite fond of chewing tobacco and had a despicable habit of needing a spittoon. His first wife, who was one of the Houghton daughters who owned Corning Glass, had long given up because of this and other habits that could turn one off.

Rushmore was an enigma; he was accomplished at most things, intelligent, well-educated and likeable. However, he gave the impression that he lived life lightly. He was general manager of new products in Corning Glass and a vice president when I knew him. He had John Lanning as his assistant and I managed to get a post in Corning Glass for David Wardale from England, when Wilmot-Breedon gave up the development of heat exchangers. I came to know the Corning people quite well, and their families.

# Automotive gas turbines

We were sorry to leave the Corning district on the way back through New York. We visited Len Raymond and his wife, who had an apartment in one of the tall buildings in the city. Len had been a past-president of Society of Automotive Engineers (SAE) and it was on a visit to Rover that I was pressed to give my first lecture to the SAE. Len Raymond was one of the guests at the dinner arranged by George Huebner, the Chrysler director responsible for gas turbine developments, and his wife Trudy who was Regent at Michigan University in January 1963. The gathering was in my honour and was quite distinguished.

George was a great showman and we had to guess the locality of the wine. Trudy had tipped me off that it was one of George's customs to ask if anyone could guess the source of the wine. Trudy had tipped me off regarding the wine for the dinner that night. So, when the time came for George to ask if anyone could guess the locality of the wine, savoring the taste in a very demonstrative way, we all took a taste and my fellow diners gave up one by one but when confronted I took a second sip. I hesitated long enough to ask what George would forfeit if my guess were right. He was big hearted and answered anything you wish to choose, within reason. So I took my time guessing. I narrowed down the country of origin. George nearly fell off his chair. Then I named the wine and he was further astonished. Lastly, I gave the name of the vineyard. He was simply amazed. I hesitated before expressing my answer, tasting the wine numerous times.

I was due to stay there that night and the following morning was looking to obtain a lift to downtown Detroit. As I waited for the train next morning, a gentleman pulled up and asked me if I wanted a lift into town. I soon joined him and he turned out to be Mayor of Bloomfield Hills. He commented on my accent and guessed I was from Europe. I became quite friendly and we exchanged cards. I promised to phone him before I left Detroit.

It was about the time of the great train robbery in Britain and the local radio station in downtown Detroit came to hear that I was from Rover. Everybody knew what a Land Rover vehicle was, but very few had heard of the Rover Company. To

# Automotive gas turbines

the average person it was the name of a dog. The Land Rover was the 'getaway' vehicle. The news of the robbery was of worldwide interest because never before had £3 million pounds been stolen from a mail train in such a well-planned 'getaway'.

So I was invited to the radio show and when the music stopped, listeners to the show were asked to 'phone in and guess the identity of the mystery guest in town. The guest was always either in the news, or associated with the news. Lots of people tried, but no one guessed that I worked in England for the makers of the 'getaway' vehicle. It must have been very good publicity for Rover.

Such was my first taste of Detroit. I returned to Detroit several times, having visited other US companies that were developing small gas turbines. I can well recall leaving Chicago in a blizzard to travel to Caterpillar, Peoria to meet Ed Johnson, and then returning to Chicago to travel to Phoenix, Arizona, to meet Charles Paul. In Phoenix it was 80°F and sunny. From there, I went to Los Angeles and caught an aircraft to Solar to see Paul Pitt. I then travelled back to Detroit to see Sam Williams. From Detroit I was due to reach Corning Glass but I managed only to reach Albany, due to bad weather. As there were others going to Ithaca, we teamed together to sport a cab for the 40 miles, each paying $25.

We reached Ithaca around three in the morning and I had to knock up the old night porter at the hotel. I was the last one to leave the cab that dropped me just outside the hotel. My team members had paid their 25 bucks to the driver. I was concerned that he would not have had a tip so I gave him $5, but he refused to take it as we had become the best of friends.

When I woke next morning it was late and I was worried that Rushmore would be waiting for me, but he had spoken to the porter on duty and was told that I had arrived in the early hours of the morning, so he knew what to expect. When I reached Sam Williams at Williams International I found I had mislaid my electric razor, so I borrowed his old one, which was better than the one I had mislaid. The weather must have followed me across the Atlantic in 1963 as there was deep snow when I reached England.

Williams International claims today (2014) to be the world leader in the development, manufacture and support of small gas turbine engines. The privately-owned company is headquartered in Commerce Township, Michigan. A second facility, located in Ogden, Utah, is the most modern and efficient gas turbine design-to-production facility in the world. Williams continues to expand its development, test, production, product support and customer service capabilities at both facilities. The result has been a steady well-planned growth since 1955 into a large, versatile organization.

Meanwhile, on my return with Sybil from the visit to the US in July 1965, we were due to leave from New York; Len Raymond and his wife Beatrice gave us a lift to the International Airport. And so began our journey back to England and the uncertainty that the arrangements had worked out for the care of our children and house. Our thoughts were satisfied when we landed at London and Roy, my brother, came to meet us. Everything had gone well; he had become very fond of Roland, the baby the son he never had. Deborah and Jonathan were well and Doreen had managed to cope with the load. We had left a lot of cash so that the children did not want for anything – and none was returned.

It had been Sybil's first taste of the US. I think she was dazzled to meet so many of my American colleagues at one time, but I am confident she enjoyed the visit.

I had to return to Rover to many engineering problems. I used to say that everybody who sought my engineering opinion had a problem of a sort; some had given it a great deal of thought in order to find a solution. Others, however, were at sea, having puzzled over the question. There was a third, much smaller group, which nevertheless could not resolve their problems, and to these few I encouraged them to think it out. So one of the phrases I often used, good-naturedly, was "Think. It could be a new experience!"

# Automotive gas turbines

The 1965 Le Mans effort had nearly eclipsed other gas turbine developments, of which there were many. The whole of Rover Company had been orchestrated to support the race project like no other. It was fascinating to feel the immense support from every quarter of the company. It was just as if the approximate complement at the time of some 11,000 employees were orchestrated to one cause. It was like wartime all over again; even the communist shop stewards went out their way to cooperate. Among the other developments under way at the time was the 'Rotax direct-cranking starter' gas turbine engine for the P1154. The development of the 2S150A for the HS748 aircraft APU was another, and the TJ125 jet engine for the 'Belgium-Emperier' promised to my colleague and fellow director John Griffith, director of sales, Rover Gas Turbines Ltd. In addition, we had a commitment to Carl Kiekhaefer for an engine to take part in the 1966 Paris six-hour boat race. These were a few developments additional to the Ministry requirements and routine R&D. All needed trouble-shooting.

In the event, the P1154 was cancelled and the P1127 would take its place, so about half the power would be needed. This requirement indicated a new engine. Happy days. I was needed in several places at once. I was well away from the politics of the day. I even had to think of who was in Number 10. Of course, it was Harold Wilson. He had become Prime Minister last year, after Sir Alec Douglas-Home.

In August, we took the children for a long weekend. Deborah was making good progress; the boys were excellent.

I had been allocated new Rover 2000 vehicles each year, as a result of the successful races with the Rover-BRM in 1963 and 1965. Rushmore bought a Rover 2000 in the US from Rover North America, where an office was reserved for me in the Chrysler Building in New York; this was across from the Pan Am building where helicopters were allowed to land on the top of the building. I was distinctly embarrassed at the number of faults with Rushmore's Rover 2000. In the end, he was forced to abandon it, allowing it to rust in one of his farm buildings.

## Automotive gas turbines

He excused the vehicle as too refined for the harsh Northern States' weather and I thought of the UK supreme test.

Occasionally, Mark Barnard used to borrow my Rover 2000. Whenever he did so, it felt as if it had been wrecked, for he managed to loosen all aspects of the vehicle. The engine was off tune, the gearbox and transmission, and for that matter the whole car seemed loose and falling to bits. We used to call that supreme test the 'Mark Barnard treatment'.

As 1965 drew to a close I reflected that it was one of the busiest years of my life. It began, as it ended in the depth of winter; one difference for me was to persuade people to spend long hours working.

I sensed that something of great importance was being discussed at the top management levels of Rover. There was a pretense or cover-up during board meetings, so I tackled the present chairman, who at that time was A. B. Smith. He gave me a knowing look and held his hand to his ear. I also asked Roy Pearce, who replaced Geoffrey Searle as managing director of Rover Gas Turbine Limited, but he knew nothing and I believed him.

Roy had come from Rolls-Royce when Geoffrey retired. Roy Pearce had retired from the Royal Air force before joining Rolls-Royce's Atomic Energy Developments. He had taken an apartment off Blossomfield Road. He was slightly balding, with grey hair. He had a military bearing and was about 5 feet 11 inches tall. I remember he was in good shape for his age. He had taken Miss Nicholls as a secretary, who worked for Geoffrey for many years. My office was just across the passage way from his. He claimed to have flown the 'Queen Bee' aircraft alongside the first unmanned radio-controlled aircraft in 1938 as an experiment. I believed his episodes in relation to the RAF. He had a grown-up young son whom he thought the world of. I felt sorry for the boy as he was slightly retarded.

Later in life I came across the lad at Twycroft Zoo. He was doing what he always wanted in life, namely to look after birds of prey, at which he was quite successful. Regretfully, Roy was

# Automotive gas turbines

thrown out of his post when Leyland took over Rover; he landed up in Falmouth where he built-up a marine business. I never saw him again after his sudden departure. I urged Donald Stokes to salvage Rover Gas Turbines Ltd by transferring it to Alvis, the group company that was most keen to take it over.

However, I soon learned why there was such mystery at the top of the Rover management. It was rumoured that Leyland Motors Ltd was negotiating a deal to swallow up Rover. The Leyland Corporation would take over the whole of the British motor industry eventually.

Harold Wilson and Tony Wedgwood Ben visited the Rover Company on the eve of Leyland's takeover. I was with the line of gas turbine cars outside the offices of top management. Harold Wilson shook my hand and said that, "He could use a slide rule." I replied, "Good for you." And he moved on. Tony Benn was chatty. He exchanged the time of the day and mentioned, "We had a mutual friend in Bob Weir". Bob was then director of the NGTE at Pyestock.

Tony Benn said that, "He looked forward to the opening of cell number five at NGTE for the testing of Concorde's engine and maybe I would allow T4 to be at the event." The Olympus 593 engine needed a special cell to be built at NGTE because of the supersonic speeds that would be under test there. Cell number five was built at Pyestock, near Farnborough, to accommodate this testing. I rang Bob Weir later to arrange for the T4 to be present at the opening. Bob was delighted.

The 593 turbojet engine was derived from the Rolls-Royce (Bristol Siddeley) Olympus engine, which powered the Vulcan. However, in the case of the 593 installed in Concorde, because of the aircraft's supersonic speed, the air intakes had to be designed in a special way. A means of sucking to fill the air intake was essential at speeds above sonic. Air, when an object is passing though it travelling at high speeds, forms into a solid bar unless there is means to pull the air back on to the boundary walls of the intakes. Then the intakes become impoverished of air and the efficiency is decreased.

# Eight
# Gas turbines in trucks

In which we receive a commission to develop a novel gas turbine engine for use in commercial vehicles.

DR BERTIE FOGG was director of engineering Leyland Corporation. I knew him slightly. He had been director of MIRA before taking up the post with Leyland Motor Corporation. He was a small man with a triangular face and greyish hair. I had not seen him for many years. In fact, the last time was at a routine Ricardo committee meeting to discuss tests on radial inflow turbines that had been supported by MIRA several years ago. I agreed to meet him at my place of work, as he was keen to see the gas turbine department at Solihull. His visit was arranged in a month's time so I did not give the matter another thought.

We had many visitors from all over the world. Most of the celebrities who came to Rover asked to see gas turbine developments. Among those were Prince Rainier and Grace Kelly, Harold Macmillan, Tony Wedgwood Benn, Colin Chapman, Peter Townsend, Donald Campbell and Frank Ifield. Frank's father had been a designer at Lucas before taking his family to Australia where he still worked for Lucas. I knew his dad when he was at Lucas in Britain and the visit was arranged through his father. The girls from the Rover press shop got to hear that Frank was visiting the gas turbine area and great crowds of girls from all over Rover congregated round my office window. They followed us both to lunch as we walked; he was quite used to this treatment wherever he went, anywhere in the world at the time. The time came for Dr. Fogg's visit.

I knew when I showed him round that it was not the main reason for his visit. He was quite pleased with what he saw and when he returned to my office he broached the real purpose of his visit. First, he explained that what he was about to say must be kept between himself and myself.

"As you may well know, negotiations are taking place to determine the future of the Rover Company as part of a larger plan for the British automobile industry," he said. He went on: "It is the intention of this government to form a combination of the motor industry in the UK. Harold Wilson's deliberation after a lot of careful review was that Leyland Corporation should play a central role.

In the context of Leyland's position, Lord Black had died recently and the board of Leyland was in turmoil as to a successor. For a time, Donald Stokes did not have the support of the whole board as the majority of members had backed another candidate. The other man who was tipped belonged to another political party and was unacceptable to the majority of Wilson's cabinet. To cut a long story short, the other board member resigned when he realised 'who was the power behind the throne', so to speak. Donald Stokes became the new leader of Leyland; he had Government backing to work towards a united UK motor industry. Now you know the essence of the plan we are about to review all aspects of engineering and manufacture across the broad spectrum of UK industry."

Many questions went through my mind. The simplest question was one of 'political pressure'. If the heir to the throne in Leyland could be overturned due to politics, what hope was what hope was there for all the Conservatives and Liberals?

I asked Dr. Fogg "If he could divulge the name of the board member who, before Donald Stokes, was tipped to run Leyland?" He declined to do so saying, "It was water under the bridge now".

I could appreciate the sense in combining purchasing, and rationalising the wasteful elements, but I was aware that there was something in the back of my mind. That night I thought hard to see what it was in my head. The only notion was the independence of the private companies that I knew so well. They got on better for being allowed to work as small teams.

# Automotive gas turbines

Britain's greatness was the result of small entities, as well as its inventive genius that must be preserved. Britain's empire was dwindling, so we could not depend on a global market. Britain's future lay in the essence of what the country was good at; it was just as if we gloried in innovation and not in production. I was firmly of the view that we should stick to advanced technology rather than try to emulate the US or Japan.

Dr. Fogg indicated that Donald Stokes was keen to apply the turbine engine to trucks and buses and he wanted my advice on the feasibility.

I talked enthusiastically in relation to their application to trucks, buses and trains. The turbine has many advantages, however, it has to be better than best diesel engine on fuel consumption and cost for it to make serious inroads into the volume market.

He asked me to compose a case for the turbine in competition with the diesel engine. He forecast that the legal limit of power to weight ratio for trucks of gross weight of 32 tons would at some time in the future be projected to rise 8 to 10 bhp/ton; that implied engines of up to 300bhp.

Dr. Fogg asked whether there was a developed engine that Leyland could install in a truck to illustrate the point of a gas turbine truck? The only engine that might be suitable was the one I had just raced at Le Mans, the 2S150R. But whereas this engine was just about applicable to cars it was no match for the diesel engine's fuel consumption. I pointed this out clearly, but he seemed set on a demonstration unit, even if it could not match diesel engine fuel consumption. The other snag I pointed out to him was that the power for best fuel consumption was only 130bhp.

Dr. Fogg was already working out what truck could be installed with the 2S150R engine. He suggested the Leyland Super Comet (photograph, p.442). When fully laden at 22 tons gross vehicle weight, it was normally powered by the Leyland 680 diesel engine of around 130bhp from memory. He also said that Donald Stokes' interest for gas turbine engines was centred on trucks and buses, and he had no interest in aviation. I stressed that Rover Gas Turbines Limited had a responsibility to the Ministry for the front-line bomber in the shape of the Vulcan. In the event that the negotiations were successful, and Rover became part of Leyland, then there must be a way of ensuring that RGT would survive. He was non-committal about the future of Rover Gas Turbines Limited and I suspected that I would have to use all my ingenuity to save the business.

He turned to another line of thought. He touched on the future but was careful to use qualified statements. "Assuming that talks between Leyland and Rover continue," he said continuing the discourse, "Would you be prepared to set up a gas turbine company here in Solihull to place the concentration into developing a gas turbine truck, say of gross weight fully laden of between 32 and 38 tons."

The idea had me reeling. On the one hand, the challenge was just what I wanted; on the other I was thinking of Rover Gas Turbines Limited and all the personnel within. There were over 300 members of Rover Gas Turbines Limited and I was the technical director. These numbers dwarfed the Rover gas turbine department that I headed. The number of people there was less than 90, most of whom were qualified engineers. Rover Gas Turbines limited (RGTL) was responsible for selling and, like any other company, had a manufacturing organisation, sales, service, publications and training. They were reliant on Alvis, which had been taken over by Rover Company, for making many of their engine parts and the balance came from all the associated industry. Assembly was at Solihull in RGTL's premises called, the 'Coseley-building' adjacent to the Rover gas turbine department.

My reply was guarded. I replied, "If RGTL could be found a place in Rover, I would welcome the opportunity." Dr. Fogg said he would regard that as my willingness to do so. And with that he shook hands to leave smiling, saying what a good day he had had. I asked him where he was living now and he said, "In the St. Annes area, on the sea front." I recalled that when he was director of MIRA he lived near Leamington Spa at Blackdown.

That summer, Carl Kiekhaefer thought he would enter the Paris 6-hour boat race that would involve the 2S150 with its annular combustion system. Three 2S150 engines were already allocated – two race engines and one spare.

Every year the Paris boat race took place in October. The River Seine was closed off to river traffic for one hour in six, but the traffic was allowed to start up after one hour. Powerboats race up and down the river reaching 70 miles per hour over a course of a mile. They spent more time out of the water than in it. The mortality rate of boats was quite high and there have been fatalities, as the slow moving traffic was allowed to move after one hour. In powerboat racing, to be thrown in the toxic river waters is to be avoided. Drivers are treated before the race just in case they are thrown into the water.

Sybil was able to come to Paris with me at the last minute, as both John Griffiths and his wife Jean could attend the race with us. I say at the last minute, because Sybil's tweed yellowish-green costume was at the cleaners when we were packing to leave.

Fortunately, I was able to telephone the owner of the cleaner's shop in Solihull and he agreed to turn out late that night and search until he found the particular garment. I had explained to him over the 'phone who I was, and that it was important to trouble him as we were leaving from the airport early next morning before the shop opened. He was an elderly man; he was most friendly and he opened the shop cheerfully.

John and I decided we would fly from the airport with each other's wives on different flights, so that if in the unlikely event of the aircraft crashing both couples shared the risk. Jean and I would travel first, and Sybil and John would leave just 30 minutes later as the aircraft were quite frequent from Heathrow. In any case, they were travelling on Air France while we were on the BEA (British European Airways) flight to Orly.

When Jean and I arrived at Orly, as we were waiting for the other two, Peter Spear emerged from the landing exit. He was just about to suggest I travel into Paris with him when he noticed Jean Griffith. Stammering, I introduced Jean and mumbled an explanation as the other two emerged. John knew Peter of course, but Peter had never met Sybil. The other two were much calmer than Jean and I. After pleasantries, Peter looked askance and said, "Do enjoy yourselves, you four," I don't think Peter ever believed me. Sybil and I had a good time in Paris during this trip. We met Kiekhaefer, unfortunately his two gas turbine powerboats sank in the Seine and were out of the race. His boats were Italian-designed Molinari class vessels.

When both boats sank, Kiekhaefer, who was a law unto himself, claimed another two boats that finished the race. We removed the two engines from the river Seine and returned them to the company. When they were cleaned out and reassembled again they were tested; they gave marginally more power than before the 'insane-test', as it was nick-named.

Mrs. Bunting, Dr. Fogg's secretary, rang me and said, "I should ring her boss at my convenience". When I rang him, he said, "Rover has concluded its negotiations. I should hear an announcement shortly." He also indicated that Leyland had located a Comet truck to be driven to the Rover works. A note would be posted off to me about this and other matters.

The note from Bertie Fogg confirmed the Leyland Comet tractor unit would be available to have the gas turbine engine installed at Rover. This would coincide with the timing for the hand-over. Also included was confirmation of three other matters. (Comet became the forerunner of trucks for BP, Castrol and Shell. See photograph, p.395).

Peter Waters had become my right hand man on gas turbine administration. He started with Rover during the end of the 1940s when he was demobolised from the Fleet Air Arm. He had been a wartime officer and had worked his way up on the administrative side. Peter had never married and looked after his mother. He often talked of his father, who died during

Peter's later years at grammar school. Mrs. Waters, we suspected was Peter's burden and the reason he never married. However, he was no oil painting. Often, old Mrs. Waters would ring round to the hospitals and the office, if he had to work over, but that was rare. She had many critical illnesses that we took with a pinch of salt. Once she was reported to be suffering with a very critical illness that we doubted, and my children, who used to frequent the local swimming baths, spotted her in the baths attired in full regalia and swimming cap.

Peter had an old Rover, one of the first P4s made. He wore a trilby hat and always the same light-blue tweed suit and brown shoes. He was slight in build, average height and sported a Guinness mound where his stomach appeared. Peter was a godsend to engineers; he could turn the most difficult and long technical report into something understandable. His was renowned for his understanding of English and how one should to express oneself. He believed in simple sentences and he often said to engineers, "Write it down as you who say it so that anyone can understand." He was a great believer in the reports written by the Air Registration Board (ARB) of the time, because of their clarity and objectivity. When our children where young, they were amazed that I always knew when Peter was streets away from visiting me. His old Rover had a habit of groaning quite loudly when it was forced to drive round a corner.

One day at Cambridge Avenue, when Peter came to pick me up to go to a meeting, he placed his briefcase in the hall at the bottom of the stairs. When we came to leave, his briefcase was missing. We looked everywhere to find it; we even looked inside his car. Then I had an idea; I looked up the stairs and two little eyes were peeping round the banister. Jonathan, who must have been four or five at the time, had taken the briefcase upstairs for a practical joke. He was always a joker. We soon recovered it and went on our way. Peter regretted that he had no children.

He would have made a good father and husband. He could turn his hand to anything. It was always a great pity that he promised his dying father that he would take care of his mother, who we thought played on that fact and kept her son captive. Peter Waters proved an immense help to me in many ways. He wrote paragraphs that I used for talks and lectures. He was a born administrator and he knew how to remain in the background. Above all, he was a man you could turn to for advice on any subject. We used to laugh about his wide range of abilities, emphasizing that he could speak on any subject for at least three minutes and then came to full the extent of his knowledge. One occasion of his immense assistance to me was during the Leyland merger of Rover and the subsequent formation of Leyland Gas Turbines Ltd.

In the autumn of 1966 we considered moving house but we had no idea where to go, except nearer the school. The Wilson government was in power and becoming stronger every day it seemed, so the stability of the country appeared to be assured in future years. On the work front, Rover Gas Turbines Limited was to be closed and its employees looked to me to save them.

Whole families turned up at our home in Cambridge Avenue, worried about their jobs within RGT. We decided that the only place for them was the back lawn where the children could play. It was there that I assured them that I would do all in my power to see that RGT survived.

Roy Pearse and John Griffiths, the other directors of RGT, had left, having resigned their directorships. John Griffiths later turned up at Centrax, largely because he knew someone close to Dickie Barr, who founded the company. Roy turned up much later in Falmouth where he founded a marine organisation.

At the last minute, I persuaded Dr. Fogg to have words with Donald Stokes in order to transfer RGT into Alvis in order to enable the company to meet its obligation to the Ministry, which by that time had become the Ministry of Technology (MinTech). I had undertaken a commitment to find all those displaced by the closure of RGT that I would find them alternative employment. At the last minute, Donald Stokes relented and RGT was transferred into Alvis and with that the majority of the jobs were saved.

However, I still had to find jobs for the balance of people who Alvis had decided it was not going to 'double up'. Some of these people found posts that were scarce to get; it depended on age and experience. It was the time of mergers and acquisitions; the word 'Rationalisation' became an excuse for sacking people, some of whom had long service and expertise of great value.

I was expected to remain as technical director of Rover Gas Turbines Limited, even though it had been transferred to Alvis in Coventry. My new post was managing director of a new company called Leyland Gas Turbines Limited (LGTL). The company had been founded but still had to be formed. Dr. Fogg and Ron Ellis of Leyland were the other two directors appointed. I was also invited to join the Corporation's weekly technical meetings at Leyland, Lancashire.

I found it difficult to contribute to the meetings as most of the matters arising concerned the design of trucks and buses. Sometimes diesel engines were discussed. Bob Frier was chief engineer of Leyland. He always gave me the impression of being a very sound engineer but he was a difficult man to get to know. There were representatives also from most of the truck companies Leyland had taken over, such as Albion in Scotland, AEC in Southall and Scammel Heavy Vehicles located at Watford, in the South. I recall Lolly Watts of Scammel, who was tall and a good engineer. I sometimes got a lift with him. I cannot remember why, as I used my own car most of the time. Lolly had a good insight into other members of what he called a 'Committee'.

Sometimes, I was the guest at the hospitality house in Leyland. The house was large and I recall, to the best of my ability, that a very wealthy northern industrialist must have owned it at one time. It was well kept and the rooms were many and all were immaculate. Donald Stokes must have stayed there many times as he had a desk in what was called the library.

At breakfast, the food was excellent and served in silver dishes. It was the only way I became acquainted with others attending the Corporation's technical meeting. They stayed the night because of the distance they had to travel. Once a week proved a great burden, especially when I stayed the night, as I was used to being a 'hands-on manager' and, in any case, we had a new gas turbine to design.

I tolerated this for three years. When Bertie Fogg left his wife, after being married to her since he was young and at university, and thus became absent from the weekly meetings, I ceased attending. Fogg's office objected, but I used the fact that I had a company to run as a good reason for my lack of attendance. At that time we had problems with the new gas turbine truck engine.

During Harold Wilson's early reign in government in the 1960s, a radical change in the education system took place as part of the new look socialism. Colleges of higher technology became full universities. In addition, new universities were established. Sussex was the first one and thereafter followed Kent, Warwick, Lancaster, East Anglia, Essex, Stirling and others.

These new universities were to play a significant part in the higher education for women. There were profound changes in other areas of education most notable of which was the criticism to abandon public and grammar schools. Local Authorities were encouraged to establish comprehensive schools. Political comparisons were made with European and American counterparts. The socialist call was for more business training. The emphasis was put on entrepreneurs and vocational training in place of classical scholars.

The number of public and grammar schools had expanded in Victorian times to educate the growing middle class capitalists, and to introduce more industrial training. The headmasters of public and grammar schools favoured classical education for a very long time. Headmasters used to look down their noses on anyone who wanted to follow a business career. The change to these sentiments came about albeit very slowly. Secondary education became theories of class as Labour made it so. Although most of those in power had, one way or another received secondary education and, in the case of the Prime Minister, had the benefit of Oxford University, yet he still spoke with a Yorkshire accent that endeared him to his socialist colleagues.

The first new house we occupied and were buying through a mortgage linked to a life policy was put up for sale. We went to see a large looking dolls house in Alderbrook Road that was being sold by Saint Martin's School through their headmistress. At first sight the property was not inviting. It seemed that every guttering was dripping with rainwater in a torrential downpour. We were lucky to get the place for a low price, although it needed a lot of attention. I decided that before we moved in to have at least the electrical power system converted by the local electricity board and central heating installed. We managed to salvage an oak fireplace place from the scrap heap when Saint Martin's School moved their lower school in Alderbrook Road. This was also installed in the main room. The floorboards in this room were one and three quarters of an inch thick and were replaced and a new vent added. When we were ready to move in I swear that no male had ever lived there. The house was built in 1903 and had been occupied by two old ladies who started Saint Martin's Boarding School for foreign diplomats' daughters of refined upbringing. The school had degenerated into taking the local daughters of 'the nouveau-riche'.

The report on the new turbine truck engine was complete. A smaller brochure was compiled together with pictures taken through the years showing the history of gas turbines and culminating in several pages of the new 350bhp engine.

Sir Walter Cawood was chief scientist of the newly formed Ministry of Technology. Tony Wedgwood-Benn was the Minister of Technology. I had requested the Ministry for £1.5 million pounds to support the new Leyland engine for trucks, as set out in the new brochure. I think John Parks, the head of Alvis, came with me. When I completed the presentation Sir Walter, who knew we had a mutual friend in Bob Weir, (director of the NGTE) smiled and said, "You should apply the circular rule against the risk of escalation". I asked him what he meant, and he said, "The circular rule is that you multiply by about twenty two over seven for escalation commonly called pi". Both John Park and I laughed with embarrassment. Sir Walter was very appreciative of the presentation and even John was impressed. He had not heard the case before.

The case for development support included a foreword and contents. The experience gained so far was outlined and illustrated with Rover photographs from Jet 1 to 1967. The power requirements of heavy vehicles were discussed in terms of future legal requirements, advocating a design area of 10 to 12 bhp/ton. Remembering that the year was 1967, the projected EEC legal limit on gross vehicle weight was 38 tons as opposed to the then existing limit in the UK of 32 tons. Considering the time from development into production, one could safely forecast the need for a compact 350bhp engine. The market was outlined in terms of the US and Europe. It was estimated that in the year 1975, the European need for heavy commercial vehicles above 32 tons GVW would be 29,000lb plus, while in the US the requirement would top 85,000lb.

There were at least 12 heavy truck and bus makers in the US and, of these four had gas turbine engine R&D programmes for trucks. So it was reasonable to estimate the Leyland could sell into this market with a developed gas turbine engine.

In Europe, Mercedes-Benz, Fiat, Deutz, MAN and Volvo all had gas turbine developments under way.

Leyland's name in the US was poor compared with that of the Rover Company, and Carl Kiekhaefer had ceased to trade with Rover on the grounds of the take-over. Leyland sold products into Cuba, despite the US trade embargo.

The case for the specific gas turbine engine power plant recommended was outlined in stages. While Stage 1's performance hardly competed with the diesel engine, by Stage 3 at 350bhp the engine was more competitive. However, Stage 6 was essential to meet the best diesel engine fuel consumption.

Comparisons of the diesel and gas turbine engine operating costs were based on information taken from major oil companies and *Commercial Motor*'s 1966 operating costs for the appropriate gross weights up to 38tons GVW. The overall comparison of power plant parameters of the diesel and gas turbine engines was tabulated showing diesel engine fuel consumption at about 10% less than the turbine engine. The factory cost was difficult to estimate in the case of the gas turbine engine as no vehicle gas turbine engine had ever been produced on a volume basis. Nevertheless an estimate was given of the factory-cost, based on the same production volume of a 350bhp diesel engine of a known type, namely a Caterpillar engine of 1966.

Also included was a comparison table showing a heavy oil tanker performance with its diesel engine removed and replaced by the proposed gas turbine engine. The proposed development programme listed instruction to proceed, based on July 1967 start date, and with volume production estimated for November 1973. The existing team that was about to join Leyland Gas Turbines Limited, the associated test equipment and the environmental and social aspects of the gas turbine engine, were also included. Lastly, a critical path analysis of events to November 1975 was shown in in a 'pull-out'.

Sometime later I heard that we would receive the £1.5 million, provided we added at least an equal amount that could be based partly on overheads applied to the basic cost of manpower and materials of the total development.

The question of whether we should accept the Ministry offer was raised with Donald Stokes. He decided that accepting Government money was onerous as the Ministry could claim their investment with interest be returned if, and when, the developed engine went into production. He gave as an example the Rolls-Royce Dart engine for which the aero-engine maker was still in debt. Thus I had the odious task of turning down the Ministry's offer, having made the application to Ministry of Technology.

Fred Morley of Rolls-Royce, who had just taken over the company's plant at Leavesdon, Hertfordshire, known as the Rolls-Royce Small Engine Division, approached me to examine his case for a new engine, the BS360. His new engine was not to replace the famous 'Gnome' but was Rolls-Royce's attempt to develop a low-cost engine. I said we would do a technical cost critique, providing he would do the same for the new Leyland engine, the 2S350R. It was agreed.

The result was interesting. Rolls-Royce placed my case for a new engine, the 2S350R, at 10 times the cost of my estimation, and this included the Leavesdon overheads. We estimated a tenth of the Rolls-Royce cost estimation, using overheads provided by Chris Peyton, executive director and financial controller of Rover Company. Admittedly Rolls-Royce had included more development engines and rigs than we had put in our development project of the 2S350R.

When we were at the meeting with Rolls-Royce at Leavesdon to analyse the overall result, we took up the opportunity of being shown round. We were amazed at Leavesdon's overheads; they were over 900%. I remember commenting at the time that for this the fire station surgery, and the flagpole with its flags must be included, to say nothing of the far-flung empires of Rolls-Royce. It was an eye-opener as to how things were done at the time by businesses outside the automotive industry in the large jet engine business. The jet engine was largely dependent on the world's civil aircraft industry as well as military business; the company was wholly in the pocket of the national whim.

The words of Stanley Hooker rang in my ears. He said to me once, "Rolls-Royce could not compete with the automotive industry, and produce a low-cost gas turbine, to save their lives".

Stanley was at Derby in the early days of the jet engine when Frank Whittle was at Lutterworth. He had the distinction of designing some of the first turbo-blowers used by Rolls-Royce and as a result figured largely in the early days of the jet engine.

Due somewhat to the internal politics of Rolls-Royce Derby, he moved to Bristol Siddeley, located at Filton, near Bristol. There he became chief engineer. When Rolls-Royce and Bristol Siddeley were merged, Stanley then became a director of Rolls-Royce and was back working in Rolls-Royce.

We were sad for many reasons to be leaving Cambridge Avenue. The first was that we had reared our family of three wonderful children. Through the children we had made lots of friends, both in the surrounding area and in relation to their schooling. The family had survived all kinds of set-backs, the biggest being our daughter's road accident. We were there for about 12 years and in that time we had witnessed intrigues, puzzling affairs, and good and bad times. I suppose this is just a cross-section of any community. Sometimes it is folly to name names, as many of our old neighbours could be still living today. However one woman, who lived several doors away from where we lived at number 29, we suspected as having an affair with her best friend's husband. He worked in a bank in Solihull and his car was parked frequently outside the woman's house frequently in late morning.

The husband worked in the Black Country at a company founded by his father. He was sometimes late leaving and as soon as he left, the best friend's husband's car appeared, or so it seemed to the women in the road who kept a careful scrutiny over what went on in the Avenue. Call it nosey or just outright jealousy in relation to the supposed sexy orgy taking place at the bottom of the road. There must be a 'Peeping Tom' in every road. A young married man walked round the block most evenings at the about the same time. He stood in the shadows outside our property looking across at the lady's shadow in the upstairs bathroom at the front of the detached house opposite. Sybil remarked many times, whenever he took up his stance, "He's at it again. I must let Pamela know not to shower so regularly and to draw the bathroom curtains". There were one or two families that had broken up for one reason or another and with surprised shock. Usually it was ones with young families.

The break-up could come as a blow. One might comment, "Who would have ever thought that of 'old-so-and-so'. They were the last ones you would have suspected." But it happens! There were the usual scandals. When one of my neighbours passed the time of day with a man at the bottom of the road enquiring what he was doing now, and knowing he had set up his own company, he got his answer, "Voluntary liquidation old boy". Another neighbour was jailed for not paying his council tax. His excuse was that he refused to support a council that was forever increasing council tax more than inflation. And so on. That was life in Cambridge Avenue in the 1960s.

We moved into the house in Alderbrook Road at the end of November. The children loved it as they now had their own rooms. There were two main rooms at the front of the house on the first floor. The one at the top of the stairs was Roland's room, at the other end of corridor from which a stairway went up to the second floor. Sybil and I occupied the room at the front; off this there was a wide hallway that could have been a dressing room when the house was first built in Edwardian times. Jonathan had the room at the back that looked out onto the back of the house adjacent to ours. Deborah was located next to Jonathan's room also looking out on the back garden. Deborah's room became the playroom, as it had large cupboards. On the second floor was a further bedroom looking out onto the front; this was at the top of the stairs which had a 'dog's-leg' turn.

Another passageway led to a smaller room that eventually we turned into another bathroom as there was already one on the first floor adjacent to Deborah's room. At the end of the passageway on the second floor was a low, thin door with a small cupboard bolt to secure it. The low doorway opened into a room with a window at the top of the house. Off that room were two other rooms, one larger than the other. A thick wooden beam ran through all the rooms at the top of the house. It reminded me of a full-sized dolls house. It was in the far room at the top of the house that I discovered coat hangers from all areas of the globe; evidence I supposed that this had been a school for diplomat's children from all corners of the world. Knowing that it was previously a girl's school, I felt as if my two sons and I were intruders into the spirit of bygone days.

One day, Donald Stokes visited Rover. He usually drove himself in one of the Corporations vehicles. He was actually in a Mini-Cooper. At this stage, negotiations were taking place with the British Motor Corporation; it was just a formality that Leyland would merge with BMC. History would show that Sir George Harriman, chairman of BMC, had taken a back seat during these talks on the grounds of ill-health.

Previously, he had helped Bill Martin-Hurst, whose son Richard was accused of being in possession of a Mini-Cooper car knowing it to be stolen, to avoid being prosecuted. There was no love lost between Martin-Hurst and Harriman, but they joined up to give evidence at Richard's trial. Richard Martin-Hurst was a tear-away at that time, which brings to mind the story of his arrival at Rover during lunchtime. It was his custom to bring his car into Rover to fill up with fuel because his dad was managing director. He banged on the hatch way, where the man who served fuel was sited, and called out, "Open up". The hatch-way was lifted and quickly closed again and a voice said, "'F--- Off', I'm having my lunch". Richard then banged the hatchway again and said, "Do you know who I am, my man". A voice shouted behind the hatchway, "I don't care if you are 'F------' King-Kong, piss off". Richard's father heard the account from Richard. It was alleged that he immediately called for the petrol attendant and made Richard apologise.

Donald Stokes, who I had not met, wanted to visit the gas turbine area some while in the future. I was asked to see him at the top offices in the office of Bill Martin-Hurst's assistant.

He was alone and we shook hands. He asked me, "What would you need to have a running truck with say a 300bhp gas turbine engine installed for exhibition at the Motor Show at Earls Court next year?" I did not hesitate and answered, "An open cheque". He was taken by surprise and said, "I will ensure that you have a budget in keeping with the task. I will visit your empire when the gas turbine truck is run for the first time. We need to have a special truck designed and styled for the purpose." We exchanged other pleasantries, and then he walked out, complaining of long corridors and doors without windows.

He was as good as his word. The budget allocated was generous, but would not be adequate to resolve the two basic problems we encountered in developing the 2S350R engine in its later stages. Nevertheless, the object of having a running gas turbine truck for exhibition at the motor show in 1968 was achieved, but the engine was under protracted development. I rarely saw Ron Ellis in the early days, but Dr. Fogg was a frequent visitor. We were both proud of Leyland Gas Turbine Ltd. The setting was neat in so much that we were all sited in one building; only the test rigs and test beds were located elsewhere in the identical position of the Rover project department.

The first engine was manufactured from a design that caused me certain misgivings. I could not rightly put my finger on what was wrong but three aspects concerned me. The engine was virtually a scale-up of the 2S150R, but with variations. One was the main casing. Whereas the previous engine had a sheet metal casing, the 2S350R was designed with a nodular iron casing that was cast for low-cost. There were basic differences in the design, in particular to accommodate cooling.

The turbines were not a scale-up but were designed by a man from Holland who studied at Deft University under an English professor I knew well. I suspected that he was new to the game but my assistant, who became chief engineer, trusted him.

Also, I had misgivings in relation to the scale-up of the Corning Glass ceramic discs or cores to 30in diameter. I mistrusted large pieces of ceramic materials. One rule of mine for designing in ceramics was to use the materials in only small pieces where possible. My misgivings were soon founded. The first engine run showed that the compressor turbine developed a small crack at the root of the blade on the convex side. That was unusual and could only be seen by means of the Zyglow ultra-violet light crack detection system.

The other sign was a defect in one of the ceramic discs on an outer diameter. It appeared to be a separation of the ceramic convolutions that had not shown up when the part was delivered. This amounted to gross weakening. Several years later the problem was traced to leaching of the lithium – the material being Lithium Alumina-Silica or abbreviated to LAS material.

The secret requirement of glass-ceramic, expressed simply, is to hold stability throughout a wide temperature range and establish near-zero expansion. The strain measured in parts per million should also remain stable throughout the working temperature range and, ideally, the gas turbine heat exchange environment. By so doing, it could remain perfectly flat, even when glowing white-hot. It exhibited a non-brittle behaviour.

The day came for Donald Stokes to visit and see the gas turbine truck (see p. 394) with the new engine installed. The truck appeared in an excellent style. It had been styled by David Bache, at that time head of Rover's styling department. He had styled the Rover-BRM.

The truck looked splendid with a flat floor and there was no bulge running through the cab, a characteristic of the diesel truck. It was truly an ergonomic design. We had installed the engine only the day before Donald Stokes visited and we kept the run-up of the gas turbine truck within the build shop as a surprise.

He visited alone. When we said we could start the engine installed in the gas turbine truck within the build-shop, and without any ventilation because the gas turbine engine was so clean, he was astonished. We were prepared for a request to have a run in the truck but he only wanted to see a start. The truck was stationary but we demonstrated acceleration up to full speed from idling while he sat in the cab as a passenger. He alighted from the gas turbine truck and was well pleased. It was just three weeks to the Earls Court Motor Show.

The engine was still running when I noticed a fine, white dust appearing from the exhaust. It settled on the back of the jacket of his dark suit. I signalled to the fitter in the cab to cut the engine. Donald was absolutely oblivious to the white glass ceramic power on his back. I nearly had an urge to brush him down but kept quiet.

When he had gone, later in the day, his secretary rang to say that he would speak to me by phone shortly. When he rang my heart was in my mouth. I thought he may have discovered the white dust, but there was no mention of it.

He was in raptures that we could run the truck inside due to its clean environmental aspect. He suggested that we arrange a demonstration at AEC Southall, on the outskirts of London and invite Tony Wedgwood-Benn, the Minister of Technology, to the proceedings. I said I would cooperate with my counterpart at AEC to make arrangements for the gas turbine truck to be diverted on the way to Earls Court.

He said, "Ensure that the truck is a runner and have it driven inside to impress Tony in relation to the aspects of the gas turbine engine cleanliness."

The scene was set. The arrangement was to give a slide show and then the gas turbine truck would be driven inside a marquee tent to illustrate its environmental cleanliness. The Minister would be invited to say a few words. The point was, it was just about possible live in the exhaust gases of the turbine, as illustrated by the use of the 1S60 gas turbine used to heat Army arctic tents directly from the exhaust gasses. The piston engine exhaust gasses would be fatal after only a few minutes.

The engine demonstrated in the build-shop for Donald Stokes was dismantled and one Corning Glass heat exchanger core was in distress. Whether it was a coincidence or not, the right-side core on other engines failed before the left-hand side. Also, under crack detection, there was the slightest sign of a crack on the convex side at the root of the one of the compressor turbine blades.

# Automotive gas turbines

My chief engineer invented a new term. He referred to the 'big-bang'. I preferred not to use the term, as the phenomenon amounted to a gradual deterioration at the start, before speeding-up to the point that the glass ceramic core failed like someone breaking a digestive biscuit.

We took no chances with the Southall demonstration and a new engine was installed in the gas turbine truck. We had taken special care that each component was inspected twice, particularly the heat exchanger cores and compressor turbine blades. Nevertheless, I had my suspicion that we had experienced one of those problems that all those working with gas turbine engines fear: a 'blade flap vibration at full speed'.

The Southall demonstration was a great success. Tony Wedgwood-Ben caught on and made an excellent talk on the virtue of gas turbine vehicles and the benefits of its environmental aspects. The gas turbine truck entered the marquee tent in quite a majestic way. To say that I was impressed is putting it mildly. Donald Stokes and Bertie Fogg were present and when Wedgwood-Benn entered he was loudly applauded.

I gave the slide presentation using too many slides. It was a problem of mine. I had to be more ruthless in the selection. I showed too many slides. The best talk I ever attended was one given by Spen King; he talked for 40 minutes and used only three slides.

The Earls Court Show was another equal success. Although I wasn't around at the time, Ivan Swatman the head of Ford's gas turbine work in the US and who was also a friend of mine, had an argument with Donald Stokes.

Donald Stokes tried to show off the gas turbine truck. Ivan of course was responsible for Ford's 'Big Red' gas turbine truck with a much larger gas turbine engine than that installed in the Leyland development tractor unit. Until that time we had concentrated on cars and Ford's developments had been on gas turbine trucks.

I was roped in to take part in a film and to sell the idea of the turbine to Castrol, BP and Shell for the company's oil tankers. I enjoyed the stint in London and Donald Stokes held a dinner, to which Sybil was invited so that she could meet Nora, Donald Stokes' his wife. I had to return from the glitter of London to the more mundane task of attempting to lead the team to solve the difficulties we encountered with the 2S350R engine.

Life in the new house was exciting and my young family was growing up. Roland was just starting school. He was quite adept at helping me dig the garden, as he wanted to do his bit. There was waste ground on one side of the house and we heard it was to be acquired by a well-known Birmingham firm, Second City. Underneath the topsoil was clay, and Roland and I had no way of removing it. I always suspected that there was a stream that ran underground.

One of the director's was living nearby. A while later, I wrote to him to request that I bought a strip of land adjacent to the side of the house. I had heard that they were about to sell off the land to a builder, named Benningman. Benningman decided to build a house on the land for his family and therein lay a story. His wife screamed down her nose when she talked, and was in the habit of screaming 'Bas', whenever she needed to attract his attention. It must have been short for 'Basil' but it was the most hideous sound.

I did manage to buy a strip of land at the side of the house. I think it was at a price of £20,000 an acre from Second City. The building of the house next door by Benningman, flooded our garden. In so doing he must have changed the water line.

His boasting was overheard by a friend of mine in a local quantity surveyor's office, to wit that he had flooded the neighbour's garden in building his house. We never took to the man and his wife and we avoided them like the plague. We were thankful when they left; they were loud people.

The headmistress, Miss Bacon, who sold us the house in Alderbrook Road on behalf of Saint Martin's School, would certainly use the name 'nouveau riche'. I liked Miss Bacon; she had such an apt way of summing up people. She often referred to the people who bought and sold antiques as 'Magpie's'. When she left the school it went down in stature.

On the work front, the head of British Rail Research in Derby approached us. He was seeking an engine for a new concept train, the Advanced Passenger Train or APT. He was an ex-Rolls-Royce man and had eliminated the Dart engine as being too thirsty on fuel. I was filled with mixed thoughts, because the years ahead spelt trouble. It was not easy to resolve basic problems in development. Especially the fundamental problems I suspected.

In the case of the compressor turbine, the only cure I could think of was to start with a fresh sheet of paper; a full blade flap in vibration was a problem magnified.

The other failure, of the heat exchanger cores, was a mystery apart from the reservations I had about large lumps of ceramic. I knew that Corning Glass was conducting work to research the LAS material, but this gave me little comfort as it could take years to track the reason for sudden failures. Fatigue failure of turbine blades through vibration is the most prolonged trouble of the gas turbine engine.

At the time, technology was scant. The stressing problem was lacking in so many areas, especially discovering more about the energy involved when a blade vibrated. Under resonant conditions, high amplitudes are possible and these imply high stresses. The first step is to obtain the natural frequency. There can many but the figure is approximately six. A blade may vibrate in several ways, giving rise to modes from which natural frequencies change. The most common is a trailing edge flap. The worst one is a full flap – a blade where the cantilever is in flutter. Others are torsional flap, and interference from disc modes. The presence of an obstruction in the airflow anywhere throughout the engine may also give rise to an exciting force. The frequency of the exciting forces is the number of obstructions in one revolution multiplied by the speed. Thus any number of static parts, whether it is one or many in the airflow, will constitute a resonance. Typical examples include struts, blades or probes.

One example is the effect of turbine nozzles on the turbine blades. This resulted in a rule of thumb. If at all possible, the turbine nozzle guiding the gas on to the turbine blade should be at least one chord away from the rotating turbine.

Today however, the determination of natural frequencies using holography is possible over any range. Computational analysis is applied to determine energy levels of any exciting force existing in the airflow.

'If only these tools were available then' must have been said so many times!

The concept of the APT advanced passenger train was of a tilting coach to accommodate high speeds. The idea was to have one turbine per coach. Ideally 600bhp was required per coach, but this was unavailable from the 2S350R. The idea of the Rolls-Royce Dart being unsuitable, because it was thirsty on fuel puzzled me. I had studied the use of turbine engines applied to the railways and formed the opinion that, as a percentage of operating cost, fuel cost was very low. It could be that the logistics of diesel fuel were difficult at that time. Stan Smith, who was head of the Railway Research Centre at Derby, was our contact point and I meant to ask him but never did. Stan invented 'Smith's Co-relation' for assessing the turbine designs.

Mike Newman was Stan's second-in-command and a frequent visitor to LGT. I went along with supplying engines to the APT, hoping that we would solve the problems of the 2S350R before we were due to deliver.

It was decided to carry out an estimation of the cost of manufacture of the 2S350R. Ron Ellis, who had become general manager of the Leyland Truck and Bus Division, had requested his accountant be responsible for the cost estimation. I was reluctant to supply drawings of an undeveloped engine and I expressed this view to Dr. Fogg and Ron Ellis, both of whom were on my board. However, I was overruled; two against one.

We struggled on with blade failure after blade failure and no action seemed to have any positive result. We tried to change the natural frequency by cutting the blades several ways, but all to no avail. Peter Higgins, our metallurgist, contacted Rolls-Royce Leavesdon and they had a thought they could count the striations to determine the force of energy of the vibration.

This was done but it was of little help; instead it only confirmed the view that that we were plagued with a basic problem. The heat exchanger cores continued to fail but failures were more difficult to forecast as the patterns of failure were more erratic than compressor turbine blade failures. The development program of the 2S350R was slipping due to these aforementioned failures. This placed all kinds of pressures on the new company.

It was once said that the worst start to a new engine development was mechanical failures. It was preferable to be developing to meet the performance by improving component efficiencies. A new engine that suffers mechanical failures slows the entire development programme.

The biggest problem I faced facing was that of creditability; we were well over budget and spending money became like using water.

I somehow managed to keep levelheaded and find work for the last member of the team from Rover Gas Turbines who might have been made redundant.

Meanwhile, I had an argument with my chief engineer, who was at sea on the compressor turbine blade failures. I thought we should design a blade of a different aspect ratio and lower the loading by scaling the 2S150R compressor-turbine. He argued that we should find the forcing vibration and eliminate it; but a year had passed and he was still looking.

The argument over the Corning Glass ceramic core failures was the converse. I was of the opinion that only Corning Glass could (would) find a solution; that the fault must be in the material. I suggested that when the material was subjected to engine conditions some change must occur. John Lanning was of the opinion that a single core might be the solution on the grounds of reliability. I countered that argument suggesting the way to solve the problem was not to juggle with the reliability factor but to solve the problem of failure.

Donald Stokes rang me from the office of the Minister of Technology, Tony Wedgwood-Benn, to say he wanted a steam-driven Mini.

Soon after, Alec Issigonis from Longbridge visited my office at Solihull requesting I use my gas turbine budget to develop a small steam engine. I objected, due to being already well committed in creating Leyland's first gas turbine truck. However, my mind focused on how I could apply existing gas turbine parts to form a small experimental steam engine. The answer was to be found in the adaptation of the combustion chamber of the single shaft 60bhp gas turbine engine, the 1S/60, now well developed and in production at Alvis, and to which I had transferred the responsibility of Rover Gas Turbines. The 1S/60 became a turboprop engine – see p. 444.

The newly formed Leyland Gas Turbine company had no engineers experienced in steam engines. So I asked Prof. John Horlock at Cambridge if he knew of anyone experienced in steam at any of the colleges.

John said there was a chap working on a steam car in his spare time; he was writing up his post-graduate thesis on machine tools. So on my next visit I met 'little Alec', as he became known, to distinguish from Alec Issigonis, who was nicknamed 'big Alec'.

I knew from the first time I met 'little Alec' that he would become an asset to my team. His vigour, enthusiasm and entrepreneurial approach were just what I was looking for to balance the team. His knowledge of steam astounded me, not only in theory but he seemed ready to roll up his sleeves and get stuck in.

He even proposed, on our first meeting, another cycle – a small gas turbine using steam raised from exhaust heat and injected into compressor discharge air before combustion as a means of enhancing thermal efficiency. This was later patented by a Japanese firm and became known as the 'Chen cycle'.

I immediately set him on, subject to his word that he would complete his post-graduate thesis. And, after finishing his work with 'big Alec', he was to join the team on gas turbine developments. To the best of my belief he never completed his thesis at Cambridge, such was his great enthusiasm for hands-on development and invention.

He complimented 'big Alec' exceedingly well. The development was banished to what was 'nick-named' the 'Kremlin'. The combination of designer-par-excellence and a true entrepreneur who knew what he was doing proved just right for the project.

It was decided to use six Mini cars, one of, which would have a development steam engine installed. One would be installed with a small gas turbine; my intention was to retrieve a Harrier unit that was designed and developed by my team at Rover. The others would be fitted with experimental transmissions, including a hydro-kinematic unit and a hydrostatic device.

All went well with the steam development on the test stand. However, the weight and volume proved prohibitive for the Mini, even excluding the condenser. The condenser was essential to the cycle for operation on its own water, thus minimising the need to 'top up' and for achieving reasonable fuel consumption.

The whole concept was founded on the premise of tolerable emissions. However, a steam engine uses fuel to create the steam. Other considerations are payload, low-mass installation and many other aspects, such as warm-up time.

The lesson, long considered on any alternative power plant for cars, trucks and buses, is that, there must be at least one overwhelming advantage. In the case of vehicles, it had to be either cost or fuel consumption; all other parameters must be equal or better.

Later, when I set up my own company, Noel Penny Turbines Ltd in Siskin Drive, Coventry in the 1970s, one of the first to work for the newly-founded gas turbine company was Alex Ritchie – of whom more anon. Unfortunately, my good friend, David Budworth, was killed in a light aircraft tragedy. I bought his company; that became known as NPT Harwich.

While on the subject of Alex, in the 1970s, the first research contract for NPT came from Caterpillar Tractors in Peoria, Illinois, USA. For some time, I travelled to Cat's research plant in Mossville, near Peoria, for one week every month to set up details to design and develop a truck gas turbine power plant to replace the direct injection Cat diesel in an 80,000lb truck.

The original project was for three years, however, we were contracted to Cat for nine years to carry out other research projects, some of which involved seconding NPT engineers to Caterpillar Inc. in the US. Alex led the team to the US on behalf of NPT; his wife Jill joined him.

I recall one occasion when Alex rang me from the US in the early hours of one morning to say that he wanted to buy a Stanley steam car, and would I loan him the money. I was in no mind to discuss the issue in the early hours, having been awakened from a deep sleep. So I suggested that I call him back at a Christian time.

The following morning I relented, after hearing that he would drive the car to Boston where a boat would transport the car to the UK. So I agreed the loan.

I heard of this venture next when the police rang me from somewhere in the US to ask if I employed someone by the name of Alex Ritchie. They had apprehended him on route to Boston in a vehicle that was unsafe to travel on US roads. I assured them that I did employ the gentleman and that he was a steam enthusiast from the UK. I asked if they would kindly allow him to proceed as a date had been arranged with a boat at Boston. Alex managed to charm the cops to allow him to complete the journey. Just outside Boston, he encountered another cop who stopped him to say that steam was coming from the car. Fortunately this cop knew the Stanley steamer car and was taken by Alex's accent, his 'Magnus Pike' manner and amazing knowledge of steam. So again Alex was allowed to continue his journey to the boat.

Completing his secondment of many years to Cat in the US, I appointed Alex to become general manager of NPT Harwich. At that time we were contracted to the Israeli Air Force to deliver production jet engines to Israel for Delilah unmanned cruise missile intended for moving targets. A colonel from the Israeli Ministry of Defence rang me to say that all the jets delivered were painted in different colours. I nearly dropped out of my chair. I said, "I would look into this and ring him back". He said, "Don't worry. All of jets have passed the production acceptance tests, but I was just curious as to why."

I rang Alex and he said, "He and Major Bashan, the resident Israeli officer, thought they would experiment with stealth paint". I was shocked to silence. Alex was the most remarkable, able and energetic engineer with whom I have ever been privileged to know.

Returning to the late 1960s, the system of higher education was still being criticised for not providing the type of education the economy required. There was nothing wrong with graduate apprenticeships; these turned out trained engineers. The enterprising universities offered courses for a master's degree spread over two or three years but with attendance at the university for a mere 21 days. Harold Wilson was a manipulator and as the unions became more aggressive and significantly militant he was driven closer to Callaghan. It was expected that he would appoint him as Foreign Secretary but it was not for long, as Wilson lost to the Conservatives in 1970.

Edward Heath became Prime Minister and was elected at a time when the unions were flexing their muscles. In the late 1960s, British industry was no longer managed by businessmen, consequently nationalisation was becoming the saving of 'lame ducks'. Shipping, British Steel, Rolls-Royce and, soon to be joined, was British Leyland. In the 1960s Labour had supported mergers, grants and loan guarantees in the name of modernising industry. Instead it had produced inefficient monoliths. Edward Heath's government stood the risk of being voted out, like Labour; it depended on the unions that were forever negotiating higher wages and striking if more money was not forthcoming.

I must have told my three children hundreds of bedtime stories. I rarely read to them because I disliked reading. Instead, I make up stories out of my head. Some of the children's favourites were the 'Cheeky Monkey' stories. This was a story about animals in the wood and the little old man who lived on the edge of the wood. There was Cheeky Monkey, Freddie Frog, Timmy Frog, Toad, Badger and the Little Old man. Two stories became told over and over again.

One story, I recall, was about Cheeky Monkey walking through the wood with Badger when they encountered a brightly painted house. Badger was all for ringing the bell that had a note on it: 'Ring this bell and get a surprise'. Cheeky Monkey had an awesome feeling, as if they had never seen this brightly painted house before. He implored Badger not to ring the bell. Badger would not be told, so he pressed the bell. A hand came out and he disappeared inside the house. Cheeky Monkey ran off to alert his other friends in the wood. They had all gathered and went to the spot in the wood where Cheeky Monkey and Badger had encountered the painted house, but there was nothing there.

Then they decided to go a long way into the wood to find Orgy, the giant, in the hope that Badger was in the giant's castle. After walking a very long way they came to the castle. The big door was open and Orgy was asleep with his head on the table. Cheeky Monkey and Freddie Frog crept in, but the giant large cat was asleep on the stairs. They heard a cry from the second floor, but the giant cat was in the way. Carefully they went up the stairs without disturbing the cat. Getting round the cat without disturbing it was a miracle. Luckily, the key was in the door where Badger was a prisoner. They rushed in, locking the door behind them. Badger was tied up and gagged but was able to make sounds through the gag.

They untied Badger, who was in great distress. The giant's cat was just waking. Hurriedly the bed sheets were tied together to lower Badger out of the window. There was a loud 'pitter-a-patter' as the giant's cat ran up the stairs to investigate the noise from the room where Badger was imprisoned. Cheeky Monkey and Freddie Frog had locked the door and were now piling anything they could move in front of the door. Badger was afraid to use the sheets to reach the ground but Cheeky Monkey urged him to escape. There was no argument and Badger almost fell from the open window into an open rainwater barrel that was full and overflowing.

Next, Freddie Frog was on his way down. By this time the giant's cat ran down the stairs thinking the giant must be waking with all this noise, but the animals of the wood were hiding in the bushes; they hoped that Cheeky Monkey would join them, escaping the giant and his cat. Cheeky Monkey removed the clutter from the door when he heard the cat go down stairs. He unlocked the door and scampered out of the front door, not waiting to see whether the giant was waking. He ran as hard as he could and did not stop until he reached the Little Old Man's cottage. He banged on the door and found Badger waiting for him in the comfort of the Little Old Man's dwelling.

There was another story, one of their favourites, which began with Cheeky Monkey walking through the wood with Timmy Frog. They heard an animal whimpering furiously. It was the man-eating lion with a large thorn in his paw. He had escaped from the local circus and had wondered off through the wood until he trod on a big thorn. He was whimpering and sobbing, sitting on a fallen tree. Cheeky Monkey and Timmy Frog peeked through the trees and looked at one another, thinking 'the lion is in a lot of pain'. Cheeky Monkey approached the animal with great caution saying, 'There, there'. Cheeky Monkey could see that the man-eating lion had a large thorn in his paw. He became suddenly brave and stroked the lion and said, 'He was going to pull out the thorn'. With the thought of pain, the lion began to whimper even more but Cheeky Monkey said, 'Don't be a big baby, it will soon be over', and with that Cheeky Monkey gave a gigantic pull and the thorn came out. The man-eating lion was calmer now as the pain had almost subsided. Soon the man-eating lion was carrying Cheeky Monkey and Timmy Frog on his back to return to the circus. At the circus they were so pleased to have the lion back safe and sound that the ring master gave all the little animals of the wood free tickets for a week and as many sweets as they could eat.

On 20 July, 1969 Neil Armstrong stepped onto the moon and made the dramatic speech 'One small step for man'. The dish built in Australia by 'NASA' was the prime source of pictures throughout the world. Apollo 11 was successful in exploring the moon.

As 1969 came to an end the 'writing on the wall' for the Socialists became apparent. Harold Wilson could not control the 'far-left element', nor could he prevent the unions from tearing the Labour movement apart. Of course, any government would have a hard task to prevent disruption. The unions seemed hell bent on destroying the economy of Britain. The Conservatives won the election in 1970 but the unions mounted a massive campaign.

The situation in Leyland Gas Turbines was hindered by the personal relationships between my chief engineer and my young secretary. I was the last one to know of his attraction to her. The chief engineer's secretary was also walking out with the second-in-command of 'Rover Research' during the lunch hour. The latter two were eventually married, despite the fact that he had to divorce himself from an unhappy marriage (to the best of my belief). There were great jokes in LGT, again unbeknown to me, that they went into the woods during the lunch hour to read 'the thoughts of Chairman Mao'.

To cap it all, the general manager of Leyland Truck and Bus Division proposed that gas turbine activities should be located at Leyland in the north of England, and not at Rover in Solihull. Naturally I was against this on the grounds that we would lose key people. Also, to relocate made no sense. I argued that when the 2S350R went into volume production it could well be manufactured at Leyland alongside the vehicles. I did not hear any further mention of relocating for several months, so I thought the proposal was dead.

Things dragged on as we tried to pin down our two basic problems namely, ceramic core failures and the blade vibration problem. Certainly it appeared to top management that we were taking too long to develop the 2S350R, as the programme slippage was now several months.

The basic strategy in any advanced development is to avoid the use of new technology and to use only well-established techniques and components if slippage is to be avoided. Even then, basic difficulties could occur so the safest design in relation to the gas turbine engine is a scale-up, employing the formal rule of true scale.

When Rolls-Royce was later forced into bankruptcy, due to problems on the RB211 fan jet engine, I gave a lecture at Sussex University dealing with established and advanced technology. I took examples of known projects and plotted the excursion from the forecast programme timing and cost. The lesson is to avoid the use of miracle materials or processes.

Advanced technology designs are proportional to the availability of expenditure, time and the quality of the team engaged to carry out the programme. The more the design involves extreme risks the more money and time is involved. The problems of the 2S350 truck engine could be resolved in time with an ever-increasing budget. I was no longer responsible for engineering but as managing director I was aware that 'the buck stopped with me', to coin an American saying. It is ironic that in time of greatest pressure, top management foibles appear in many ways.

I was requested to report to the main board at their monthly meeting at Berkley Square House in London to explain why the LGT budget was overspent. It was the Rover system of working. The various departmental budgets started the year with a reserve such that, in case of overrun, there was an indication at the end of the financial year that the budget forecast was too pessimistic or optimistic depending on overspend or the opposite. I tried to explain this to the board, chaired by Donald Stokes. My boss, Dr. Fogg, remained quiet, seemingly in awe of what I said. In the end the chairman said, "Noel, go back to running your company". With that I said, "Good morning gentlemen" and left.

There was another sign that British Leyland was taking too long to get rationalised (a word that was much used in the 1960s). There were numerous engines of similar power within the group, but individual companies would insist that their engine should not be compromised by rationalisation. There was no one man strong enough in British Leyland to argue otherwise.

This was 'the writing on the wall'. I felt in my heart of hearts that this spelt the dream of a successful compounding of British companies, but it was a myth for the sake of ruthlessness. One ingredient was instilled into British companies, small, medium and big, that gave them independence and freedom. It was that there was an almost fundamental issue in their make-up, dating back to the industrial revolution, which called them to 'stay as we are'. It was almost a kind of revulsion against change. It was manifest across the whole of British industry in the late 1960s; there were few exceptions.

One observation focused on the difference between the home and world markets. The concentration on both home and world markets had yet to come for most industries. 'Lame-ducks' were the call of the day. Anything with 'British' attached to its name seemed doomed. The Department of Trade and Industry was forever pumping cash into British Steel, the national coal industry and the nationalised railways. Surely British Leyland was not going 'down-the-pan'?

Wilson's endeavour was to take Britain into the EEC as a means of regeneration. President de Gaulle had other ideas. He vetoed Britain's application out of spite.

Edward Heath had formed a new analysis of the country's economics and formed a new set of policies. However, it failed due to insufficient investment in the potential economic successes and sadly there was a lapse into throwing good money into 'lame-ducks', such as steel, coal, rail and then the automotive industry. In 1969 the number of stoppages peaked at 3,116 and the days lost to strikes amounted to 6,846.

On the domestic front, all was well in the new house. Many things had been done to change the character of the interior in name of easier cleaning, understandably. We decided to change back to the more Edwardian style, albeit slowly. I had central heating installed and designed two new more modern ring mains for the electrical power points. These I arranged for the electricity board to install. As for the lighting, I had to leave as it was, due to the nature of the system. It was a source of concern to me that the rubber coating had perished after all this time.

On the work front, the prospect of moving gas turbine developments to Leyland reared its head again with renewed fervor. The rumours were that along with the other two high fliers in BL, namely Mike Dunn, and another at Cowley, it was anticipated that I should be considered for the post of the Corporation's chief of research.

Although this was only a rumour, I was not a bit interested in moving. My first love was developing gas turbine engines that were slowly becoming a design issue. No longer, despite two basic problems on the 2S350R, was it a development task but a design one. Computer science was evolving so rapidly that calculations on the slide rule that once took weeks could be accomplished in less than a day.

The length of time to receive approval of financial capital became almost too long, and the only aspect that saved the day was the fact that we were working on the Rover site. I complained to Dr. Fogg and he told me, "He would see what he could do to get Ron Ellis's signed approval". A note came from Ron Ellis spelling out that we were spending too much. I replied that we were well within the budget. He replied that taking the programme slippage into account we were well overspent. He did not raise the issue of approval of capital but I was confident that he was bent on playing politics.

He said, "He and Bertie Fogg were due to visit," but they never did. Mrs. Bunting, Dr. Fogg's secretary, rang me and said that, "Her boss had to enter hospital at Preston." I eventually visited Preston to see him, and he admitted that the doctors had discovered a lump on the lower part of his leg, and they were forced to operate. As a result, he would be out of action for some time. This placed me at the mercy of Ron Ellis, who visited and announced that it was his intention to move Leyland Gas Turbines Ltd to Lancashire. He said it was "too bad" if we lost key personnel.

Of course, I opposed the move and I went to see A.B. Smith, in confidence. I explained to A.B., who was a good friend and had become chairman of Rover Gas Turbines board, before the relocation to Alvis in Coventry. A.B. listened carefully to what I had to say. His comment surprised me. He said, "The Corporation was reneging on many of its promises and he was thinking of getting out. He could see the whole lot becoming a bed of thorns in the side of any government." It was up to me as what action I took, but he swore me to confidence in relation to our conversation. I returned to my office, a matter of several minutes' walk away. On the way it offered the opportunity to think of what I would do. I decided that if Ron persisted in raising the question of relocation to Leyland I would threaten to resign. I thought I would receive the support of every person in LGT. Of course, they did not know what had transpired.

Months passed and it became more difficult to obtain authorisation for even the approved budget. I took every opportunity to voice my dissatisfaction over monetary sanction, especially in respect of the approved budget. I checked with the heads other companies only to find that it was widespread. Others knew more than I did, and some said, "There was a shortfall with cash flow within BL." The main board of the group was at crisis level.

Things came to ahead when Ron Ellis rang me and asked, "What was I doing about relocation to Leyland?" I replied, "Nothing," and he said, "I would be sorry" then slammed down the phone.

I decided to appeal to Donald Stokes and after some delay I heard from his secretary that I was required to attend a meeting at Berkley Square House. When I turned up for the meeting, both Ron Ellis and Bertie Fogg were waiting for the same meeting. I acknowledged them and it was more than they could do to pass the time of day with me. I had not seen Dr. Fogg since visiting him at Preston hospital. Both men looked put out and nervous. When the meeting convened, Donald Stokes asked what it was all about. The others were withdrawn, so I was forced to explain that Ron Ellis wanted to relocate gas turbine developments from Solihull to Leyland and, as a result, we would lose most of the key people who had made possible the former successes.

"This is the first I have heard this proposition," said Donald Stokes, adding. "Understandably, Noel has a problem with this and it is up to you three to find an agreement." Then the meeting was over. When the other two were out of Donald's office, Ron Ellis said to me, "You will be sorry for this, dragging me down all this way." So I turned and left them.

The last straw came when I received a note, circulated to the heads of all companies in the group, stating that New Year's Day fell on a Thursday that year. The use of paper must have been enormous. This indicated to me that regretfully BL was going down the slope fast and A.B. Smith was right.

I was determined to resign, but I wanted to leave the team intact at Solihull for the sake all my colleagues and friends. I felt certain misgivings about my second-in-command. He had adopted a stance of self-righteousness and I doubted whether he was the right personality to head the team. This was by no means my only problem. My concern also was for the development of the engine that I would be leaving behind with two fundamental problems. It was against my nature to renege, especially with engineering difficulties.

I suppose I was born with a fair degree of tenacity and the will to follow issues to their bitter end. This side of me rebelled against resignation, as I could not face a situation of an unfinished scenario. This caused me to hesitate and think out my course of action.

I pretended that I was the master of my own destiny. I sought the advice of John Parkes, managing director of Alvis, into whose willing hands I had transferred Rover Gas Turbines Ltd. John explained that I must be a 'good soldier' when I implied that I had considered resigning my directorships. I have always recalled his words. I found them strange. I was still technical director of Rover Gas Turbines Ltd and I visited Alvis as such. Whether John Parkes gave this advice, because of the potential absence of my knowledge from the Alvis gas turbine developments, remain a mystery to me.

Years later, when Arthur Varney came to work for me, he confided that John Parkes, who was his managing director, had misgivings about my intended resignation and that he was as 'thick as thieves' with the top directors of Leyland. John had mentioned to Donald Stokes about our conversation, that was in confidence, and should have been treated it as such. Arthur Varney, of course, was chief engineer of Alvis and had grown to dislike John Parkes, whom he mistrusted.

My ardent wish was to find a way out of the basic problems besetting the 2S350R engine. It seemed to me that my second-in-command had lost all heart; admittedly his affair with my young secretary did not help matters.

I appealed to Corning Glass to heighten their research. I sought the help of Rolls-Royce to find a solution of the blade vibration problem. Rolls-Royce came up with the following solutions: design a thin ring for the outer diameter of the blades would increase the stress and change the design of the blades altogether, using an increased prime number. This is what I had suggested 18 months ago. It could still be worth taking this step. The drawback could be the 'lead-time' in procuring yet another new precision casting that was estimated at six months. Originally, we could have solved the problem had we moved when I first proposed this modification.

It was usual to call a meeting of Sybil and the children in the event that vital decisions were to be made affecting their future. It was the custom to hold these meetings on the stairs. I called them together to announce that I was thinking of resigning from BL. There were many questions arising from the discussion. The big issue was what would take the place of going to out to work. Would I get another post? This was fairly easily answered as I would set-up my own constancy until I could build a new gas turbine development company.

Having spoken to Sybil and the children, I had their agreement to choose the path I favoured most. Deborah, of all the children, was old enough to consider the different options. The boys were less circumspect and if things went wrong would go along with whatever fate had in store. I thought of all the friends I would leave behind and a thought occurred to me that many would join me and take the risk of job security. I was determined to succeed with whatever was the outcome. My regret was I would leave behind two basic engineering problems.

In 1970, British Rail took delivery of the first engine, and my selling job turned out to be a success with Shell, Castrol and BP, all of which bought gas turbine tankers as an experiment. Early in 1971, I tendered my resignation to Dr. Fogg. I had previously explained to the key men in LGT the step I was about to take and the reason for the necessity of taking radical action.

Donald Stokes came to see me and persuaded me that it was not vital that I should resign. If I stayed on he would change things. After a full day's discussion I asked the question, "Would he get my fellow director's to sanction Leyland Gas Turbine Limited at Solihull with the budget cash already approved?" He hesitated and I thought 'Who was in charge of the Corporation'. Looking back, I doubt if he knew about the politics of preventing the flow of cash from the already approved budget. His hesitation about moving LGT settled the issue for me and my fate was sealed.

When I left my office a week later, my departure hit the headlines. Stories abounded. A typical story was 'a backroom boy leaves BL'. I was a director of two of the Corporation's limited companies and the contracts stipulated that I was restricted for one year in relation to setting up a new gas turbine company.

I took legal advice with respect to the contract and the outcome was another meeting at Berkley Square House with Donald Stokes, whom I respected. He ruled that as long as I acted for six months as a consultant I would be released from the contract. When I met him he asked me, "What was all this I read in the *'Daily Telegraph'* in relation to the legal term *Sub-judice*." I answered, "The term means under judicial consideration, it was the case at that time" He said, "I know what the legal meaning is. I was wondering what was *'sub-judice'*. The contract?" I explained that I had taken legal advice in relation to the contract. Now that we had agreement I went away in a good frame of mind.

Three organisations took me on as a consultant: Carl Kiekhaefer in America, and Firth Vickers and Lucas in the UK.

Added together, my equivalent of a year's salary was double that which I had been earning in BL. I sold my Leyland shares at 53p a share and a week later the same shares had dropped to less than 6p.

As 1972 emerged, the miners' strike began. The vice-president and the secretary of the National Union of Mineworkers (NUM) were communists. The most threatening aspect of the strike was the 'flying pickets'. The strike coined a new phase 'the winter of discontent'.

I had an approach from General Motors (GM) followed by a further one from Caterpillar. GM requested that I take-up residence in the US. I thought about this long and hard. Eventually I wrote to my contact, Ty Nagey, and pitched my salary at the equivalent of £150,000 a year. I assumed that even GM would find this excessive. GM proposed that a contract would be awarded to Williams Research Corporation (WRC) for a gas turbine car development and that Sam Williams would take me on.

Sybil and I went over and decided that it was not for us. So I took another tack with Sam I suggested that we start up in the UK on a 50/50 basis. WRC would finance the start-up and within the first 18 months the company would just about pay its own way. An agreement was drawn-up accordingly between WRC and Noel Penny. I searched for premises at Stratford upon Avon. Mason Road seemed to be the core of the industrial area. I looked at several buildings along that road but none appeared suitable.

About this time Joe Grandfield had taken over from L. E. Johnson at Caterpillar's gas turbine department based in Mossville, within Caterpillar's Research Centre. Joe came to England to make contact with me. He already had a team working in Caterpillar on small gas turbine developments and I became attracted to working with Joe, who suggested a proposition. It was more of an invitation to work with his team in a way that was easy for me to operate.

I liked Joe's set-up. It was a mixture of young professionals already within Caterpillar, working on the development of a small gas turbine. The engine was known as the 700. The team was led by Brom (Abraham) Zadoks. Additionally, Joe had brought with him Bill Eichberg and Jessie Wiggins from Pratt & Whitney. The other senior engineers were Mell Eide, Harry Dawson, Bill Hoftiezer, Don Krull, and others. They were a good team and I got on well with them, thanks to Joe.

Joe was of Irish decent and he was known for his rages. When he was like that he could rip the phone away from the wall. We were good friends and I visited once every month for a one week at a time.

Joe was anxious that I was not out of pocket and as a result he was keen to chase-up my frequent payments, including my

# Automotive gas turbines

airfare. Joe and Nancy set up together; Nancy was, I believe his secretary when he was chief engineer at Pratt & Whitney in Hartford, Connecticut. This was the research arm of Pratt & Whitney. Joe seldom talked about his former wife, with whom I believe he had children. We always had breakfast together and visited the stores around Peoria.

When first I visited the US I stayed at a modern hotel that was decorated inside with what only the American's thought imitated an English castle. The name escapes me but I know it had lodge at the end. I took pleasure in going home sometimes with Brum. They were a happy family; his wife was named Mia. Mia and Brum had two young sons, Geff and Rick. I played baseball with them and always brought them large bars of Cadbury's chocolate that I wrapped in tin foil to last the journey.

Joe had a cigar almost continuously in his mouth. When he ate breakfast he was the only one I knew who could eat while chewing on a cigar. He had one of the original Ford Mustang cars, with a long bonnet culminating with a horse emblem rearing-up. He was a good host, looking after my welfare in all things. Sometimes I stayed with Joe and Nancy. Nancy did every crossword puzzle imaginable and Joe used to buy crossword books. I had not seen one before.

One hot sunny Saturday, Nancy, Joe and I went to the local wine shop to get my favourite dry vermouth and other wines. Joe had the air conditioning in the car on full blast. Joe and I went into the shop while Nancy stayed in the car. We were some time in the shop and Nancy joined us. The car engine was running to maintain the air conditioning on full blast. When we came out of the shop in the heat of the day we found both car doors locked with the engine running and the keys of course in the ignition. Joe started to whine uttering, "Why did you do that Nancy?" He repeated it over and over again, raising his voice as he did so.

Not wishing to see Nancy humiliated in this way I ran into the shop to obtain a wire coat hanger. I did not know if a wire

# Automotive gas turbines

coat hanger would work on the Mustang but it certainly worked on my Rover 2000. I went out of the shop straightening out the wire. I pushed the wire into the door trim and flipped open the door 'push-down'. Joe and Nancy were so amazed they were speechless. Joe stopped his ranting.

I enjoyed my frequent visits to Peoria and following this it was decided to approach the board of directors to see whether they would sanction a major project with a new company. Joe and Rex Robinson, the head of research and Joe's boss, were supportive, providing the directors decided it was a good defensive measure to safeguard the diesel's vee-eight engine sales. At least we would give Caterpillar Inc. a measure of the benefits of a gas turbine truck.

A meeting with Mr. Nauman, the president of Caterpillar, was arranged. Joe came with me and it was agreed that Mr. Nauman would have a discussion with Rex Robinson before placing a motion to the board of directors of Caterpillar Inc.

It was decided also that a specification for a flat-rated 450bhp gas turbine be compiled by Joe's gas turbine department (photographs of parts for Phase 3 engine, p.452). The flat rating was on an 80°F day and 5,000ft altitude. This equated to a sea-level performance at normal temperature of 580bhp. I was determined to not fall into the same trap as the 2S350R Leyland engine and pitched the performance conditions generously. So I arranged the pressure ratio at 5 to 1 (2S350R was 3.9) and the air mass flow of 5lb/s whereas the 2S350R was 4. Admittedly, the power of the 2S350R was less at 350bhp.

The specification was completed and I noted that the vehicle selected was an International Harvester Tri-Star with a fully laden weight, or GVW, of 80,000lb. Also, the engine had to fit within the frame rails of the vehicle; for the International that meant a standard of 28in. Most diesel engines were installed into the vehicle with a large bulge in the cab.

One of the virtues offered by the gas turbine engine was a walk-through driver's cab. The gas turbine engine also was one-third of the weight of the diesel engine and more compact for roughly the same power. With the heat exchanger, the gas turbine was competitive on fuel. The noise level with the gas turbine engine was less and there were no vibrations. The gas turbine also could, with the heat exchanger, meet the forward noise legislation in the US. The diesel engine had difficulty in doing so, unless it was made significantly more expensive.

I felt it was the noise legislation that decided the Caterpillar board of directors to reluctantly give the go-ahead of a major overseas research contract. The way Joe explained it was that if the programme were unsuccessful, then the Caterpillar board would be blameless. Joe had an uncanny determination that the engine would be a success. He almost became my mentor and a great friend. The initial contract was for a total of £3 million sterling and was of three year's duration. Joe encouraged me to go for an advance payment of £250,000 to get a head start, and to avoid the risk of running out of cash at the start of the programme.

At that time, that I was part way through negotiations with Tony Sheldon for the site in Siskin Drive that Harry Ferguson had built for Harry Ferguson Research (HFR). Tony had married Harry Ferguson's daughter prior to his death. Tony had delegated the task of the negotiation to John Peacock, a tall but quite thin, pale man. John had been a prisoner in Germany during the war. Negotiation with John proved more difficult than with Tony, but John Peacock was the man who forced Ford to pay-up when they infringed Ferguson's inventions related to tractors. Henry Ford 1 worked with Harry Ferguson during his early years in the States.

Sybil came with me for the final negotiation in which I agreed to take on all the shop-floor workers, the HFR activities and continue on the site for approximately 12 months, or longer if they failed to find new premises.

The Caterpillar agreement to go ahead with the R&D contract arrived just in time as I hoped to receive the advance payment prior to handing over the cheque to HFR. I had already taken on six new employees. One was of these Bob

# Automotive gas turbines

Chevis who later was to become a great asset to my new business. Another was a good, loyal designer/draughtsman in the shape of Ted Griffiths, who left Leyland about the same time. I needed a secretary and Jill Wilks (photograph, p.452), who had been working for Bill Martin-Hurst agreed to join me and worked for a time at the house in Alderbrook Avenue.

I installed one of the early Hewlett Packard electronic calculators in Alderbrook in order to carry out quick performance calculations. This was the forerunner of the laptop computer. My work as a consultant was paying off and every time I had to visit Caterpillar my time and airfare were paid for out of Joe's budget. I felt better than I had ever done since I left Leyland; things now were slotting into place.

I had the prospect of buying my own company and its workforce from a man for whom I had a great respect. The late Harry Ferguson was a difficult man who did not suffer fools gladly, but I certainly looked up to him. He was an entrepreneur and a trailblazer. He was also an outstanding engineer. I was sorry that he was not present to be in on the final negotiation.

I had previously been to see A. V. (Val) Clever, the head of Rolls-Royce Rocket Division, on the outskirts of Coventry. It was on the point of closing down, as the country no longer had a use for rocket developments. Rolls-Royce had a hand in Bloodhound, Thunderbird, Blue Steel and Black Knight. Fortunately the Rolls-Royce facility was closing as I was just starting Noel Penny Turbines Ltd., as my new business was called. I took many of their engineers. One key person was Reg Lowe who had been chief engineer under Val. Reg was a very stern chap, an ex-RAF pilot who used to fly Mosquitos. I well remember Reg, who did very good work for NPT during the formative years, explaining that many Mosquitos crashed on landing due to the high landing speed. Among others that I signed on for NPT from Rolls-Royce was Bob Petipher, who became an outstanding development engineer. Bob was responsible for testing of some of NPT engines off-site.

# Nine
# Noel Penny Turbines Ltd

In which at last I start my own company, geared to the design and manufacture of small gas turbine engines.

MY COMPANY, Noel Penny Turbines Ltd, or NPT, was incorporated on 12 July 1972. The chairman I invited to join the board was Edward Ponsford. Eddy had two sons and resided in Littlehampton. We visited there occasionally as a family. The other director I invited to join the board was Viscount Melgund, the eldest son of the fifth Earl of Minto. My wife Sybil became a director and company secretary.

We had already signed the Caterpillar contract to develop a gas turbine engine for heavy vehicles, as already mentioned. *The Engineer* magazine ran an article on the incorporation and my face appeared in the published edition. 'Cash flow' was 'the end all' with me. I kept a simple account of all my expenditure, even household expenditure; even more important was the money received daily. It was as if my motto was cash flow, 'the God of survival'.

Returning to the Caterpillar contract, I decided that a good discipline was to compile quarterly reports, a practice that ran throughout all NPT R&D projects. There were two reports, one financial and the other technical. The financial report included the budget. Shown against the budget line was drawn a graphical against time and that was the actual expenditure. Monthly billing was also shown. This included manpower and overheads.

The billing system involved a list of the cost of materials. Added to which was the general administrative overhead. A profit of 10% was applied at the end.

The second quarterly document was a record of all technical aspects and was a most comprehensive book, rather like report.

Even the programme and specification were commented on in order to illustrate progress. A team leader headed the Caterpillar project. This was Chris Bramley, from my old team. The project was given a number, with much debate as to the significance of this number. Many wanted to call it Project 101. I recalled the tragic circumstances of British airship 101 and we decided to call the Caterpillar engine Project 201 (see illustration p.446). Each R&D subsequent project had a number as distinct from future NPT engines that were classified according to their use.

The start of any development in those days commenced with a network or Gant-chart. Usually the first item was analytical, or concept design, according to the specification. As soon as possible into the project, long-lead items must be given attention and material where applicable ordered. Long-lead items must receive early attention if delay is to be avoided. Usually concept design can identify these items. For example a precision cast turbine or stator could take six months from order to delivery, and so on. A similar treatment involves the ordering of special materials, such as light alloys; this is especially so if they cast, these can be subject to long delivery times. High-speed bearings are another long-lead item.

Expertise is a vital part of any development. To have trained people in all disciplines was a great asset. I was fortunate to employ Rolls-Royce people as well as some of my old team of trained personnel, right from the beginning of the Caterpillar project. It was not just trained engineers in design, development, or the specialist functions of performance and aerodynamics but also positions of procurement or buying and programming.

I was fortunate to employ a programmer, David Wilks. He was married to the secretary who joined me at the start of the company. I sent him to Caterpillar for a few days to become trained in their systems. David came back with a valuable book of Caterpillar paper work that I used to model the systems of the new company.

I contacted PriceWaterhouse and arranged for them to become the auditor. Bland Payne, later taken over, covered insurance matters, while the initial travel agent was Lunn-Poly. I started with two banks, the Midland and NatWest. Solicitors were Skillington & Brown that have now been renamed.

The first meeting of the board of directors of Noel Penny Turbines Limited took place on 20th of July at Alderbrook Road. John Brown a solicitor, also a friend, and Douglas Sandry from PriceWaterhouse were present also. Sybil and I were there and the chairman was appointed. Viscount Melgund was invited to be a director, both accepted.

The board decided that NPT should adopt the 31st December as the company's financial year-end. Therefore, as we were incorporated early in August, the first accounts included nearly 18 months of trading. I felt that the company was unique and highly successful, so we had a printed booklet to commemorate the first 18 months' trading. The booklet contained a foreword, the chairman's report, the report of the directors, a managing director's report, brief accounts, the company's assets and finances and an item about 'how the gas turbine works'.

The foreword took the form of the following: "The first 17 months of operation involved hard work for all of us and we felt that you would all welcome the opportunity of reading at first hand just what your efforts have achieved. Apart from the simplified presentation of the audited accounts and financial status of the company, we have included the year-end reports from the chairman, Mr. Edwin Ponsford and our managing director, Noel Penny. Finally, in response to the intense interest, which our turbine projects arouse, we have also reproduced the diagrams that show, in elementary terms, just how these simple yet efficient machines work. The first trading period indicated a sales turnover of over £1 million. A power crisis hit the whole country but NPT was one company that did not stop work. By foresight, we installed a Caterpillar 250 KVA stand-by electrical generating set."

Meanwhile, Britain suffered various miners' strike in 1972. The irony was that coal was a key fuel and although its importance was declining, the number of miners had fallen by nearly one million to approximately 300,000 in just 15 years. Coal remained still a vital fuel for power stations.'

Ted Heath's government decided to float the pound and return to an incomes policy. The government was forced to bring in legislation to support the wage freeze. Britain, having suffered a wage freeze, then faced an oil embargo at the end of 1973. This was due to the Yom Kippur War between Israel and the Arab States. Whole industries supported the miners and the country ground to a halt. In 1974, Harold Wilson, the leader of the Labour party, took over from Heath's Conservative government.

Ted Heath was a good man but he had lost the political battle and never led the Conservatives again. He was forced to make way for his rival, Margaret Thatcher. She was destined to fix the National Union of Miners (NUM) using radical new policies. Margaret Thatcher believed that most men with whom she came into contact within the Conservative party should have remained at school. Tom Trenchard, who was Minister of Defence and a good friend, was the son of Lord Trenchard, made famous for the naming the RAF. Tom once told me why he was sacked from the post of Defence. He had a tendency to debate with Mrs. Thatcher. In her company he attempted to correct her on a defence matter, he started by saying, "I do not wish to get my grandmother to suck eggs". From that moment he knew he had overstepped the mark with Margaret. He was sacked a day later.

Wilson appointed Callaghan to the post of Foreign Secretary when he became Prime Minister in 1974. Callaghan was less inclined to dither than Wilson. Callaghan was more decisive and waded in, right or wrong, even at the risk of party unity. In contrast, Harold Wilson seemed to be out of political touch, merely standing on the sidelines. Later in the 1970s Callaghan was given the post of Prime Minister.

## Automotive gas turbines

Returning to NPT and the home front, I was feeling like I never had it so good when suddenly the news came through that a man I admired so much had been killed tragically when he was landing at Norwich. David Budworth had many skills and one of his pursuits was flying. He had a private pilot's license and was flying a second-hand light aircraft that he had bought the day he crashed at Norwich. We had to wait a long time for the Air Registration Board's (ARB) report into the crash.

The ARB report stated he had no chance but it took issue with the mechanic who certified the aircraft as being fit to fly. One of the chains working the flap was rusty. When David attempted to land the chain fractured. Consequently the flap was in the wrong position for landing. David was in the Walsall area and had persuaded the mechanic to pass the light aircraft in order to get back to Norwich in daylight. Although I had great belief in David's own genius, he was a bit light of people with expertise and resource. When he was tragically killed all the assets became liabilities. I decided to travel to the meeting of creditors with Douglas Sandry of PriceWaterhouse to determine what could be done.

Two of the prime creditors at the meeting were the Eastern Electricity Board and Flight Refuelling. Condolences were expressed to Mrs. Budworth, who had five sons. It became apparent that both Flight Refuelling and the Electricity Board needed their projects completing. I offered to consider taking on the projects and make an offer for the company without taking on the company's liabilities. The condition was that NPT would be paid for completing the projects, the sums to be negotiated. All creditors agreed; the two prime ones subject to negotiations.

At the Board meeting on 31st December 1974, I was authorised to make an offer for the complete assets of David Budworth Limited of the order of £55,000 with no liabilities. This was accepted.

# Automotive gas turbines

I appointed David Noble as general manager of Harwich with specific instruction to ensure the company was profitable. The name was changed to Noel Penny Turbines Limited. I was against this but gave way to PriceWaterhouse. They argued was if we retained the name of David Budworth, suppliers of the old firm would expect us to pay the Budworth debts.

Harwich soon became profitable and turned out to be a great asset, though one difficulty was its distance from Coventry. Its services could be described as electronics, engineering of small jets and a lost-wax casting capability. The three-story 'Naval house' was listed and it was alleged that Samuel Peeps had written his diaries in the house. It was just round the corner from the workshops. The workshops were adjacent to the sea. We used to stay at the local public house.

In the late 1970s, when we received an order from Yanmar, which had its headquarters in Osaka, Japan, we were responsible for training their young engineers in small gas turbine technology. The company's managing director visited Harwich and stayed in the local pub; its toilets were along the corridor from the rooms. All doors had Yale locks and when Mr. Motoyoshi, the managing director of the Yanmar's research department, got up in the early morning to visit the toilet he inadvertently locked himself out of his room. There was quite a rumpus when the landlord of the pub found a small stark intruder dressed only in a lion cloth in the bar downstairs. The visitors from Japan had a habit of looking up the recent record of the population of a town or city.

Mr. Motoyoshi asked, in a casual way, "What was the population of Coventry?" I guessed that it was in the region of 400,000. He said, 'Would you like a bet on it'. Of course, he knew the correct figure, so I lost. He used the same treatment to David Noble in Harwich and he lost too.

Early in 1975, Viscount Melgund became the Earl of Minto on the death of his father. At the board meeting in February 1975 it was decided that Harwich Division sales would be based on the two main projects of 'CEGB' and 'Fight Refuelling' until further engineering technology planned to bring the engines involved in line with the standards of NPT.

Both quality control and manufacturing, following the initial expansion when NPT took over from HFR, needed time to evolve. The payroll in 1975 had grown by at least 25%, keeping pace with new areas of the company. Fabrication and weld shops had been installed. A metallurgical laboratory and crack detection facility were also operating. (Noel Penny included a record of early days of the company in the NPT appendix of his original *My Story*.)

Project 201, stage one, for Caterpillar proved a success with the exception of the Corning Glass ceramic heat exchangers that were plagued with the same failures as experienced at British Leyland. The only difference was that the ceramic cores gave a longer life before failure. The average time before failure was approaching one hundred hours. Corning Glass indicated that this must be due to 'quality control' but I had my doubts, as the basic problem had yet to be discovered. The final resolution was some way off, so Ford Motor Company and NPT agreed to review the complete historic data of ceramic core failures, both on rigs and during engine running. We agreed that British Leyland should join the review, so an invitation was sent to BL.

British Leyland by this time had incorporated a bigger core and experienced catastrophic failures. The result of the three-day session with Ford, BL and NPT indicated that endurance running with kerosene and not diesel gave longer life. This was a red herring and tended to hide the real basic reason for failures.

Later, Corning Glass discovered the solution when they explored the chemical composition of the material. The lithium was being leached out and the material had become susceptible to erratic changes to the material's physical properties, particularly in the brittle behaviour. Corning Glass searched to find a way to lock-in the lithium into the ceramic.

The fact that engines running with kerosene gave a different failure pattern was explained by diesel fuel reacting more severely with the material and accelerating the failure. Corning Glass found that an ion exchange improved the life during engine running conditions. The US company introduced this basic solution and extended running was achieved. However, Caterpillar were dubious and suggested NPT should examine the stationary steel heat exchanger, known as a recuperator as distinct from the regenerator. Many design layouts were schemed with the steel heat exchanger. Marginally the weight and effectiveness would increase and the engine would be more of a box shape.

The breakthrough came when 0.003in-thick sheet was introduced. It was possible with one tool to make the 'wavy fin' approximately three times the area on the gas side than on the air side. This enabled the gas-side pressure drop to be similar to the air-side. Wavy fin also enabled the boundary layer to be energised for improved heat transfer. It was decided to carry out the design detail of the most compact scheme.

Meanwhile, the Harwich plant was operating quite profitability and it was decided to expand the electronics section.

In mid-August 1974 we took the children to the US, flying via New York and Elmira to Corning. Here we visited John Lanning and his wife Nancy. While we were there we also went to meet Rushmore Mariner.

Deborah, who was the eldest of our three children, thought that Rushmore was quite odd as he had a habit of chewing. Little did I let on that he chewed tobacco! On the same trip we also met John's mother, whose name escapes me. We hired a Chevrolet shooting brake. This was a popular General Motors vehicle and Jonathan was intrigued that all the windows were automatic, even the rear window in the tailgate. He was a very useful lad and when we came to leave Corning he had great fun in packing our luggage in the vehicle.

From there we motored to Albany where we were due to stay one night at the local Holiday Inn. The children loved it, but Sybil and I found it quite tatty. We were on the way to Niagara-on-the-Lake in Canada. Crossing the US border into Canada enthralled the children. Niagara Falls is best seen from the Canadian side. The two rivers meet having flowed from Lake Erie into Lake Ontario, dividing at Goat Island into two falls.

The Horseshoe Falls on the Canadian side are 176 feet high and 2,215 feet wide at the crest. We stayed on the Canadian side, near to the 'Falls'. The children loved seeing the spectacle. "Why is called a Horseshoe, daddy?" "Because it is shaped like a horse's shoe," I answered Roland. It became clearer when the boat we took to view the spectacle approached the crest of the waterfall. We could see the water was falling in a curve. The boat appeared to be tug-like and curved round the open horseshoe. We were all dressed in yellow sou'westers. As we curved, the spray was saturating and we were thankful for the yellow covering, at least it kept most of the body dry. Only our faces were open to the spray.

We crossed back into the US to motor back to Albany where the shooting brake was left with the hire firm. Then we flew to New York. I remember thinking about 'bird-strike' at Albany when a flock of birds happened to circle the airport.

In New York City we were staying as a family at the Westbury Hotel. We were tired out and hungry so we all tucked into dinner. Next morning I was awakened with the telephone ringing at the crack of dawn. It was Sid Hill. He had been to Alderbrook, for which we had left the keys with him, and had managed to open the post. Deborah had gained eleven 'O' Levels, one was a converted GCE for mathematics. It was the good news of the trip. I thanked Sid for the excellent news for which he was overjoyed. We all celebrated that day while we enjoyed New York City.

Deborah was allowed to join the sixth form at Solihull school; it had introduced girls to what was a formerly boys' school. Jonathan and Roland were already there.

Turning to the country and politics, the Liberal party was shaky and David Steel worked hard to revive what had been a good cause under the leadership of Jo Grimond. Nevertheless, the party was under some stress due to Thorpe's exposed homosexuality; this eventually forced him to resign. His career ended when he was arrested on charge of conspiring to murder the man claiming to be his former lover.

Alex Ritchie had taken a vacation in Israel and while there the partner at the head of a firm called TAT was in touch with him. The partner's name was Ostersetzer. He requested an invitation to visit NPT. An invitation was duly sent to Mr. Ostersetzer. Some months later he visited. He was short and looked like an Israeli. I got on with him very well and he said, "The Israeli Air Force (IAF) was searching the world for a small turbojet of about 150lb net thrust". I mentioned the David Budworth small jet developed for Flight Refuelling. He said, "That the 'MoD' requirement was that the engine should become the property of the Israeli Air Force."

So I indicated that in eyes of NPT this would constitute a dedicated engine required to be funded wholly by the customer. He thought the IAF would agree. So I sketched out the engine on a scrap of paper, using the well-tried compressor and turbine from my Rover days. He was amazed and years after would mention that this is how it all started.

We heard several months later that the Israeli Ministry of Defence had accepted the principle of NPT working through TAT to design and develop a dedicated turbojet engine. I was invited to visit Israel for the first time. I decided to take with me Rod Carr, who I had engaged as a consultant.

Rod was an abbreviation for Rodney. He was a handsome, good-looking designer that led the Rotax team when Rover had the task of drawing and detailing the gas turbine starter engine under my guidance. When the Rotax team became part of

# Automotive gas turbines

Lucas, which had owned Rotax for many years, the unit was moved to the North of England. Rod decided not to go with them. He did however move to Haversage near Sheffield. His marriage was a tumultuous affair, a real love/hate relationship.

I recall that when, as managing director of Leyland Gas Turbines Limited, he visited me and related that his wife has stabbed him with a kitchen knife; he was living apart in a boarding house. I was in touch with Sam Williams and found him a design post with Sam. Sam reported that he was a good designer but very temperamental; as a result his usefulness as a designer was negated. It was Rod's first experience of the States and little things became magnified.

I remember that he rang me one evening to say that his socks had become stuck in the laundry of the hotel and what should he do. I said, "I would buy him new ones if he went out and bought new pairs of socks". He seemed quite at peace with my instruction. The job with Sam lasted several weeks and when he returned to this country I set him on working for LGT as a consultant. He followed me to NPT; here he was useful in setting up concept schemes prior to design. Rod adopted me as a benefactor.

When Rod and I arrived in Tel Aviv we stayed at the Hilton. As we were at the desk of the hotel, booking in, a US gentleman was also doing the same and rubbed shoulders. He introduced himself and said that he would like to have dinner with me when we rested after the journey.

Rod and I met him for dinner that night and he had with him, Rolf Wyer, a colleague. He turned out to be Bill Rutherford from San Diego, head of the Pacific Group of Teledyne. After pleasantries we got down to business. Bill was curious why we were visiting Israel. I explained that I had founded NPT in the Midlands, UK, and we were canvassing work for the company. I mentioned we were there at the invitation of TAT an Israeli firm.

Bill Rutherford said, "His daughter was at the London School of Economics and did I know David Plaistow, managing director of Rolls-Royce Motors?" I knew, of course, that David Plaistow was managing director of Rolls-Royce Motors but I had never met him. Answering Bill I said, "I know of him". Bill was most affable and easy to get on with. Rolf and Rod were lost to the conversation Bill and I were having, so they held their own discussions. It was an excellent dinner and I learned a lot. Apparently Dr. Henry Singleton, of Litton Industries, had founded the Teledyne Corporation and Bill had become vice-president and president of the Pacific Group of Teledyne Inc. Teledyne bought a proliferation of companies, including Ryan Aeronautics, Continental Aviation, Continental Motors and many other firms.

Bill Rutherford was vice-president of the electronic division of Ryan and an experienced navigator. He was frequently in Washington, peddling Ryan's Doppler radar navigation systems. Competing with Bill for military navigation systems business was his friend, Henry Singleton of Litton Industries. Early in 1961 Bill replaced Ed Uhl as vice-president operations, assuming management responsibilities for all Ryan product lines, both aerospace and electronics. During his first year at Teledyne, he became responsible for all 'Black' programs. He was responsible for the well-known unmanned aircraft 'Firebee'.

He ended the dinner writing and sketching on a napkin the outline of a small gas turbine engine he wanted NPT to design for Teledyne Mobile, producers of piston engines for light planes. The connection the company had with Rolls-Royce Motors concerned a licence to maintain spares for Continental piston engines for light aircraft.

This first trip to Israel resulted in NPT designing both the NPT 151 turbojet for the IAF, and a turboprop engine, the TP500, that later became known throughout NPT as the Project 145 for Teledyne.

I saw Mr. Slohmo Ostersetter at TAT next morning and he showed Rod and I around the TAT company. I was surprised by the heat exchanger section where wavy stainless steel sheet was being manufactured. TAT believed both brazing and welding were satisfactory. I pointed out my experiences at Rover and the problems of brazing thin sheet. The penetration of the surface of thin sheet by brazing made any thickness less than say 0.005in dubious, because of gross weakening of the mechanical strength.

I asked Mr. Ostersetter whether I could meet representatives of the Ministry of Defence (MoD) during this trip and he said, "A meeting had been arranged". He offered to take us to dinner that night at our hotel, as the meeting with the Israeli MoD was in Tel Aviv. In the course of my first visit to Israel I got to know Slohmo quite well. He entertained us royally. The MoD meeting confirmed that if NPT worked through TAT, any funds paid to NPT for the project of creating a dedicated engine would not include funding for TAT. I queried this with Slohmo later and he said: "Providing TAT was requested to supply parts, however small, for the NPT design of the 150lb turbojet that was more than expectation".

The design of the 151 for Israel started before the Israeli's released the funding. This practice was dangerous and was against anything I had instilled into my team, economically. I requested that the Israeli's make an advanced payment for work already committed, but they argued that NPT had taken on the task entirely of their own volition. All in all, I found that Israeli is the one nation not prepared to give any ground in negotiations. I wrote a report for the Chief of Staff in 1982. In this, I said the average Israeli is educated and will beg, borrow or even steal to save the nation of Israel. This is not said in any critical way, but by my observation it is quite true.

It was a while later that I heard from Bill Rutherford. Meanwhile, the Israeli commitment went ahead.

On the political front, the Scottish Nationalist Party (SNP) became a thorn in the side of the Labour party and threatened the tiny majority late in 1974. The fear was that Plaid Cymru and the SNP would become more popular than the Labour party in both Wales and Scotland.

In 1975, it was decided to adopt the steel recuperator heat exchanger in Caterpillar's Project 201. The Stage One performance had been achieved but endurance testing remained plagued with failures of the Corning Glass ceramic cores.

Design of the steel heat exchanger was complete and it was decided that Stage Three would mark the introduction of the recuperator.

Chris Bramley, who had been the Project 201 leader from the start, resigned. I believe that Chris, who was under great strain, felt that he was being undermined due to failures of the ceramic heat exchanger. This was not so. I was very sorry to see Chris leave and tried to get him to stay on with a promise of promotion. He was determined however to join Ruston Gas Turbines in Lincoln, which he did. I racked my brains to replace Chris, but it turned out to be difficult. For a time, Reg Lowe took over on a temporary basis while a permanent project leader was appointed.

In 1975, Stage Three of the Caterpillar engine was built using steel recuperator. On the first test run, 559bhp was achieved with a specific fuel consumption of 0.385lb/bhp.hr

This marked an achievement. The first ever engine run on Project 201 had taken place just a year previously on 17 June using the ceramic regenerator. Then, 549bhp was achieved together with a specific fuel consumption of 0.39lb/bhp.hr.

Plans were considered for vehicle trials, but prior to this it was planned to carry out a 500-hour cyclic endurance test. Caterpillar selected an International Harvester Tri-Star as the heavy truck in which the Stage Three (201 gas turbine engine) would be installed. At 80,000lb gross vehicle weight and with a five-shift automatic transmission the gas turbine truck could be compared with its diesel counterpart. The Tri-Star was powered by a Caterpillar direct injection flat-rated 450hp diesel engine. Vehicle trials were planned for late 1976.

During the early years of NPT, I drove to Coventry daily, except during holidays and when I visited elsewhere. During the period prior to the site at Siskin Drive becoming available, I used my office at Alderbrook Road. When I resigned my directorships in Leyland and Rover I bought the Rover 2000, which was now not getting any younger. I contacted Bernard Smith in relation to the purchase of a new Range Rover.

I paid £2,900 for a new Range Rover that at the time was a 'Desert Yellow'. I recall that the registration trade plates started with a 'P' The family loved it and the children soon nicknamed it 'Jumbo'. I remember it did not have a fuel reserve tank; instead a green light would come on when the fuel level was getting low. I gauged it one day in a brave moment. Fortunately I found a fuel filling station before I ran out of fuel. I suppose that I clocked 25 miles from the time the green light came on fully on the level road. Depending on how it was driven, the Range Rover averaged less than 20 miles per gallon. I only used it in bad weather and for trips out to the country with my family. Roland became quite concerned when the green light came on to indicate low fuel level. He kept on until I was compelled to find the nearest filling station.

The journey to NPT from my home in Solihull was quite sensitive to peak-time traffic in both directions. Five minutes late in the morning made the journey time at least 15 minutes longer. I remember thinking many times, when motoring along to work, that some people worked in the direction I was travelling, also started from the direction of Birmingham. If the people travelling from Coventry in the Birmingham direction were to swap places with their counterparts then the majority of traffic would not be as intense and sensitive to peak timing.

It was usual for the children to give members of the family nicknames. Many of these were for life. Deborah's nickname was 'Daisy', Sybil's nickname was 'Elf' and my nickname was 'Bear'. Jonathan's was 'Rabbit' and Roland's was 'Kitten neck'. Looking back, these were rarely used, unless in conversation.

In 1973/74 there was a big oil price increase. Britain's standing in Europe was at low ebb. The economy was in an atrocious mess. President Ford and Jimmy Carter tried to lift the prestige of both America and Britain. Nixon had resigned and Germany's Chancellor Helmut Schmidt and the French President Valerie Giscard d'Estaing reconfirmed the Franco-German axis in the EC. Domestically Britain, as a nation, seemed to be 'going to the dogs'. However, NPT was on a high and Margaret Thatcher was battling with Edward Heath for the leadership of the Tory political party.

During 1975, Jonathan was due to take 'O' Levels and I attempted a similar system to that which I tried prior with Deborah and her 'O' Levels. This was in relation to revising the weakest subjects that demanded more time swatting, until all subjects are up to roughly the same standard. Jonathan was slightly less interested in revising so we had to persuade him that it was worthwhile by pointing to Deborah's achievement. Meanwhile, Deborah was sitting her 'Advanced-Levels' (A Level) examinations during May and June of the year in which Jonathan took his 'Ordinary-Levels'. Jonathan did quite well in his 'O' levels; he had become the captain of the school's shooting club and was getting on well at school, especially in technology.

On the work front, during 1975 the board gave me approval to purchase the site, buildings and freehold at Southam. I had a desire to start a new business in engineering models but soon had to learn that it was not as easy as I thought. There was one thing above all I had learnt the hard way in life.

Ted Matthews, who was responsible for my upbringing over the wartime years, had a simple comment he often repeated. He said, "You have to learn the bloody hard way in this life." He often swore and he was the first man in my life I encountered who could swear in the middle of a word. I was heavily committed and something had to give way. The lesson in life is to try to keep within one's limitations. Whenever I attempted to exceed my limitations, calling on immense reserves only evident in a crisis, I usually landed in trouble.

NPT progressed and the profits of the company were ploughed back, as they always were. As the company grew, it became more efficient at handling major R&D projects of varying complexity. The three main areas of NPT's activity became more self-explanatory and clear as time went on. My main desire for the company was to design, develop and market its own engines. I felt that projects of varying size, as well as sub-contract for other companies, were good ways to earn profits, as long as these were compatible with the main core business.

Nevertheless the real commitment lay in the urge to have a product. I suppose this was my reason for starting the company in the first place, although it was something one had to work towards painstakingly, building the organisation on a broad front.

In this respect, I am not counting David Budworth engines. For David was a concept genius in his own right. David Budworth's engines required the engineering expertise of NPT to produce the reliability that customers expected. As I have mentioned, there were two main customers when NPT took over David Budworth's company: the Central Electricity Generating Board, or CEGB, and Flight Refuelling.

Both requirements were for relatively short-life engines for quite different applications. One was an air producer of high mass flow and low pressure; the other was a small turbojet that powered an Unmanned Ariel Vehicle, or UAV.

In the 1970s, the CEGB had to transport huge generators weighing in the order of 600 tons on the road. These were essential in the construction of power stations. The generators had to be moved from the site of manufacture to the power station and NPT came to their aid. When the Romans built most of the British roads the bridges needed strengthening to sustain such loads.

The development was in its early stages when David Budworth was killed. But the solution lay in an ingenious invention to adopt the principle of load relief by evenly distributing a very heavy load over a large area. By so doing a 600-ton load could be reduced to 100 tons, providing the vehicle was long enough and had a hover skirt to prevent leakage of compressed air.

However, there were problems that had to be overcome. The route by road had to be carefully selected beforehand. For example, we discovered manhole covers were thrown like tiddlywinks; so all man-hole-covers had to be bolted down. The compressed air was likely to lift even the heaviest of man-hole-cover.

The hover skirt had to be sealed against the road and this caused the other main problem. We soon found a solution however in Ferodo brake lining material, despite its friction coefficient. The vehicle train was quite long (a third of a mile) and slow moving. The air producers had to be mounted on the moving vehicle and explains why a gas turbine engine – the NPT100 – was adopted, namely because of its low weight.

With careful management, over 1,000 bridge crossings were negotiated successfully in the UK by working this system on UK roads organized by the police.

The small turbojet was equally successful and several carousel Ariel launches were demonstrated in Scotland with Flight Refuelling. The small turbojet was engineered into a teaching set whereby students could be taught to extract performance calculations regardless of the small size.

The most significant migration to the UK before 1950 was by the Irish. Both the Irish, and the Jews before them, moved into run-down areas of London and northern towns. By 1960 there was a more rapid rise of immigrants and by 1968 Enoch Powell warned of the dangers in opening British borders to all immigrants wishing to settle in this country. Successive governments have been unable to control the problem since.

During the 1970s, a new issue arose. Families of immigrants already established here began to raise their voices. Legislation was established to allow family members to enter Britain in an attempt to control marriages of convenience, and use it as a vehicle to enter for immigration. The difficulty of 'bogus' asylum seekers was yet to arise but already the 'writing was on

# Automotive gas turbines

the wall'. The government later was driven to insist that potential asylum seekers obtained visas prior to their arrival in Britain.

In 1976, Jonathan won a place in Canada with the air cadets. Regretfully he damaged his ankle in the pentathlon at Solihull School. As a result, he was unable to take up the trip to Canada. I took him to see Mr. Rose, the surgeon in Earlsdon who successfully treated Nancy Grandfield with a fractured hip. He advised that only time would be the healer; regretfully Jonathan should not entertain the idea of a trip to Canada. He was very disappointed. So that year as a family we went to Jersey, accompanied by Deborah's boyfriend Gareth Edwards, a rugby player. He was not to be confused with the professional, who played for Wales. Jonathan was on crutches, something he disliked intensely. For most of the time we were in Jersey he was like a bear with a sore head. Deborah had gained her 'A Levels' examination in 1976 and decided to take a foundation course at Solihull College of Technology, prior to entrance to Coventry Polytechnic to take an arts degree.

On my return from Jersey, it was off to the office and a loaded desk. I deplored going away for this reason. However, the company moved from strength to strength. During those early years it progressed on a broad front in preparation for designing our own product engines.

The prospect came to do another major project for Caterpillar. This prompted me to consider expanding the company by building on the Coventry site. Plans for the new building, in the form of a build-and-test unit, were drawn up. When completed this became the boardroom, as Caterpillar finally decided to carry out development of the engine in-house. Caterpillar completed the design of a 3 to 5 MW gas turbine. The idea was to create three frame sizes for power generation operation. We were disappointed not to have another research project with Caterpillar, but we kept in touch with their people through our team seconded there.

# Automotive gas turbines

One of my team, John Clarke, came to see me; he was a key man. I used to have a sort of premonition before any of my team asked to see me on private matters. Sometimes it was wrong, but more often than not it was right. Years of experience gave me grounds for these premonitions; usually whenever someone was seeking an audience privately it was to move.

John Clarke was a good man. He was quietly spoken and always appeared immaculately dressed. I had immense respect for him. He was of average height and build. He had a dark complexion that was tinged with the slightest of grey hairs. I suppose he would be regarded as a quiet, handsome man as his skin was in excellent condition.

He was a civil servant with the NGTE) before joining NPT and this is where I first met him. Bob Weir was director there and was a good friend of mine since being Director-General of the Ministry. Bob was always enthusiastic about all aspects of engineering. Certainly, there was no exception to his interest in the case of John Clarke's invention.

Unlike a gas turbine engine, the cycle of his engine was more akin to that of a piston engine. I believe the reason it was developed at NGTE was simply that its inventor worked there. It was a rotary engine, rather like the Wankel rotary. I went along with the slant-axis-rotary (SAR) although many engineers regarded its claim to fame as bogus. It was said to have free-body motion. This meant that it be mounted in a vacuum to wobble in a certain way; neglecting inertia then free body motion was assumed.

These characteristics were fine in theory, but the 'SAR system' had to operate under other conditions. Free-body motion was achieved when the revolutions about its own axis had a magic ratio with the speed of the procession or wobble. Amounting to frequency of tipping of the rotor in relation to the speed about its own axis, just as a spinning coin when allowed to wobble will rest on a flat surface. The National Research Development Corporation (NRDC) was the owner of the invention, it being invented by a civil servant. I never accepted John's SAR invention had an advantage. To me, it had the fundamental snag of any rotary (piston) engine; the combustion shape was not ideal. The Wankel was no exception.

I managed to convince John that he should join NPT when the directorship of NGTE changed. He joined NPT in 1973, largely as a result of his engine being regarded as a 'red–herring' by the new director of NGTE. Later, Caterpillar showed an interest in John's SAR invention. So I was able to do a deal with NRDC to sell the device to Caterpillar. NPT never made any money from the transaction. John Clarke compiled several research assignments for Caterpillar, so this was compensation enough. I think John always thought that NPT made a large sum of money on the deal but when Caterpillar paid the contract money it was sent straight to NRDC.

Returning to John Clarke's presence in my office to broach a private matter, opening in his usual voice, he said, "I must appreciate that he had his heart set on the SAR system. Since it had been transferred to Caterpillar, his centre of mass had moved and he was thinking of taking a post with them in the Research Center at Mossville".

I recovered after what seemed a very long time. John Clarke was a key man and he was leading NPT into the computer age. His simple statement filled me with astonishment. I had several questions and I did not know if John knew the answers. My mind was raising all kinds of questions; it was literally working faster than one of my small engines.

My first question, whether Joe Grandfield had known this, I could answer for myself. My next question was to ask John whether he was committed yet. He said "He was". He had discussed this with his family and although they had considered the 'for and against' arguments it rested with his decision. He had accepted the post so it was a *fait accompli*. I asked him how much notice he was giving and he replied that if it was alright with me that he would like to stay until all arrangements were made and it could take up to six months. The earliest he anticipated was three months.

Unfortunately at this point my secretary rang to say I was due in a meeting. So I asked John to excuse me while I recovered from shock. John Clarke had advanced my company quite considerably during the time since he joined NPT; he was one man to whom I could always turn to ventilate new ideas. He would constitute an immense loss.

When I was next able to see John he still had to apply for a 'green card'. A 'green card' was an emigration requirement for people who wanted to work in the US with a view to becoming an American citizen. It turned out that John was to be in the post for several months at NPT.

I had given a lecture at Lucas Research for Bill Arrol, who was in charge of the Research Centre. Whilst there, I was shown round by a young bright engineer named Ives, who lived at Knowle. Eventually I hired him to replace John Clarke. He turned out to be a 21-day wonder and eventually went back to his post in Lucas. I knew Joe Righton, who was a main board director of Lucas. Joe had hired me as a consultant while I was working out my contract with British Leyland. Joe had been very keen on Ives and it left a nasty taste in his mouth that he thought I had enticed Ives away from Lucas. All was not lost as his bright boy rejoined Lucas.

In 1976 work began on a family of engines that became forerunner of the Teledyne engine project and called the High Specific Power project. It was an attempt to develop NPT's own family of small gas turbine engines inspired by John Clarke.

John was head of research and the concept was to increase the power of the engine by increasing air mass flow and pressure ratio *pro-rata*. By so doing, another more powerful engine could be based on the smaller member of the family and the resultant development was reduced.

Only the concept and planning phases were completed of the High Specific Power project, as the Teledyne contract was

# Automotive gas turbines

adopted, after two initial funded reports. Of these two reports, one for the SB137, was completed in May 1978; the other for the SB138, was completed in July in the same year. The SB138 report outlined a programme for a 450bhp turboprop engine.

The contract was signed in San Diego in the fall of 1978. Sybil and the three children joined me after the signing. Bill Rutherford laid on an escort and transport into Mexico. Our return flight from San Diego to the UK was marred by Deborah, who was nervous of the flight. She was reading a novel about a flight that was due to crash and this, she imagined, would happen to our aircraft. I cursed the boys for giving Deborah such a book. They found it quite funny that their sister was so worried.

It was the start of the named Project 145. Project 145 was a development of a turboprop gas turbine engine, albeit of more horsepower than my previous developments. I had previously received the Air Registration Certification as technical director of Rover Gas Turbines Limited for the smallest turbo prop in the world, this despite the fact that I had little to do with the engineering of the installations in three aircraft, the Currie Wot, the Auster and the Chipmunk. The latter two aircraft were powered by the TP90; the Currie Wot had the TP60 engine installed. Douglas Llewllyn was a maverick and Frank Lord, head of the ARB, was more of a stickler for the system. I had systems and Douglas was a more cut-and-try merchant, however he was amazingly successful and a first class designer.

His understanding of the single-shaft engine coupled to the feathering propeller was a joy for me as a professional engineer. We use to have many debates about the fundamental basis of marrying the prop to the single shaft gas turbine engine. One characteristic of the single-shaft gas turbine engine was that the cruise speed was higher than that of take-off. The single-shaft engine was susceptible to a rise in temperature if maximum power was exceeded.

We never experienced such a case, although I suspected that Viv Belemy, who was a wartime pilot, sometimes went over maximum temperature doing his aerobatics.

We soon got into the swing of the engine design based on Teledyne's specification. Teledyne adopted Jeff Fawn, ex-managing director of Rolls-Royce Ltd after the crash. I spoke to my friends at Rolls-Royce and they related a story of Mr. Fawn going round to each office saying, "Get rid of all your reports." It was not black and white that it was 'Fawn'. I prefer to deal with engineering experts and Jeff was no expert. In fact, he once argued that it was impossible to design an 8 to 1 pressure-ratio centrifugal compressor with the adiabatic efficiency of the specification. He was proved wrong.

Returning to the specification, the single-shaft engine proposed was to operate at a speed of approximately 56,000rev/min and an inlet temperature of 1300°K.

Principal features of the engine were that the performance should be competitive with the typical piston engine of similar power. In respect of the equivalent piston engine, the integrity, reliability ruggedness and life should be at least on a par.

It is interesting to note the steps NPT took for research projects of this kind. With the instruction to proceed, the preliminary design was undertaken in accordance with the customer's specification. The next job required detail design, with all long-lead items being established. Then rig designs would be started. There is a time in the programme when all long-lead items must be ordered, as with all parts and components on order. When parts start to arrive and are inspected, sub-assembly build can commence. The next vital milestone is the first engine run. Three initial engines are built, one to establish mechanical integrity, one to establish performance and the other is a slave engine in order to maintain the programme. All being well, a 300-hour endurance test is commenced, followed by a 1000-hour endurance test. By this time, if the programme proceeds satisfactorily, the sixth engine is built and target two engines can be delivered to the customer.

I have never known an engine programme to meet these targets. Something usually happens to cause slippage to the project. The Teledyne Project P145 was no exception. On the engine strip during the first engine run the auxiliary gears showed distress. The gears were manufactured in the US at one of Teledyne's own companies, 'Neosho'. The excuse the Teledyne people gave was that NPT inspection was at fault for accepting the gears in the first place. The gear teeth had serious wear; had running continued a serious failure would result.

The incident was so serious that I called in a gear expert, Ray Hicks, and agreed to send Ray into Teledyne's gear firm to investigate. I told Ray to leave no stone unturned and to examine every aspect of manufacture. Ray was familiar with gear design and manufacture. His survey soon produced results. The angle plate on which the gears were bolted and teeth were cut was out of true to an extent that it would certainly account for the problem. Ray suspected sabotage; there was a great deal of jealousy from CAE, the engine manufacturer of Teledyne, and it was easily possible that the gear firm could be persuaded. While I had trouble convincing Teledyne I would not order any further gears from 'Neosho', my point was finally accepted.

Back tracking for a moment to early 1977, when I went to China during Mao's time and became marooned in the British Embassy because of the actions of the Red Guard. I had agreed to lecture at Peking University by invitation of the Machinery Group. At that time, China had strict rules governing all trade and the Machinery Group was one official body for transacting trade with mainland China. When I was released, I joined Ian Grant in Japan; Ian was NPT's sales manager at the time.

Ian had received a request from a company called Yanmar, one of the world's makers of auxiliary diesel engines. Yanmar had an office in Tokyo. This meeting turned out to be quite successful, although we did not know it at the time. Later, I returned to mainland China during the time of the Gang of Four. This time I was welcomed with open arms and actually sold them one or two engines. The tour was most exhausting. I started my lectures at an unearthly hour in the morning and this

# Automotive gas turbines

continued until 10 in the evening. I was allowed to have my own interpreter and she made sure I had a most impeccable Peking accent. Every morning that I appeared in front of the class, the audience clapped their heads off saying "I had the most excellent Peking accent".

When I left China, all my class came to the airport to see me off; most of them had tears in their eyes. They were a friendly people. I must have picked up a bug on this trip; it lasted several months. I was on my way to the US via Los Angeles and I slept all the way during the trip until I reached Peoria.

Then I became ill with a kind of pneumonia. It plagued me for many weeks and turned into an all-over stiffness.

I went to the see the Birmingham Symphony Concert with Sybil and experienced difficulty in clapping. It was so strange that I told people I had suffered from the 'Goby Desert lurgi'. It took many months to diminish.

After a bad start, Project 145 with Teledyne progressed quite well. David Moss, who was responsible for combustion and stress, invented a forked injection point that was nicknamed the cobra's head because it had two prongs. The pre-vaporizer worked quite well. The only snag was the second turbine had a slight drop in its total-to-total efficiency; this however was made up by an increase in compressor efficiency.

There was great competition between Bob Chevis and David Moss. David was the weaker in constitution, but not in ability. However, there was nothing to choose between both individuals in this respect. Due to this I nurtured David. I allowed two technical heads to operate over many years, but I knew well that the only real management solution was to appoint one or the other to lead. It was one time that I was unable to be decisive about which one to choose.

I knew well too that I stood the risk of appointing one or the other, but only retaining the winner. So I vacillated, and as a result this was one of the only times I turned my back on making a firm decision. David eventually suffered a nervous breakdown, but he was too good to lose so I did everything in my power to bring him back to full health, and I succeeded.

Although Bob was good at concept planning and performance I had misgivings he could not be always trusted. When my back was turned, watch out. On the other hand, David Moss was difficult to read but was full of integrity, so I had no qualms. In my subconscious mind this probably was why I vacillated over appointing one over the other. 'A coward lives to fight another day.'

The year 1978 came to an end, but the problems of the Teledyne project persisted. There was a scare, invented by Homer Wood, who was a trusted consultant of Teledyne. He challenged the calculations in relation to the stress of the first stage turbine. Fortunately, Caterpillar supported the stress criterion and Homer had to back down.

It was usual to hold a quarterly project meeting in NPT and I laid on an event in the House of Lords. The Earl of Minto, (Gibby short for Gilbert), officiated. A party in the House of Lords must not to be held for commercial reasons; a member of the House of Lords must invite overseas guests as friends. A luxury bus was laid on from Coventry. Unbeknown to me, Reginald Keetly was being critical of the leader of the project for personal reasons; this behind the backs of the management.

Next morning, I sensed that the visitors from Teledyne had something on their minds. Bill Rutherford, the most senior of the visitors, spoke up and said, "It is no good beating around the bush. Your designer, who shall be nameless, was critical of the way the project was being run". I knew who it was. Reg used to ring me up most evenings; he was always apologetic.

He had a distinct tendency to move out of line. I had no alternative but to change the project leader. However, I took my time over appointing a new one. Bob Chevis was the obvious choice, but Bob had to manage his own department, performance or concept design. The gear train worked well having changed the source of supply, though engine performance showed a deficit in overall turbine efficiency, partly made up by the compressor efficiency.

The next quarterly meeting was planned to be held in the US at Bisley, in North Michigan. It was at this quarterly meeting that Bob Chevis announced that the 8 to 1 pressure ratio compressor had delivered its predicted efficiency

We were expected to play golf and, later in the evening, to play pool. Mrs. Rutherford had her leg pulled, until her husband stepped in and showed a side I had not witnessed before in him.

Caterpillar's 201 Project automotive engine was renamed G45. The G45 engine benefitted from developments that had taken place at NPT since starting in 1973. The initial phase involved the design and development of three combustion systems, two different forms of heat exchanger, and two types of radial compressors – one with sweptback vanes and the other with straight blades. Numerous details were made to align the design to cater for the major change of main components.

Off-the-shelf proven design systems were to benefit NPT's own engines. Twelve 201 engines were built and tested. About 4,000 hours of endurance and performance were carried out; about half of these were at simulated vehicle conditions. Two engines were installed in the International Harvester Tri-Star truck of 80,000lb fully-laden weight. A special Caterpillar team carried out production planning on a wide range of volumes of 100 to 50,000 gas turbine engines per year.

Turning to the politics of the day, by the 1979s, the SNP had become a declining group of politicians. It was searching for its true identity. Although the leanings of many of its members were left-wing, however the party's roots were anti-Labour, competing with its Liberal ideals. These divisions were visible.

When I returned from the journey to Japan I was looking forward to a homecoming. I was confronted with some difficulty. When I had rested, Sybil explained that she had seen Hilda, my sister, during my absence in Japan. I was decidedly pleased about that, but I soon became disturbed because of Sybil's demeanor. It was as if I was under interrogation.

# Ten

# Downfall

**In which not only the business I created began to stumble, but my personal life also.**

AS the company progressed in the early 1980s, my problems seemed to worsen. At home, Sybil was becoming such that I was at a loss as to know what was wrong. Her doctor prescribed some drugs but after a while she was in such a daze she gave them a miss. I went along with her to see all kinds of specialists and doctors. Finally, we landed up with an expert Dr. Meldrum had recommended. He was a very eminent psychologist. He listened to what had turned into a slanging match between Sybil and myself. Dr. Meldrum gave me the result: Sybil had an incurable obsession. When the malaise was triggered, she was in another world of unreality.

Over the years up to 1984, I left home three times. Once, when I was taken to Sidney Hill's in Alcester, by Roland and a friend; on another occasion, I stayed at a hotel and finally, I took an apartment near Roy and Doreen. Life was sheer hell and I lost sleep. I was beginning to turn away from my work.

On the political front, trouble was brewing with the SNP, which had been beset by tensions since 1979. It was beginning to expand and as a result it would have to oppose Labour in its heartland along the Clyde. Devolution became the new quest.

On the work front, my malaise with Sybil was taking its toll. The company had progressed significantly and was well on the way to producing its own engines. However, the next four years were hell. I lost sleep and I think that if I had stayed there I would be dead by now. I gave up many times and, as I mentioned, I left three times before finally leaving in December 1984.

It seemed that there was no incentive to work anymore. Had I drifted into some kind of ennuis? I floundered; it was the start of the company's downfall.

I had been extremely active until then. During the nightmare, Sybil used to rifle through my briefcase and sometimes she had a habit of standing at my desk and, in a quiet voice, would say, "You are a wicked man". I wrote several poems during the problem and I posted them on the board in the breakfast room where we ate. They were always torn down.

The crunch came when the Meldrums, who were both doctors, husband and wife, declared that Sybil was a subject of incurable obsession and that I should get out, or she was capable of doing me harm. I queried this as I was aware that 'incurable' meant 'forever'.

When I left in December 1984, Jonathan my son and Arthur Varney found me an apartment with the Winters' house in Leamington Spa.

Bill Winters was a short man, just about 5 feet and three inches. I knew him while I was at Rover. He had been sacked as production director for lying to the managing director, Bill Martin-Hurst. He should have been working in Cardiff but instead was watching a cricket match at Edgbaston. Rover was a paternal company it was unusual to dismiss anyone for anything other than a very serious offence.

Bill Winters had been at Rover for only a short time and thus was an unknown ingredient. One had to have worked at Rover for a good many years to be accepted as a 'Rover man'. His wife was an authority on dogs. She bred Cavalier King Charles spaniels. His wife kept many dogs in a cellar. With the door open to a room one was expected to cross to reach the entrance to my flat, not to mention a door that opened out on to a flight of stairs, the stench on entering the room at the bottom of the staircase was awful. I was always ashamed to invite anyone to my 'new home', because of the smell.

I was told that a cleaning lady visited many times during the week and it required an extra payment to enable the flat to be cleaned. Not aware of domestic staff, I paid up but I soon noticed that the flat was no better for dust than when she was absent, which was more often than not.

The flat was quite respectable. It had a small toilet facing the door at the top of the stairs. The corridor faced a kitchen. There was another corridor at right angles to the one at top of the stairway. Off this, there was a bathroom on the left, and a bedroom was facing on the right, which I never used. Where the corridor ended there was a door to another bedroom, which was the one I used. One turned right through a door to a rather medium-sized lounge, poorly furnished with a television set and a mirror over what was once a fireplace.

Slowly, NPT engines evolved. At first, the development contracts were useful in relation to our own engine-development. Then I underestimated the cost of engine development, which was tantamount to 'rubbing-out-in-hardware', of which Frank Whittle was guilty. Despite this, progress was being made in technology. It enabled whole engines to be designed 'over-the-board' by computer, especially if applied to changing the breed with updating.

The problem I faced was enabling Harwich to become subject to the same disciplines that existed at headquarters in Coventry. None of my engineering executives had the responsibility to ensure that Harwich conformed to the rest of the company, in respect of quality and systematic approach. Even David Noble, who was the first manager of Harwich and did so well ensuring profitability, was of the opinion this plant was uncontrollable. It was an example of something needing strong management that I was in no position to give control.

It came to a head when I visited Boeing Military Aeroplane Company, or BMAC, in the US. Peter Rackham came to my room in the hotel late one night; he was responsible for electronics at Harwich. He complained that the man I placed in charge of Harwich was having an affair with the cleaning lady. He could be seen every night in his company car committing an indecent act with the woman who was married. I wrote to the manager laying out what Peter had said. It was my usual practice to call management meetings every week and the manager was due to attend. He was not in attendance, which I wrongly accepted was an admission of his guilt.

Then I was told the reason for his absence. His son was tragically killed the week before the meeting, and he was at the funeral. It was the usual reason: he did not inform my secretary. I imagined this was due to distress. However, I wrote to him saying I was very sorry to hear of his sad loss, not raising any other comment in the letter.

However, it was proved that the manager I had appointed to be responsible for Harwich had been indiscrete, when the husband of women wrote to me. The fat hit the fan, Peter Rackham was right. I sacked the man immediately and he took the company to the unfair dismissal tribunal. I defended the case myself and the barrister I hired, said afterwards, my performance was OK, but I went round a bit to nail the point down.

Readers may experience difficulty in understanding the company. Noel Penny Turbines Limited (NPT) was founded on my belief in small gas turbine technology. It was more a case of a 'happening' that subcontract work materialised as one of the main core items of NPT. The other two main company pursuits were R & D for other companies, and our own designs of engines, mainly derived from projects we were doing for others.

It all started with the truck gas turbine engine for Caterpillar Tractor, called Project 201, and on which the company was founded. Project 201 had been active from 1972 to 1978. Subcontract work had been operative from the time the company started also. Mainly derived from UK customers, it also was a useful manufacturing facility to produce engines for others and on the back of this it enabled the company to build-up other sections, such as inspection, fabrication and a testing capability. NPT was an approved firm to carry out work for Her Majesty's Government establishments. It also carried out assignments testing piston engines, mainly for the Rover Company.

# Automotive gas turbines

The one project that followed the gas turbine truck engine, Project 201, proved to be the turboprop engine for Teledyne in the US, known as Project 145. This project covered design, development and certification of a low-cost 520bhp turboprop engine. P201 spurned G45, an NPT engine, and P145 turned into a family of HSP, or High Specific Power engines supported by the DTI, the Department of Trade and Industry.

In 1979, a demonstration at White-sands was conducted of an NPT151 turbojet installed in a Boeing 'poor man's cruise missile'. This led to the design of the NPT171 for the Boeing Military Aeroplane Company (BMAC). The NPT171 was an engine specially designed for Boeing.

In 1980, we designed the Yanmar engine under Project 169. This engine design was intended for standby and continuous duty as a 500KVA electrical generator (photograph, p.453).

Suddenly, all enthusiasm left me during Sybil's illness. I was about to give up when my former girlfriend came into my life. I left Alderbrook Road in December 1984 and made contact with her in November 1985, but the least written the better.

My heart was no longer in my work. It is one of life's problems that all hope is lost when there is nothing to work for.

My two eldest children had graduated; it only left Roland to do so. Roland seemed set on his degree. He needed less attention than Deborah and Jonathan. I knew he would always do well. His tutor came to see me and said he was on course for a first class honors degree with a star. I had always been keen to provide the best education for them; they were good children.

It was at that time that I began to fill several notebooks by working until the hours of the night merged into the hours of the early morning. It was as if something was guiding my notes. I filled notebook after notebook, writing about my parents and researching the background of the story.

However, I could mention that Joan Spurgeon was an ancestor of the man who wrote the Forsyte Saga, John Galsworthy. Joan's mother was a Galsworthy. There was a large family of Galsworthy and her uncle happened to be the boy who sat next to me at school. James Galsworthy was, for a time, in Hertford Hill, the open-air hospital at Warwick, for TB. After he was released, I fed him with Liquorice wood to get the roses back into his cheeks.

At that time, when I travelled back from the US, I thought about the most remarkable man I ever knew. It was Ted, who could swear in the middle of a word. Despite Ted's swearing he was good-natured and far-sighted. He had many skills. A born touch with anything that grew out of the soil, he could remember their Latin names and identify almost any plant or flower. He had an assortment of pruning knives. His knowledge of first-aid was astounding; that could be judged only by his success. Although not obvious in his appearance, he had a gentle healing touch. He could lance a boil or abscess without causing pain. In fact, he could have been a skilled surgeon. He had many stock phrases that caused much laughter, 'Sort out that bugger' and 'You will bloody-well do something, you will'. 'Look at that bugger Friary' – for Violet, my sister, he used to call Friary. When he came across anyone of a haughty nature it was 'Mr. Bloody Simpson or Miss Bloody Primrose.

He was brought up in Great Tew and his young father was head gardener at the big house. His father was killed in the first boatload to go to France in the First World War. Great Tew was alleged to be the prettiest village in England. Ted was brought up by a family not far from Aunt Min, the mother of Betty Slater, who married my eldest brother. Ted's grandfather lived at Woodstock and was killed while trying to stop a bolting horse.

Aunt Min, a distant relative, was dark, almost like a gypsy. She could look from the front door of her cottage and see in the distance the old stocks in front of the school. Across from Aunt Min's cottage a path overgrown with foliage led to the big house where Ted's father was head gardener. During the First World War, Oxfordshire depended on agricultural trades.

The story goes how Ted's mother, Katie, was so distraught after receiving the telegram informing her of her young husband's death in France, that she wandered off. She was a parlour maid at the big house in Great Tew. Some say she was looking for his eldest brother, he would know what to do. She thought he had taken a job of war work at Banbury. Katie reached Banbury after being picked up by a farmer in a horse and trap.

The farmer sensed this young girl was distressed, and drove to his farm. But when he returned from the house with his wife, Katie had moved on. Katie stumbled into the shunting siding of the Banbury railway yard and, in her distressed state, was knocked down by a shunting railway wagon.

She went in and out of sleep in Banbury hospital. A white-coated doctor told her to rest, having shone a pencil-like torch into her eyes. Katie drifted in and out of consciousness. After a few days she started to improve but she had lost her memory. She proved an asset to the hospital with her domestic training. She was determined to become a Red Cross nurse and go to France. Katie was in the right place to receive training.

She trained for a year, and doctors were amazed with how well she progressed, but the yearning to return to France was like a magnet. Finally she was accepted. In those days, a young person with such determination was easily accepted as a skilled nurse; they were few on the ground. The Irish rebellion had taken some of the nurses away from France. Katie was sent to France at the end of 1916. The battle of Verdun became a massed killing, part of the wearing-down process, later known as attrition that became the tactic of most generals. A new type of gas, phosgene, was used by the Germans but had little effect.

Katie was thrown in at the deep end and the doctors were overworked. Once, Katie was ordered by a high-ranking doctor to treat a gassed man by putting a field dressing on his leg. Katie was forced to help the gassed soldier breathe although the doctor had thought there was nothing to be done to help the poor man.

# Automotive gas turbines

Katie soon became a favorite with doctors and patients alike. She made herself work until she nearly dropped. No sacrifice was too much for her, although she did not forget her young husband. She was dedicated to nursing the wounded.

By a trick of fate, she nursed a young captain who was critically wounded at Verdun. Little did she know the young captain was a relative of the people who owned the big house in Great Tew. With her perseverance, he developed, albeit slowly, an improvement from his bad wounds. After many weeks' improvement and Katie's nursing, the young captain was on the mend.

From this point on I will not burden the reader with the narrow escapes that Katie had. During the night and day the field hospital was subject to constant shelling from the Germans lines that were not far away, as the crow flies.

Indeed Great Tew, where Katie had worked, was just only about 260 miles away from the fighting, as the crow flies. Aunt Min wondered, as she saw two crows in the trees near her cottage, if they had flown from the battlefield. "Where was Katie, I suppose she was at the bottom of the lake in Woodstock?" Aunt Min said to herself.

There was another twist of fate. Ted Matthews, Katie's son, was married to my eldest sister, Violet, and Aunt Min's daughter, Betty, was married to my eldest brother Jack.

Going back to November 1985, it was at that time that I was overweight but by various means I began to lose weight, albeit slowly, so my clothes became ill-fitting. I found a retired tailor who could alter clothes that sometimes were recoverable; so one could have two sizes in one suit within reason.

After many months I was down to my weight when I was 16 years of age; I felt a lot healthier.

I regularly used to visit Sybil in my home at Alderbrook Road to see if she was well. I think she resented my visits. Sometimes she locked all the doors and I had to wait for hours to get free. I sympathize with anyone who is not free. Freedom is one thing I treasure most.

Once, when I gathered a suit, she was greatly disturbed and ran after me as I left the front door. Getting into my Range Rover, she caught up with me and tugged the driver's door of the vehicle. My foot became trapped on the clutch pedal and the Range Rover moved forward into the neighbor's fence. Where the fence had been erected there was a deep step; I had great difficulty getting the vehicle back up the step. Finally, I managed to reverse, although the vehicle complained. The fence was a complete write-off for its full length.

I did not know my neighbour very well, but he came out because of the sheer noise and saw his shattered fence. He muttered, "You must pay for the damage." By this time, my Range Rover was in its rightful parking place. I explained that my foot had become trapped under the pedal and the brake was off. I don't think he believed me. I think he thought I was drunk, but I assured him that I would make good the damage.

Usually, there was a fracas whenever I visited Alderbrook Road. Many items I took there would land up missing. My visits became less and less until finally, after a year, they stopped altogether.

I have paid most of the expenses for Alderbrook ever since. The cost to date after 21 years has been somewhere close to £750,000. I have not been earning since 2001. It is a good thing I was earning before that.

For a reason I will not discuss, I slowly built up my confidence and belief in myself. I had lost all confidence and was becoming obsessed with Sybil's illness. No one should go through what I went through, especially someone you love immensely and who has helped shape our children's good character. To hear them swear that reality is just a myth, really is the worse trial of life.

During the worse nightmare of Sybil's illness I appealed to many people and wrote long letters seeking help, for example to her brother and my children. And during the five years of hell I turned to writing poems. It was just as if some hidden force was working my pen.

Turning to the politics of the day, Thatcherism had ruled since the early 1970s. It allowed ailing industries to go to the wall. It was a policy that kept the pound high hedging against inflation. Later, Tom Trenchard, the son of the famous Lord Trenchard who instituted the RAF, at the time was Defence Minister, told me a story of a meeting with Margaret Thatcher. He had started wrongly, "That he did not wish to teach his grandmother to suck eggs". Then he realized that what he had said. He was sacked soon after. Margaret was not called 'The Iron Lady' for nothing. Thatcherism also was imposed on Scotland. Mrs. Thatcher later introduced the Poll Tax and this proved to be her political undoing. She opposed, John Major as her successor.

Turning to the work front, my team helped me a lot as I was still travelling the world. The Teledyne Project 145 was due to end in 1985. It continued for a further two years and was a most remarkable design that involved variable geometry on the inlet as well as a variable diffuser called the GPG. Other projects were a 401 for Meteor in Italy and an engine for the IAI, the 904.

About this time I was negotiating a joint project with Allison in the US for a link to LASOM. Williams International regretfully has linked with BMW. There were five Nations in LASOM: France, Germany, Spain, USA and the UK. It was eventually called CASOM, which involved only the USA, UK, Germany and France. Spain withdrew. The initials CASOM stood for Conventional-Aerial-Stand-Off-Missile.

The projects were a good leaping-off point for our own technology. Updated, NPT were developing their engine. The applications of these were wide: expendable jets, automotive engines with heat exchangers, and combined heat and power engines called industrial engines. The smallest compressor development involved the design and development of a vapour turbo-compressor for a domestic heat pump. This was for a UK company. It was less than one inch in diameter and would have amounted to six to one on air.

Another development involved low-inertia, wide-flow turbochargers. That brought in the development of ceramics. Indeed, NPT joined the ceramics development of the Department of Trade and Industry.

I have always had a wish to apply ceramics to the gas turbine engine. I am confident that ceramics will play an ever-increasing role. The LAS system of glass ceramics and used by Corning Glass captured my imagination. Parts per million strain was introduced, reducing the brittle behaviour of the material. But the application of high-temperature ceramics with some strain has proved a lifetime search.

The key to the gas turbine engine is to discover materials that have high-temperature capability, enabling these materials to handle higher and higher temperatures. Ceramics are of low weight compared with metals and therefore intrinsically offer lower stress. My one wish is that I could invent a ceramic material that would permit a long-life turbine.

I came near to this when I invented a thin, shell-like turbine of loose blades. Unfortunately, the turbine design failed when the 'Nimonic' caps were welded to the construction by means of high voltage electron-beam-welding. The caps were to be cooled in a recess.

I lectured regularly and was called upon to make talks abroad. Among the countries where the lectures were formally presented included Germany, the US, Russia, France, Italy, Portugal and all over Asia, including countries such as mainland China, Taiwan, Japan, South Korea, Singapore and Australia. I gave many presentations in the US, practically in every state.

Returning to the story of Ted, Katie was asked by the young officer she nursed, to return to England with him. Katie said that she would return with him on the boat train and see that he reached London safely; however, she said she would return to France. She was at Waterloo station and was on her way back to

France when, little did she know, but her husband's brother was on the boat train rejoining his regiment at the front at Ypres.

Katie was welcomed back by the doctors who knew her. Many of the doctors had moved nearer to the front at Ypres. The battles hitherto of the Somme and Verdun were quite different from those at Ypres. The constant shelling at Ypres had an adverse effect on the drainage system that was close to the surface. Ypres soon became a quagmire of slime, mud, dead men and horses. Sometimes it was many feet deep, allowing troops to become lost in the mud and drown.

Ted's brother was sent to Ypres and suffered from the day and night bombardments. He had terrible wounds and was also shell-shocked.

There was another twist of fate for Katie. Unbeknown to her, she nursed her dead husband's brother. He had been sent to the field hospital on the outskirts of Cambrai (in the Hauts-de-France region) where the first major tank battle took place on late November 1917.

Ted's uncle was recovering from his wounds and became shell-shocked when the field hospital came under attack from the German artillery. He survived, but once more was injured in the shelling. Katie was unhurt; however, there were many other casualties in the field hospital.

Flags were flying on Armistice Day, which was the eleventh of November 1918. People danced in the streets and Charles Penny took young Violet in his arms to see the marches in Broadgate. Although only four, she felt comfortable in her father's arms; she carried a little Union Jack.

Katie, her job finished at the field hospital, remained for several more months as the chronically-wounded could not be moved.

Between 1985 and 1994 manufacturing output fell by 23 per cent. Scottish capitalism had taken a knock and this gave rise to the SNP. The payoff came for the SNP in 1988 when Jim Sillars won the safe Labour seat of Glasgow Govan. The Poll Tax was a further advantage to the SNP. The SNP, the Scottish National Party, allowed Scotland to take a lead in a tax boycott. The anger of Labour's Scottish MPs over the imposition of Thatcherite policies drove the party to adopt the nationalist concept of the SNP mandate.

On the work front, in 1987 I became concerned that the number of people responsible to me had grown to the point that it was too many. I had seven executives reporting directly to me and this, together with the consultants, was taking more of my time. I worked on Saturday mornings, allowing me some time with Bob Chevis.

Bob was head of engineering and concept. This was a focal point in order to undertake preliminary design work. But much of his time was taken in proposing and tendering for new work.

Bob was short in build and had been at Rolls-Royce in the performance section, working on the Viper turbojet. The Viper was one of Rolls-Royce's early engines and, while it was an expendable engine, it had at one time the longest life of any of the Rolls-Royce stable. He came among the first batch of staff I hired. I believe he obtained his masters qualification at Cranfield, after serving in the Fleet Air Arm. I recall he showed me the work he had done on a family of gas turbine engines during his first interview in Alderbrook Road. I was significantly impressed at our first meeting. His resemblance to a famous football player was uncanny.

Ian Grant was deputy managing director and executive director, marketing and sales. He is difficult to describe, but slightly protruding teeth was one of his characteristics. He was fairly tall, slender but sturdy. He was inclined to slight baldness. I suppose that I hired him from National Research Development Corporation (NRDC) when I was negotiating the Slant Axis Rotary (SAR) engine designed by John Clarke. He used to discuss the time he worked with Frank Whittle on the turbine drill at Bristol. His wife nearly worked for me when we seconded Ian to Teledyne's plant in Mobile, in the US. I was anxious that Ian should be moved to set up an office in Rome.

# Automotive gas turbines

Work had started on a new engine designated the NPT 4006. The engine was a new departure. Its target rating was 3MW ISO Rating and was designed with a thermal efficiency of at least 30 per cent. David Moss, director of operations, was the design instigator. This was a peculiar project.

It came as a result of a man by the name of Felix Pole; his friend in Italy was named Ricardo Raciti. He was managing director of a firm that had a name like B.A.T.

Ian Grant had made several trips to Rome to see the gentleman. At first I imagined it was another of Ian's 'irons in the fire', but it soon gathered immense problems.

Ian had other responsibilities. The sales staff reported to him through an assistant and the general manager of the Hinckley Manufacturing Plant, Roger Tomlison, also reported to Ian. In the event that Ian was traveling, my son Jonathan officiated.

In 1988, Lars Malmrup, who I knew from Volvo, came to see me and after lunch he suggested that Volvo could be interested in taking a slice of the company to see the unmanned turbojets developed into volume production. I later discussed it with Jonathan, who was non-committal. I read into this that in his opinion a decided agreement.

Also, Mark Metcalfe who was at the time executive director finance, suggested that if I listened to Felix Pole about taking the Company public, at that time it was called 'Listing', he would resign. I was not about to have my hand forced. I regretted what Mark was suggesting, but apart from that he was doing a good job.

I knew well that the company's tail was wagging the dog and some changes had to take place. The options were to 'go-public', sell the company, or merge. Which one was the right one? I chose that latter – to my cost.

Despite that, Mark Metcalfe resigned. He explained that a firm that manufactured hospital beds had accepted him; he was moving to a position as a director on the board with a share in the profits. I had no intention competing with his new firm. The long line of financial staff that had left would be an embarrassment to anyone who ran any company. It indicated that there was something that was financially wrong with the firm, despite its engineering background.

Since NPT had been incorporated, we had had Mike Morris, Geoffrey Mullet, Austin Pilkington and now Mark Metcalfe; all had left the company for various reasons.

I became tired of seeking another financial controller. So I gritted my teeth and paid an agency. Recruitment agency Page was supposed to be the best firm to locate financial staff. I was faced with a short list of two, I had a feeling that the Scots or Irish made the best financial staff and neither of the two candidates was of the other national identities. Alan Cross seemed a high-powered candidate, although he was unemployed at the time but he lived locally. I chose Alan Cross and he eventually became managing director, a post my son turned down on the grounds that he needed more experience.

Discussions progressed and Jonathan and I planned to visit Volvo in Trollhättan, Sweden, to meet the company's top management. The meeting went well and Jonathan excelled himself. The Volvo management must have felt I was peculiar, because I was not up my usual standard. Nevertheless, the legal arrangement was drawn up and we were requested to visit Gothenburg to hold a meeting with a Mr. Tryggve Wahlin, the legal consultant of Philip Lemans Advokalbyra. Ted Jeynes and Arthur Anderson of Edge & Ellison, our Birmingham solicitor's accompanied us to meet Tryggve in Gothenburg.

We were surprised that we had to lead Tryggve through the details of the Volvo proposal, but the meeting went off satisfactorily as we warmed to our host. We were flying from Heathrow and my two colleagues dropped me off at the farm in Marton, where I had an apartment. It was during the poll tax and I calculated that I was paying the poll tax in addition to the rent. So I looked round for a property that I could buy, so putting the rent I was paying towards a mortgage.

Jonathan and Arthur Varney helped me to look round for a suitable property. Both had found me an apartment owned by the Winters when I left Alderbrook. My son Jonathan and Arthur helped to locate the property I live in at the moment.

The Volvo agreement progressed and as it, so did the paperwork grew. I was not expecting to see the 'due diligence document' but the solicitor's insisted.

The year was 1989. Margaret Thatcher's biggest mistake turned out to be the introduction of the poll tax. When John Major came to power in 1990 he immediately dropped the poll tax. It was the year when there were two changes from Margaret's policy: the other was to develop a better basis for Tory voters in particular and indeed voters in general.

We at NPT had taken on Project 225 for GEC at Lincoln; that was the old Rushton plant. The managing director of the site was Calvin Bray, who became responsible for the power division of GEC. The other plant was in Rugby. Like all UK manufacturing companies at the time, the latest buzzword was rationalization. That resulted in the arrival of the whizz kids who were just out of college re-organizing industry. We used to say: Whizz kids burnt themselves out in two or three years, just when the damage was done. They usually were encouraged by the politicians of the day to change the UK's lame ducks companies that were supposedly out of date.

Some had grown up like old trees, with few management skills. That brought about an industrial thirst for 'small is beautiful' and entrepreneurship. The two examples were the steel industry; Leyland was becoming another. It caused me to get out. A month after I did, my shares would be worthless; had I not of sold them several weeks before. The theory was good but the practice was hopeless.

Management has to evolve and unless there is a ready market for the product the industry is lost. It was this that impressed me about Caterpillar. The company was highly effective, with a ready world market. The management was superb and they had guidelines; for instance, they knew the ratio between man-hours, as they called it, and materials, to which was applied overheads and general administration. One knew right away if the ratio between these elements was right for a particular industry. However, the 'God of success' was in having a world market, but all the other elements of operating, if in balance, the business would probably be successful. Change must be tempered with evolution.

That is why a country, used to low-cost labour working on every-day goods, must experience evolution to migrate to technology and higher value-added commodities. It is easier these days to bridge the gap with electronic goods. Britain is a nation of inventors; there is no ready access to world markets. That is why we are losing our manufacturing industry. It was acceptable when we started to fuel the industrial revolution. Most companies rested on their laurels without constant change or improvement.

I was still grappling with the organization, in theory. Good management suggests that in a well-regulated company, no fewer than five people should report to the top man. I saw an opportunity of appointing a new 'Queen Bee', another word for a managing director. I argued with myself that I was to being bogged down with signing for toilet rolls; my bent was innovation and engineering.

My executive of engineering, Bob Chevis, I thought objected to too much interference in his work; he was inclined to come to my office only when I could solve a problem better than he, which was rare. Whether this was my attitude to people, who relied on my solving problems, I do not know. I sent many good engineers away with the thought, "Think. It may be a new experience."

Not many good people are able to think when faced with difficult problems. Thank God, I have been blessed with the tenacity to think about a problem until the solution has appeared. This was a problematic cross to bear for one who takes on the thought that one can resolve any problem.

However, there are some problems that cannot be solved. They tend to be personal ones, I would give anything if I could solve Sybil's incurable obsession. This, above all, was a bitter blow. It made me realize that I was a mere mortal, just like everyone else. I deplored what I had done to others in the name of good management. The worst crime was to project your personality onto others, but the appreciation of this did not cause me to change.

David Moss, executive director operations, in fact overlapped Bob Chevis, the executive director engineering. David had a larger staff than Bob. David Moss was responsible for stress, fuel systems and mechanical engineering; Bob, on the other hand, was responsible for initial design, meaning concept, performance and proposals. I also had an executive manager of quality control.

My son Jonathan headed manufacturing and was a main board director. That left David Noble, who was executive director special projects, who was responsible for the Harwich Division. His work covered that of advising directors of the feasibility of the good working of projects under discussion before proposals were made. The two main consultants were Mac, short for Macdonald, and Spen King, both reporting to me, although David Moss had one consultant on combustion and Bob had one on compressors. That excluded my secretary who received communication and guarded unwanted telephone calls.

This was the company structure of Noel Penny Turbines Limited. It had changed much from 1972. The company had four plants, although Southam was used as a storage plant. It surprised me why the small-to-large firms did not maintain storage plants; they were a godsend. I used to pull people's legs with the remark that if anyone became surplus to requirement "They should second them to Southam." It became a well-worn phrase.

The Volvo agreement was signed and with it I appointed Alan Cross to become managing director of NPT. I considered that by becoming chairman my responsibilities would diminish. In fact, the reverse was true. Instead of executives, who I switched to be responsible to Alan Cross, they came to see me even more.

That was apart from David Noble, with whom I had taken issue over Harwich. When I had been dressing the man down, he had collapsed with a suspected heart attack. I called on the first aid man from the works and he advised me to call an ambulance. Despite having a monitor installed in the David's bedroom for many weeks, it recorded a healthy heart; he was at home for three months.

It became known that David was having an affair with a married lady, also employed at NPT as a secretary. He used to take her out at lunchtime and the other girls had great fun with this. The fat hit the fan when we used to have a lady look after the switchboard to take calls while the regular lady telephonist had her lunch. That is one difficulty with an international company, people expect to telephone at any time.

Mrs. Noble, who had become suspicious, telephone during lunchtime and was told that David had taken Mrs. Needham out to lunch; and that this was his usual practice.

Eventually Mrs. Needham was divorced. It was a great shock for Mr. Needham too. David left his wife and married Mrs. Needham. That was not the end of it, because David had two daughters who went to the house where David and Mrs. Needham were living to deliver Christmas cards. Discovering the keys in the company car, the girls left the cards in the car and took the keys into safekeeping. When David returned, he found the cards in the car but there were no keys; he became furious. He soon discovered the culprits and remonstrated with his daughters. He told them that taking the keys was an act of stealing, since the car was not his own. One daughter had the presence of mind to write to me and ask if she did wrong. I wrote back and thanked her, saying the company that owned the car would let the matter drop. Such is life. They both asked me to attend their wedding and make a speech, which I did.

## Automotive gas turbines

That year NPT supplied two jet engines for the Leopard to fly at Farnborough. The expendable turbojets were half the power needed for the aircraft and very noisy. Nevertheless, the test pilot handled the aircraft very well and it proved a great success. NPT regretfully stopped trading before we were able to design and develop the fanjet having twice that power. I had a notion that the military engine we were developing with Allison in the US in some way could be made suitable for the Leopard.

The engine was a 'hotrod', even for its purpose. It was a single-shaft fanjet for LARSOM. What would be ideal for the Leopard was a much higher bypass ratio engine offering lower fuel consumption. Later, Williams International became interested but requested more than the designer of the Leopard could afford. Such was the problem with NPT. I was used to doing engine development at low cost. I had been brought up in a vehicle environment where cost is vital. The two crucial issues with cars are fuel consumption and cost. I was of the opinion that this could be read across to development cost. I was proved wrong; I was too used to doing things on a shoestring.

Margaret Thatcher was forced out of government because of the poll tax and her ultra-strong leadership; as a result Tory voters experienced euphoria. A lesson for the strong leaders of life must be to temper leadership with humility.

Ian Grant requested that I visit Portugal many times during 1989. It all started with a government-owned company, LNETI, based in Lisbon. It promised to develop a fluidized bed combustion system for burning waste. Dr. Ibrahim Gulyurtlu was on the verge of demonstrating the system but prevaricated at the last minute. I felt his real purpose was to manoeuvre NPT in partnership. Eventually we were put in touch with Colonel Oliveira Marques and his assistant Dr. Ramos, both of the government. We nearly sold them NPT turbojets for unmanned aircraft. My impressions of Lisbon were mixed; some parts were like London and others enjoyable for scenery. (See illustrations on p.449 and p.450 for NPT turbojet families).

About that time we considered starting an office in Rome to cover the NPT 4006 project. We were thinking of setting up near to Cosenza in Calabria as a start-up in competition with Piaggo, also of Italy. I consulted PriceWaterhouse (PW), our auditors, and the junior partner working with NPT at that time came to see me (I have just forgotten his name). If I located to an office in Rome, they suggested that NPT should consider occupying the same building as PW. A rent could be arranged later, but it was important that two salaries should be agreed for anyone who was domiciled there. The thought of paying two salaries I found appalling.

It was agreed that Ian would go to Rome and be in charge of the office. Albert Panatti, who spoke fluent Italian, would go along there too. This was a good arrangement as Ian had Albert reporting to him as one of his sales staff. We had the prospect of hiring Philip (I think his surname was Shore or something like that) who was made redundant by his previous firm; he also spoke Italian having had an Italian wife. Philip was divorced and had remarried an English girl.

That Ian was a good negotiator appealed to me when I had the opportunity of convincing him to join NPT. Ian would receive his normal salary as in the UK and would also receive an Italian salary. In addition, NPT would pay his airfare to allow him to commute every now and again; it was as loose as that. When I had to arrange the terms for his staff it was left to Ian. I hoped that he would not use his own remuneration as a guide, which he did. The Rome office was set up. At first the money I received from Volvo was expended on NPT Italia.

I was determined to sell the Italian engine to Ansaldo SpA in Genova; its size was larger than anything NPT had designed and developed previously. I arranged a meeting with their chairman.

David Moss was well down the road with the design of the NPT 4006. Volvo had a parallel development to which they had partners and I considered the Volvo participation in their engine too risky. Phil Wash, who was part of Bob Chevis's team, had carried out a performance study and rejected the Volvo engine for the NPT 4006 gas turbine engine. The Volvo gas turbine engine, if I may refer to it, was complicated to say the least. In theory, it was capable of over 20 to 1, but the temperatures were up to the limit of materials of the day. To reach optimum performance, the temperatures were only obtainable by the use of materials that could at best be described as 'Miracle Materials'.

I carried a life-size print of the NPT 4006 engine in cross-section which I rolled out for the chairman of Ansaldo SpA. Genova in Italy is a city on the Mediterranean Sea.

Christopher Columbus sailed from Genova when he discovered the USA. I used tell the following story: When he set out he did not know where he was going, and when he arrived he did not where he was, and when he returned he did not know where he had been; and he did it on borrowed money.

The Rome office was created and I visited four times, once to meet our Ambassador. Italy has two Ambassadors – one for the Vatican, which is a separate country, and one for the country. The other visits involved formalities.

We decided to hold a board meeting in Rome with the two newly-appointed Volvo directors: Per Erik Mohlin and Lars Malmrup.

Per-Erik Mohlin was born in 1946 and had an MSc from the Swedish Royal Institute of Technology. He was a former chief executive officer of Volvo Aero and Volvo Car Corporation and executive vice president of AB Volvo. He was also chairman of Almi Invest, and a board member of Swedish Space Corporation and SEB Funds.

The two men were due to meet our Ambassador and all went well until after the meeting of the directors. Lars Malmrup, in particular, asked questions about the project we were pursuing in Italy. I explained that it was to create an electrical generating engine of 3MW. He questioned on what basis the project was funded. I explained that the funding would come from the Italian Ministry, mainly for the purpose of

## Automotive gas turbines

creating new jobs in Calibria. I emphasized that so far no cash had been received and that the total expenditure had been a drain on the company's cash flow. It was unusual to be faced with such an interrogation and Per-Erik Mohlin was distinctly embarrassed.

There was another directors' board meeting in Trollhättan at which all the English directors were present, with the exclusion of John Butcher MP, and the Earl of Minto who was excused due to having an operation.

The chief executive officer of Volvo Marine attended the dinner, kindly held at the private home of Arne Wittlov, deputy to Per-Erik Mohlin. We discussed the marine application during dinner and I explained that the main snags were fuel consumption and cost. However, the gas turbine was catching up but we still had some way to go to compete with Volvo's diesels.

Spencer added to the conversation by pointing out that some diesel engines were prone to using more oil than fuel and mariners all over the world are faced with oil pollution.

The chief executive of Volvo Marine asked: "How much do your engines cost?" I gave was an oblique answer. "It all depends on market demand," I replied. If the small gas turbine engines were made in production volumes equivalent to those of the piston engine, then I dare say the cost would be the same. But as yet no gas turbine has ever been manufactured at the same rate; the cost of tooling also must be amortized.

I extolled the other virtues of the marine gas turbine engine, pointing out that our latest turbojet could be easily turned into a marine engine, simply by adding a free-power turbine and producing an engine one-third of the weight compared with the competition. Furthermore, the small gas turbine used practically no oil and could be installed easily. It was almost vibration-less compared with the thump, thump noise of the diesel. Also, the power turbine had a multiplying ability and therefore could be installed without a transmission. It could use a standard SAE flange. It would be easy to fit a firewall protection against the biggest risk at sea, namely, fire. The rest of dinner went off perfectly. The following morning we had the board meeting.

Lars was quiet and I put this down to the fact that he only had Per-Erik to impress; also he was outnumbered by English-speaking members of the board. However, Lars could speak English well. I announced that Ian Chichester-Miles was due to show his Leopard at Turin and there were rumours that the Duke of Kent's plans would take in the Turin Exhibition. NPT Italia planned to be present.

Also, I wanted to introduce Per-Erik to the financial controller of Fiat, Corso Ferrucci, should his diary allow him to attend the Turin Exhibition.

Lars Malmrup decided to visit our Rome office with his own Volvo agent, who was an Italian. The purpose was to review the NPT 4006 Project. I was bound to be in Rome, both to defend the project and the Italian government's undertaking to finance the jobs in Calibria.

There was another reason why I should be in attendance in Rome. The project was to be supported by a contract whose details I never remember seeing. I took Ian's word for the fact he had seen the contract. The company's cash flow was strained. I no longer was in charge, but I thought that auditors PriceWaterhouse had approved the previous year's accounts. I also thought that provisions attributable to the Italian government were just a formality; they would pay in the end. I consoled myself that the order book seemed quite healthy.

I was due to attend the ceremony of the signing of the contract, which was overdue. The discussions had taken place with a man in the Minister's office, following Mr. Riccardo Raciti's introduction of Ian Grant to the Minister. Mr. Raciti was managing director of an Italian company B.A.T. I naturally assumed it was the follow-on of the agreement with Mr. Raciti but later I discovered it was not.

The meeting in Rome with Lars and his Italian agent proved to be rather an interrogation. However, someone came in to interrupt. "Maggie has resigned," we were told. I could not believe it. I had met her only once. The myth that Tories are loyal to their leaders is a load of nonsense. Many leaders right back to Robert Peel have been disposed of in a bad way without the words 'Thank You'. There has been always a political opponent within the same party. Ted Heath was Margaret's enemy. It made Lars Malmrup's interrogation seem almost like child's play.

The political rebellion to overthrow Margaret Thatcher, once it caught on, was rapid; but it started slowly. John Major was confirmed Prime Minister. As he had worked his way up through the ranks over many years, his lack of experience began to tell. Major was personally associated with the European Exchange Rate Mechanism, or the ERM, and instructed Norman Lamont on Black Wednesday – that day in September when the country was forced to withdraw from the ERM. John Major led the Tories to an amazing victory in 1992, winning 336 seats; he never looked as strong as he did in April 1992.

I felt punch drunk following my return from Italy. All things in life seemed to be going wrong. In life, when difficulties mount up one seems to be looking for the next bad thing to happen. In those days, Maggie's resignation was always at the back of my mind. I had to grin and bear it. I well remember the words of my father. He said: "If there is no room to turn; stop her and go astern." It may be one of the first rules at sea, but it is the same as saying 'God give me enough strength to ignore those things I cannot control'.

I felt better after a good night's sleep. As soon as I reached the office in Coventry life was pandemonium. The post had to be cleared. Even after just a few days there was an amazing amount of post. I consoled myself that it was the luck of the draw.

Bob Chevis knocked on the door. He was compiling a proposal for a turbocharger design and he asked for my agreement in hiring a Professor at Cranfield University. Although I was against academics undertaking outside work, I readily agreed because I knew Bob would not ask idly. He asked me how my trip had been. In particular, he referred to the discussion with Lars. I mentioned that Lars had been uncertain the Italian Government might provide the funds for the NPT 4006 project; namely that they could renege. The whole question was subject to investigation. He confirmed that the Turin Exhibition was going ahead. He was due to be there with Chichester-Miles, showing off the Leopard aircraft. Bob had a soft spot for Ian C-M and when I knew he was due to visit I invited Bob to take lunch with us.

Bob came to see me and said, in confidence, that he wanted me to 'put-my-hand-on-my-bible'; that what he was about to say could not be repeated to anyone else. I had no bible there and I wondered what was on his mind. I was nearly flippant but, sensing Bob was quite serious, I just listened. He said, "I fear I have cancer". I was dumbstruck as to know what to say. When I recovered, I asked all the questions, such as have the doctors diagnosed the cancer. He said not, in so many words. But he was sure he had the disease.

I told him that until the doctor declared he had cancer, I would not take it for granted. I knew he had been seeing the doctors for some time; he was under strict orders to time himself every time he was due to visit the toilet. In fact he had a time switch for indicating the interval between his last call to the toilet. I pacified him by assuring him his secret was safe with me, remembering my mother.

My son found a new home in Moreall Meadows, where I live now. The day I was due to move into my new house, I was away in the US, securing a loan from Williams International. My work with Sam, who agreed the loan to supplement cash flow, was done.

I was due to play golf with Barbra Williams and her friend, an author who was visiting from New York. I was with Alan Cross, my managing director. The trip over there was delayed due to a hold-up in New York.

The airport in New York was called Kennedy; it had been changed from Idle-Wild after John Kennedy, the assassinated

President of the USA. (Ed – the airport was dedicated as New York International Airport in 1948 but was more commonly called Idlewilde Airport until 1963 where it was renamed John F. Kennedy International Airport or, more loosely, known as JFK.) I recalled that when I went there first as a young engineer that it took me 14 hours to travel from London to New York. We stopped at Shannon and Newfoundland. The flight was by Lockheed Constellation, an aircraft powered by piston engines. The cab driver who picked me up was black. As it was my first experience of New York he gave me some very favorable tips on life in New York City. He said, "Count your knuckles when you shake hands boy; that is the City."

The Turin Exhibition duly took place and the Leopard aircraft was there in all its glory. It looked splendid and the Duke of Kent came to speak to Ian and me. A display board with NPT Italia was in pride of place.

However, in Britain, things were going from bad to worse. The result in the General Election gave John Major authority over the Tory party. This authority however was quickly destroyed on 'Black Wednesday', the day in September when Britain was forced to withdraw from the ERM or the 'European Snake', as it was sometimes called. Norman Lamont, the Chancellor, and the Prime Minister both insisted throughout the summer of 1992 it was essential to join the ERM for the sake of economic stability and recovery from recession. The result was the opposite, as the country lapsed into a deeper recession.

We at NPT were caught between two big organizations and the only thing we could do was to cease trading, despite the large order book. 'Cash-flow', the God of survival, had caught me out. To see a wonderful company that I built up go bust was devastating.

I had assured Bob that in the event of the future of NPT ever becoming questionable, he must put his family first – that was said some months back when Bob swore me to secrecy over the matter of his health. I stressed that Bob must go to see

# Automotive gas turbines

Lars Malmrup of Volvo. I never considered the day would come when Bob would need to do so.

I moved into the new house and held several meetings there for NPT employees who were now facing the loss of jobs.

I recalled the days of Rover Gas Turbines Limited where I had the whole of the company without jobs on my back lawn; I promised to find them all suitable posts. Fortunately, Rolls-Royce took most of them, even shop-floor-workers. Although I say myself, NPT workers and staff were in great demand.

It was the apprentices who caused me most concern; I wanted to find locations for them. NPT was proud of its training record for apprentices. Arthur Varney took them on and was so concerned about their training that I had many arguments with him about taking them to fit into major posts of promotion.

I kept in touch with my NPT employees by arranging every Tuesday a lunchtime meeting at the 'Old Mill' in nearby Baginton. The discussions were interesting; most were about pensions. I had ensured them that the pension money was indeed safe; I had trusting the money to 'Evershed Pension Trust Limited'. I tried to save NPT by investing my small pension scheme into the company, but was in vain.

Alan Cross, the managing director of NPT, will bear witness to the number of moves I made to try to save the firm. I was the last one to leave. It was a terrible for me to leave the place where I had spent the last 21 years. I thought of Sybil. When we started, there was no anti-feeling, only support and trust for everything I set my heart on doing. I wondered where I went wrong. I left the job of building my children's character to Sybil. She must have done an excellent job shaping their young lives.

I was only there at bedtime to tell thousands of stories out of my head to my young children. I can see their little eyes rolling in my sleep with 'Cheeky-Monkey' and the little animals of the wood. I gravitated (moved on) to telling to my eldest grandson, Timothy, most of the stories, but in fact he was so imaginative he added endings of his own.

In my opinion, the political state of the country, from 'Black Wednesday' in September 1992 onwards, finished John Major. The Tory party divided from that point on. As a result, John Redwood made an attempt for the leadership. John Major tried to hold the party together but the rift was too great. His status before 'Black Wednesday' had never been higher; however that he argued for the ERM all that year was an indictment too far. Even his chancellor Norman Lamont turned against him when he was sacked. It was a Godsend for the Labour Party that Tony Blair turned the party into New Labour, oddly becoming a puppet to the USA in his rule against all his principles. Apart from that, he was the best Prime Minister the country had had since Mrs. Thatcher.

In my opinion, there have been three great Prime Ministers in the 20th century: Churchill was one; the others were Margaret Thatcher and Tony Blair. In time, people will eventually turn to support Tony Blair, despite how Iraq is turning out thanks to hostilities aided and abetted by Iran, Syria and other middle-eastern countries.

Sam, Gene and Gregg visited London in March 1993. Jonathan and I went to see them. It was a disappointing trip for Jonathan, who took the minutes. The party explained that only I would be offered a position that would be worked out by Gene. Jonathan would be offered a post later when the company was in a better position.

I felt duty bound to compensate Jonathan. I had taken delivery of a new Range Rover, one styled by Spen. It was all in black and it carried Spen's initial on it – CSK. It was in exchange for a used 2S150R engine that was a spare for the Rover-BRM.

The head of Gaydon had arranged the deal. I explained to Jonathan if he could sell it he could keep half the cash. I regretted losing the new Range Rover, especially as it was a collector's item and styled by my friend Spen.

Jonathan soon secured a new position. He was a good lad and I went through sorrow that I was partly responsible for him being unemployed.

## Eleven

# International advisor

In which I make the acquaintance of Sam Williams and become part of Williams International.

I FLEW to Williams International and when I arrived, John Shepherdson, Sam's bodyguard, picked me up in Sam's car. I was not new to Williams International. I had been there many times. In fact, I helped them out on the Lockheed APU, which was failing rather badly. The engine had two centrifugal compressors and the overhang was too much. The engine failed in a Lockheed aircraft; this was due to it hitting the first critical (whirling) speed, which was very near to the maximum. I advocated a damper I had used with success on the Rover 2S140 in the T4 car. The modification was a great success but the project engineer moved his job. I was sorry about that as I made a friend of Jim; he and his wife entertained me many times. Sam was eternally grateful.

My time at Williams International was a good one. My title was 'International Advisor' I was introduced to Pat, Gene Klein's secretary, who soon became a good friend. She remained so until Pat retired.

I became responsible for traveling all over Asia; it was called then the Pacific Rim. Sam explained that as a company, Williams International must get a foothold in Asia and it was my job above all others to get that done. While I was there I was issued with a code of conduct. Williams International was a high-minded outfit. When I joined the company, the manufacturing unit was in Utah, at Ogden near to Salt-Lake City where most Mormons reside in the United States. In response to why he went to Ogden, Sam was quick to reply that the work ethic was good. And then, after a pause, he said the offer to build the plant there was also good. I went there many times for meetings.

The whole company was proud of its work on the fanjet engine that powered the cruise missile. The Williams International engine, known as the WR19, triggered the beginning of the development of the world's smallest turbofan. This engine was for the Williams jet belt; it was so light that a man could carry the engine on his back as a backpack. One essential feature of the engine was low weight; another its low fuel consumption.

In the 1960s, the US government appointed a scientific advisory committee to see whether a power plant for a long-range cruise missile was possible. The findings of the committee were that such an engine was unlikely to possess sufficiently low weight and be efficient enough to do the job. And even if such an engine were possible, it was likely to be of prohibitive cost.

The large engine companies agreed but Williams International went ahead with its private venture (PV) development of a small low-weight, efficient turbofan engine. The company called it the WR19, WR for Williams Research, which was the company's name before being changed to Williams International.

In battlefield surveillance during World War 1, balloons and aircraft were used to gather information from behind enemy lines. These would be used for data on troop dispositions as well as serving as forward artillery observations posts. Today, gathering intelligence remains, in combination with other systems, one of the aeroplane's most significant and fascinating functions.

In 2002, I gave a Dedicated Lecture to the Royal Aeronautical Society in London describing cruise missile engines such as the Williams International F107 and the NPT 171 for B.M.A.C. The paper was entitled *Propulsion Developments For UAV's and Cruise Missiles* and presented on 10 April 2002 in a session called Aerospace - A 2020 vision.

Norman Lamont waged a personal vendetta against John Major following his dismissal in May 1993. His ceaseless abuse

## Automotive gas turbines

portrayed Major as a weak and indecisive politician. It set the scene in 1997 for the rise of New Labour by Tony Blair.

In 1994 I toured Asia, calling at Taiwan and afterwards Hong Kong. I went to Taiwan because I knew that Taiwan Aerospace supported Williams International's competitors, namely Allied Signal. NPT had carried out a project for the China Institute of Technology (CIST). I have often wondered what the 'S' stood for; maybe it is the Chinese for something associated with military developments. Anyhow, NPT had successfully designed a turbo-pump for the Institute. I have photographs taken of the team of Taiwan visitors. The Taiwan Agency for Overseas Investment in Taipei and responsible for funding, is located at Keelung Road in the same building as Taiwan Aerospace.

I nearly lived there and once, when Sam and Gene were present, Gene made a presentation of Williams International. I also visited Geneva when knew a director of the Agency would be visiting. I soon appreciated that Taiwan would be a long haul so I reviewed my contacts in South Korea. My considered opinion was this would be more productive. At NPT, I had trained young engineers from the Agency for Defense Development (ADD) of which at the head that time was Dr. Wi Hun Kang. In fact, NPT had the honour of designing the first small jet engine for South Korea. This was the 251; it was a dedicated engine.

My first meeting was with Tony Kim of Samsung; he was a friend who agreed to visit Walled Lake and meet Dr. Sam Williams. The meeting was a success in so much as it convinced Sam that South Korea would be a much better bet for Williams International than Taiwan as a base from which to obtain a foothold in the Pacific Rim. From that point there started many years of negotiation with Samsung Aerospace.

A three-shaft engine was designed by the engineering team at Williams International for regional aircraft use. This was the FJ77; it had a thrust rating in the FJ77-1 version starting at 4,000 to 5,500lb. The FJ77-2 version would see the thrust lift to between 5,500 and 7500lb. Finally, for the FJ77-3 the thrust would reach 7,500 to 9,000lb. That was the family of engines. The thrust was later increased to power the 130-seat regional aircraft version; this was up from the 70-seat version powered by the FJ77-3. The FJ77-1 engine was equivalent to a 19-seat aircraft.

The purpose of the FJ77 engine family was to replace turboprop-powered aircraft. Furthermore, the sales pitch was 'three engines share a common core'. Regional aircraft in some parts of the world, particularly Asia, are the growing segment of the civilian market due to the speed, comfort and range expectation of the customer. With greater profit levels for the airlines, the engine technology could expand to other fields of application.

Turbine engines scale; it is always desirable to scale to a larger engine than to a small one. Although the selling point of small gas turbine engines is that smaller engines can generate a greater power-to-weight ratio than is possible with large engines, they also have lower cost and generally assure reliability. When dimensions are halved weight is cut by one eighth, but power is cut by only one quarter.

I travelled to Seoul, the capital of South Korea, to the headquarters of Samsung. Tony Kim previously had set up my meeting in the Hilton Hotel with several Samsung senior engineers (NPT had been engaged with in project for Samsung when my company ceased trading). When I outlined the Williams FJ77 to the Samsung people they were taciturn to say the least. Tony told me later that Korean people were known for their deadpan reaction. I asked Tony, who was tall, handsome and an assistant to the chairman of Samsung, how he thought the meeting had gone. He replied, "It is just a start but there are many years ahead." I remember thinking then that I did not have many years to prove to Sam that I had influence with Asians.

My first impression of Williams International was the pleasure I gained from meeting people. I was shown round the whole establishment and I travelled to Washington to meet my boss. He was the quasi-head of marketing; his card spelt out 'Business Development and Washington Operations'. I had met Ray Preston previously in Walled Lake, when Sam held his lunch in his office. Many Williams International staff tried to avoid Sam's lunch time whenever they were invited; they were so bland. Sam believed in no salt; he had a thing about fitness. I suppose that made up for his poor sight. Gene used to have soup whenever he was invited, which was every day when present on the site.

Ray liked people, so it was a pleasure to see him again. He was above average height, slender but well built. He was, I believe, a retired general. Ronald Schwedland reported to him, but being away in Washington he was rather remote. Williams International was an open-door company, so one reported equally to Sam or Gene. My friendship with Sam never got in my way of going directly to him. It was home-from-home at Williams International, for the people were very friendly. Ray had already sent me a comprehensive document suggesting what I could do for Williams International.

I had known Ron Pampreen by reputation; he had been at Allied Signal and was well known in the gas turbine world. His Williams International card spelt out Advanced Technology. Carl Schiller was director Government Field Operations when I joined Williams International. He was friendly and took me to see George and Trudy. George was Sam's old boss when they both were employed by Chrysler; they lived at Ann Arbor, Michigan. It was a short distance for Trudy to travel; she was Regent at a local University. I had to get used to all different names at Williams International. Carl gave a presentation and kept referring to a name that was new to me. I asked what the name meant and the others attending Carl's talk were amazed.

Bruce Crew was the liaison link with Rolls-Royce. Bruce was dark and well built. Undoubtedly he had been briefed by Sam. He left me with the FJ44 sheets of engineering certification. The FJ44 was a new engine that was finding favour with aircraft manufacturers such as Cessna and Swearingen for smaller jet aircraft. Rolls-Royce had a joint company with Williams International on the FJ44 engine called Williams-Rolls Inc. I got a distinct impression that if one listened to the story of the FJ44 fanjet engine project one received a different slant, depending whether you were listening to the Williams International or the Rolls-Royce version. I happened to speak to both at the Farnborough Air Show.

There was an Englishman at Williams International, an ex-Rolls-Royce man, by the name of David Carr. His shares were rising and he headed the operations; under him was the director of material, Celeste Busch. He later became vice president of purchasing. David was short and always gave the impression he was running, because he had such a red face. Edwardian's would have said this was due to drink. David had a lot to do with the Ogden manufacturing plant. In charge of the Ogden plant was Lloyd McCaffrey. Lloyd helped me when I took the Samsung engineers round Williams International's manufacturing at unit Ogden. Lloyd was another well-built man, about the same height as David Carr. He always wore a brown tweed suit. Tommy McMillion was Williams International's expert on precision casting and I was enthralled to see all the latest equipment installed and working. Tommy was a slender man – as thin as Lloyd was thickset.

Equipment at Ogden was the very latest; for example there were automatic machines cutting compressors. I noticed many of the machines were imported from Switzerland; they were made by the leading firm 'Rigid'. They were completely programmed with only the flank-mill cutters needing to be replaced after one compressor. The machinery took my breath away; even machines to remove sharp edges were automatic. In the past I longed that NPT could afford such equipment. I was told that the machinery was installed by the US Air Force for the complete production of cruise missile engines in volume, and I could well believe it.

I well recall when I travelled to US as a young engineer, the cab driver who gave me some useful tips on my first visit to New York. America was always the true land of opportunity; it was my second home. It is the land where, if one works, anything is possible. I well remember that my first office in the US was in the Chrysler Building. This was opposite to where helicopters would land on the top of the Pam Am building, until one was blown over the edge and then the practice was stopped. My next office was in down-town Peoria, Illinois when NPT was working for Caterpillar; it incorporated Noel Penny Inc. I recall travelling on Concorde and having three short meetings in one day, one in New York, one in Detroit with Sam Williams regarding the design of 'after-burners'. The last meeting was in Peoria with Joe Grandfield.

Back home in the UK, the political defeat of Neil Kinnock in 1992 heralded the rise of the Labour party. John Smith became leader after Kinnock's defeat; John was a man of the right. 'Black Wednesday', tax increases and divisions over Europe were just the opportunities that John Smith could exploit. However, when John Smith died suddenly in 1994, Gordon Brown and Tony Blair both thought they could lead the party. The latter won and created New Labour. Both men had a hand in it following John Smith's right-wing doctrine of controlling the Unions.

I soon settled down to business with Williams International. It was like a 'breath-of-spring' to be working again with such an excellent company. I had targeted Samsung of the Asia Pacific Rim. Tony Kim was as good as his word, despite the caution he gave me. The caution was that the total cost of the FJ77 family might be too much for Samsung Aerospace to swallow. He thought that we could do better if the sums of money in the commitment were reduced; that is, the building blocks of commitment should be reduced. I was spending my time travelling between Seoul and Europe in 1994.

I had a peculiar feeling that the Tories would not recover easily. The 1990s had been a remarkable decade for politics. It started with the spurning of Margaret Thatcher, then a surprise that John Major was Prime Minister. After which, Major had a victory in the election where he was proclaimed the strongest. Then followed in 1992 the rejection of the Maastricht Treaty by the Danes; and in September of that year came 'Black Wednesday'. John Smith, who died in 1994, had turned to the right in his thinking. In 1995 Major fought and won the fight over John Redwood. Norman Lamont continued to blacken Major's name personally. Neil Kinnock was on the slippery slope. Ken Livingstone put his name up for London mayor; if he did so he would be banished from New Labour, according to Tony Blair. Blair had an overwhelming win for the new leadership of the party and Frank Dobson resigned to take up the election for the mayor of London. Having lost to Tony Blair in the contest for the leadership, Roy Hattersley attacked Tony Blair on his new policies shaped on Clinton's rule.

After the General Election in 1997, which was a 'walk-over' for Tony Blair's New Labour, John Major resigned and William Hague became the new leader of the Conservatives. Kenneth Clark was once again abandoned, due to his pro-European views and the fact that many Tories considered him too old. Hague followed right-wing policies and Kinnock became unelectable in the light of pointed press attacks. Kinnock was eventually banished to the European Parliament. Tony Blair and Gordon Brown joined the euro as an economic issue rather than a political issue. Gordon Brown defined five economic criteria and how it all will end is a matter for further story telling.

My work with Williams International progressed and I was over-confident that I would succeed with Samsung Aerospace to the point of becoming tenacious about the issue. I was determined to see the contract for the FJ77 family project succeed, despite Tony Kim's worry that the cost was too high for Samsung to swallow. I varied my responsibility with items such as covering air shows and early talks with CASOM and the UK Ministry of Defence. CASOM stands for Conventional Ariel Stand-Off Missile.

The gas turbine engine is excellent at the concept and design stages as a means of creating a family of engines. The rate at which technology improves is rapid; this is particularly so as the gas turbine engine has only scratched the surface as being the youngest power plant in existence.

The Williams International P8300 Family series was applied to CASOM and is designed to meet the requirements of low-cost applications, as distinct from other applications guided by end-user customer specifications. The family, similar to the NPT family of engines, was designed with three nominal engines. One engine developed 10,000lb thrust, and another 13,000lb and a third member of the family created 15,000lb thrust; they were all within similar dimensions.

As the 'state-of-the-art' improves at the concept at the design stage, so it is possible to increase the pressure ratio, air mass flow and speed within limits, and extract more power. By so doing, three sizes of engines are possible for the same manufacturing techniques. The engine was similar to the NPT 1204 in so much as it was a single-shaft fanjet engine. The by-pass ratio was around unity so as to keep the frontal area compact, like that of a military engine.

I was required to see David Fabish when I visited the Paris Air Show. David was a Jew from Rumania; he lived in a suburb of Paris with his chain-smoking wife. He was also an aircraft designer. A short man with a ruddy complexion and slightly rotund, David gave the impression that he knew how to design aircraft. However, my knowledge stopped short of finding out whether he did or not. He was very critical of Swearingen aircraft in which the Williams International FJ44 was installed.

I will back-track for a moment to Ted, my guardian's mother Katie. Katie was having a nightmare. In the nightmare, Katie was tortured by thoughts of the wounded soldiers she knew and who had survived the trenches of World War 1, but who also lived a nightmare as long as they would live. It was a nightmare they knew would never leave them. For years later they would scream out every night when reliving the life of battle. Over and over again they would see the looks in the eyes of men and boys who were always pleading in that fatal moment. They were always asking: "Why kill? You are just like me. So why must you

# Automotive gas turbines

kill me. We could be friends. I am a mother's son. I have a wife and children. Please let me go back to them. Don't have blood on your hands – my blood." Katie and Ted realized that many men would never harden themselves to be innocent of those pleading eyes.

Returning to the Williams International saga, I was due at the Korean Air Show. The Korean Aerospace Industries Association (KAIA) had invited me to give a lecture there. The KAIA pretended to be independent of any company, but I had a feeling that it never quite made it, and the last large company with whom they were associated won the day. This particular year proved to be a big event; aircraft were on display as well as taking part in flying displays. I had been seeing Mr. Yeo, assistant to the president of Samsung Aerospace. He had a brisk manner and was unpleasant to deal with. He tended to poo-poo the idea of a regional aircraft. I learned at the show he was due to be moved and there was to be a new president of Samsung Aerospace, a Mr. Yoo. I had seen Tony Kim's old boss, a Mr. Seok Ho Noh, a vice president of Samsung. He was a perfect gentleman and I had explained my intention of getting Samsung to sign a contract for a family of fanjet engines named FJ77. He was quite interested to hear that the company responsible for powering all the cruise missiles in America should be in Seoul. I was encouraged, to say the least, when he said he would bring it to the attention of the vice-chairman of Samsung, Dae-Won Lee.

Later I heard from the office of Dae-Won Lee that he pointed out to me that the KAIA was due to execute the commuter aircraft project in a bid to improve domestic availability. The development project would take place from 1994 to 1998. It was already 1995, so I asked if the project had already started. The answer was 'no'; the delay was on account of deciding the specification.

The number of seats ranged from 50 to 100 and the maximum speed was in the region of 530km/h. The aircraft's range was pitched at 2,100km with a maximum take-off weight was 19,000kg. I asked what the favoured power plant was. He

# Automotive gas turbines

replied that at the moment it was a turboprop but he queried whether I had in mind a fanjet unit.

I outlined the proposal for the FJ77 and he asked me the cost, on the basis of 50-50 training Samsung's people. The cost was in the region of half a billion US dollars. He did not blow through his teeth at the cost. I said I would send him my study of South Korea, which I did and he was eternally grateful. I was very encouraged from this conversation; it did much for my morale. That night, I rang Sam Williams at home, having first a good conversation with Barbara, Sam's wife. I talked to Sam in a very optimistic way but I soon learned that the South Korean way was to take high cost seriously, in line with the rest of the world.

The crunch came when Sam, Gene, Gregg and I visited before one of the air shows. Sam had already written out a note stating Williams International had won the regional aircraft project. Prior to that, Dae-Won Lee, the vice-chairman, had visited Walled Lake to announce that Samsung must go ahead with the project. Everyone in Williams International believed the project would receive the go-ahead but the president of Samsung Aerospace, Dae-Won Lee and others met in a meeting in the morning before the air show and from this I assumed that the project would be started. However, the president of Samsung Aerospace, Mr. Yoo, explained that the cost of doing the FJ77 was more than one year's income for Samsung Aerospace. Dae-Won Lee kept very quiet. Finally, Mr. Yoo went to great lengths to explain that Samsung Aerospace had to decline, unless there was help from Samsung Corporation. Again, Dae-Won Lee kept quiet. Sam and Gene left the meeting in disgust and returned to Walled Lake. Only Gregg and I stayed on for a banquet in the evening that was hosted by Dae-Won Lee. Gregg could not hold back his disappointment. He walked up to Dae-Won Lee. In no uncertain voice Gregg told Dae-Won Lee what he thought of South Korean promises.

Bill Bar, an American employed by Samsung to review the FJ77 project, was a good friend. He repeated over and over again that Samsung Aerospace did not have the funds to support the FJ77 proposal but, despite this, I was not about to listen. In the end Bill gave up. He visited Walled Lake twice with the Samsung team. Bill said he would like to work with Gregg. I lost touch with him when I knew the FJ77 fanjet was but another dream.

I had invited John Copling to give his opinion of the three-shaft design of the FJ77. He was the design director of Rolls-Royce when the three-shaft design was current. Sam was reluctant to visit and John was sent by nominal airline service to Ogden. John told me later that had the project succeeded with Samsung, many details would have to be changed.

The one outstanding feature was the absence of active clearance control, which meant that the performance would be difficult to meet. The design, as it stands, had only passive clearance control and most large gas turbine engines now have an automatic means of controlling clearances. With three-shaft engines this is a must because the aspect ratio of the turbine blades are nearer to those of the small turbine designs. (The aspect ratio of a turbine is the height or depth of the blade divided by the chord. The chord is the distance between the trailing and leading edge). Small engines tend to have turbine blades that are stubbier than large engines.

Meanwhile, there were many other faults in the design too numerous and too technical to relate.

I had made many good friends in Samsung and once again I was loathe be out of touch. I turned my attention to CASOM, the Swedish requirement for a military trainer using the FJ44 engine and air shows. George Rouk and I attended meetings with DASA in Germany. I was due to put Williams International's case for CASOM to DASA. George was decidedly shaky when using computers in Europe. The two-day DASA meeting was near Munich at the foot of the Alps.

I returned via the DASA plant in Munich and met with a German who was known to me when I had the responsibility for NPT. The trip in Germany was highly successful. I felt I was on home ground. I was also required to visit London to

## Automotive gas turbines

support other competitors of DASA and at the same time was due to present the case for the Williams International P8300 to the MoD and political military leaders of the liberal party. I attended the dinner at night and left my card with Menzies Campbell, later knighted as leader of the Liberal Party. He was then the military leader under Paddy Ashdown, who was the party leader.

The FJ44 turned into a trainer engine for Sweden through SAAB. I went many times to Sweden in order to negotiate the FJ44 trainer engine. One notable flying visit was with the Cessna light aircraft powered by the FJ44. The aircraft was flying into Broman airfield at Stockholm. Broman was a military airfield and was dogged by noise. I imagined the Cessna did more to convince the Swedes, because of the noise levels of the military airfield at Stockholm, than any amount of hard selling could do. The noise level was significantly lower than any aircraft landing or taking-off. We had access to Goran Berg, the Karlebo agent for Rolls-Royce under the FJ44 agreement. John Hellstrom, purchasing agent of SAAB, was the contact point. Many of the staff at SAAB were the same people I dealt with in NPT.

One time at Farnborough, Erik Prisell, who was head of the propulsion department of the Ministry of Defence in Sweden, and who I knew very well, came to me and said, "I must listen to SAAB", once the trainer aircraft had been adopted by the Sweden, subject to the blade platform cracking problem which had materialised on the FJ44 engine.

When I reported SAAB's concern over this kind of turbine cracking to Williams International, I received a hostile reaction. My experience of this type of platform cracking at Rover suggested it could risk developing into a complete turbine failure. The attitude in Williams International was that action would be taken by the appropriate engineering people.

These were the only strong words with Williams International, apart from when Dickie Barr visited with his son. I took issue with Dick over the speed of the FJ44 low-pressure turbine. Dick said the turbine was too slow in RPM. The question was that Centrax thought the engine could be a useful combined heat and power engine. I had set up the meeting with two reluctant partners, if Dick's heart was not in anything it would not be adopted by Centrax. It was the only time I experienced Dick and Robert on the same side but my trips to Centrax were rare in those days.

We had a meeting prior to this when I was with Ron Schwedland, a good man who went round the world selling the FJ44. Ron had been at Pratt & Whitley where he trained as a good development engineer. I never discovered why Ron did not rise to the vice president position. When I queried this with Sam, I think the answer was he was too valuable in selling the FJ44 to become a vice president of Williams International. He is doing the same work at the moment, maybe he will be rewarded with the VP's office.

Dickie Barr founded Centrax in 1948; he is a well-built, tall man. His handsome face reminds one of past years when he must had the ladies swooning. He is a good and inventive engineer one could always respect. He has been very good to me and is one of my closest friends. In fact, he could be an excellent 'technical director'.

Sir Stanley Hooker once said to me that if anyone ever leaves engineering to become a managing director then he is lost; signing toilet roll requisitions is not for me. Dickie's inventiveness is amazing. He once told me that he schemed out an axial compressor, much to the annoyance of Frank Whittle. Frank was a well-known supporter of the centrifugal compressor.

One day, when Frank was visiting Power-Jets, Dickie's boss called over the intercom that Frank had arrived and would like to see how far Dickie Barr had got. Dickie was working on the drawing board and called out, "The bloody fool wants a centrifugal compressor but I have drawn it as an axial."

He thought he had been in touch with his boss only, however, the intercom could be heard also by Frank Whittle.

Frank walked up to Dickie's board and said, "So you think I am a bloody fool, Barr." Dick was overcome with embarrassment and took on a tan as if he had been sunbathing.

The name Centrax was made from the two main compressor types: one being the 'Centr'-ifugal and the other being 'Ax'-ial. Although the word centrifugal commands more letters than the axial it is in recognition of Frank Whittle. Dickie's partner, who started the company with him, was drowned in a boating accident; his son worked at Centrax for some time.

While I was globe-trotting, I had someone look after the house – I paid here from Noel Penny Associates, a business I formed when I left NPT. It was in effect an international consultancy business to cover lectures to learned societies worldwide, for example Williams International, Centrax and others.

I ended my consultant's work with Williams a year later in 2002. I heard that my former girlfriend, Joan, died on Sunday, 21 November, 2004. Her friend Claudia from Germany was with her when she died. Joan's funeral was on the 4 December that year. I was coming up to my $78^{th}$ birthday.

I ceased trading with Noel Penny Associates two years later in December 2006. As I write this it is now March 2007. This Christmas I will be 82. I clean the house and shop when I have to. I also write 'My Story'. I go to see my children, who I love, and I think the world of my five grandchildren. Timothy was the first born, then Seraphina and Michael who are both of the same age; then there is Ophelia and lastly Elinor.

There are many coincidences in the Penny family. Roland, my son, has traced the family back to the seventh century. He found another Elinor Penny, who was alive in the late 1700s.

I have hoped with all my heart that I could have saved Sybil from entering the strange 'World-of-Compulsion'. And, if I could, I would have saved Joan from cancer. But apart from those two events, I would not change anything if I could.

Looking back over the years, the most amazing decade was the 1930s. When the decade began, the number of people out of work peaked after the 'Wall-Street' crash. Britain was insulated from the worst of it by the gold standard. Germany, by some miracle, escaped from high inflation. However, a new dictator had written *Mein Kampf* when he was imprisoned, and that changed everything.

As soon as I was old enough to read, I read the English version of *Mein Kampf*. Using the first word as an anagram, the word Mien means in English demeanour or air or aura. Never was Hitler true to the screed that he wrote while he was locked-up. If only the world could have taken him literally, the Second World War might have been avoided. It was the biggest mistake in my lifetime.

Not to spread the word, if one could go round again keeping the identical circumstances, that might have been said millions of times.

Out of a family of 11 there are only four of us left as I write 'My Story', three sisters and myself. We were a happy family and the whole family reached old age; only Roy, my close brother, who was born during Coventry's carnival, did not make it past 55. I always remember that the young die first. He was brought up with me; at times we became inseparable. He was the joker and grew to be a better man than me. His caring about people was uncanny.

He had three lovely daughters, Julia, Cheryl and Janine. He was a staunch socialist, unlike his upbringing; my father was a staunch Tory. Whenever I went to see him he would always welcome me smoking a small pipe and then, as if by magic, he would start to fill a very large pipe. He was a good man and we would sit in silence, content that we were together as we had been in childhood.

He tried many times to settle the trouble between Sybil and me. Doreen, his pretty wife, is dead now. She almost lived till 20 years after Roy. She was a dance skating instructor in her younger years. She kept dolls and she had strong views and was discerning.

Discerning, determination and tenacity are all qualities that illustrate wise and shrewd knowing. One cannot get someone to believe something they do not want to believe. Someone said the worst that you can impose on others is personality. Perhaps the worse treatment of others is that one is indifferent.

Now 'My Story' is coming to an end I am lost for words. I dedicate my life story to my children and grandchildren. May each one reach fulfillment as my life has been.

**Noel Penny was born Noel Robert Penny.**

# Automotive gas turbines

# Appendices

BETWEEN 1970 and 1980 *The Engineer* weekly news magazine under my editorship published numerous articles about the automotive gas turbine. These articles were spurred on not only by developments within Leyland Gas Turbines Ltd, the hub of much of the activity in the field at that time, but also by developments in the field of engineering materials, again most notably ceramics.

Engineering ceramics, in the form of silicon nitride and principally at the Atomic Energy Authority at Harwell, but also in the private sector began to emerge.

Not only did the idea of ceramic – not metal heat exchangers – surface but the prospect of making other components also from ceramics.

Why ceramics? Ceramics offered the possibility of running engines hotter and with increased temperatures it followed that higher efficiencies might be achieved. If the gas turbine was to offer any chance of competing with the traditional diesel engine it had to at least match it of specific fuel consumption whatever other benefits it might offer besides lighter weight, fewer moving parts and increased distanced between routine maintenance.

The modern jet engine, which has proved to become an extremely reliable unit, is however manufactured largely from high-strength metals and composite materials.

But in the white hot heat of technology that pervaded the 1970s there began to emerge the prospect of an all ceramic gas turbine, for which the term 'crockery engine' were coined.

George J Huebner of Chrysler Corporation, and the 'father of the automotive gas turbine' said of ceramics: "Ceramics are OK for stationary parts, but foe anything else they are too expensive, the production yield is too low."

# 1 Alvis plc

ALVIS Car and Engineering Company Ltd., a British manufacturing company in Coventry from 1919 to 1967, had its origins in cycle making, as did many companies of the period. Later, in addition to automobiles designed for the civilian market, the company also produced racing cars, aircraft engines, armoured cars and other armoured fighting vehicles.

The original company, T.G. John and Company Ltd., was founded in 1919 by Thomas George John (1880–1946). Its first products included stationary engines, carburettors and motor-scooters. On 14 December 1921 the company officially changed its name to The Alvis Car and Engineering Company Ltd. Geoffrey de Freville (1883–1965) designed the first Alvis engine and was responsible also for the company name.

Rover took a controlling interest in Alvis, maker of hand-built luxury cars and military vehicles in 1965 and a Rover-designed mid-engined V8 coupé prototype (P6BS) was rumoured as the new Alvis model. However, with the takeover by British Leyland this was shelved. By the time the TF 21 was launched in 1966, (available in both saloon and drop-head form and with either manual or automatic gearbox), the model was showing its age despite a top speed of 127mile/h – the fastest Alvis produced. With only 109 sold and with problems already in UK car manufacturing, production ceased in 1967.

In 1968, a management buyout of the car operations was finalised and all the Alvis car design plans, customer records, stock of parts and remaining employees were transferred to Red Triangle. Alvis became part of **British Leyland** and in 1982 sold to United Scientific Holdings, which renamed itself **Alvis plc**.

In 2009, Red Triangle negotiated the legal transfer of the Alvis car trademarks. The following year, the company announced the 4.3-litre Short Chassis **tourer** would be available with manual or automatic transmission. All Alvis records remain intact at the company's **Kenilworth** headquarters together with a large stock of period parts.

## 2 Austin Motor Company

WORK on a gas turbine engine at the Austin Motor Company began in April 1952 at East Works, Longbridge. Dr. Weaving, as Austin's chief gas turbine engineer, led the team to examine the feasibility of building a gas turbine engine for fitment to a car.

By that time, Rover had demonstrated its Jet 1 in March 1950. Rover's engines later developed 230bhp at 26,000rev/min and were fitted to the rear of a standard P4 chassis.

Some reports suggest Austin's original project began in 1949 with a total of six employees intending to fit a gas turbine for in a car. Dr. John Harold Weaving, as an Austin ex-apprentice, had obtained a London University BSc degree and later moved to Cambridge University to undertake research on internal-combustion engines. On returning to Longbridge in 1946 he became superintendent in East Works Research.

Leonard Lord, in charge of Austin, had no wish to be seen falling behind in modern thinking, and it was important to show the public that the company had technical expertise.

One main component of a jet engine is the compressor, so to build Austin's first engines Dr. Weaving used a unit from a Spitfire Merlin engine. From this the team produced their version of a gas turbine suitable for a car.

In August 1954, after bench testing and curing one of the main problems, namely dealing with turbine blades that had to cope with the centrifugal forces whilst spinning at 23,000rev/min, the engine's optimum speed, results showed the engine could develop an output of 125bhp.

The engine was installed in a modified Austin Sheerline. It was necessary to lengthen the bonnet of the car to accommodate the air intake silencers at the front. These reduced the whistling sound normally associated with a gas turbine. It was this noise, when the car was driven on public roads, which often made heads look upwards expecting a jet aircraft to pass overhead. But inside the car the noise level was regarded as acceptable.

In the Austin gas turbine, air was drawn through the front grill and compressed before passing to the heat exchanger where it was heated by the hot exhaust. This reduced the amount of fuel needed to be burnt to reach the required entry temperature to the turbine. The hot gases then passed through the power turbine to rotate the shaft that was coupled to a Hobbs gearbox and the rear axle. The exhaust gases then passed through the heat exchanger before been discharged to the atmosphere. Grills in the bonnet allowed heat from the engine to escape.

The Hobbs gearbox was an early type of automatic so when the input shaft was turning at 600rev/min the car would slowly move forward.

The car was duly registered and given the registration number TUR 1. It was ready for road test in August 1954. Anxious moments occurred the first time the car was taken out as drivers had to nurse the car along; it was slow to respond to the throttle pedal. Although capable of about 70 mile/hour the gas turbine car could average only 4.5 miles to the gallon.

Slight leaks of fuel or oil could be just one hazard of driving the car. With high under-bonnet temperatures, it was quite possible to have a fire under the bonnet.

Lord felt obliged to show both Press and Public that the company was at the forefront of developing cars of the future. So it was decided to show how progress was developing at the Austin Golden Jubilee (1905-1955) on Saturday July 9th 1955. During the ensuing cavalcade, the modified Austin Sheerline purred past the crowds. Sadly this proved to be no publicity stunt. At low road speeds, power was so low the car had to be towed up the steep ramp from East Works onto the main road.

A couple of years later, Dr. Weaving came to recognise that while gas turbines were fine for aircraft, they were impractical for cars. The main problems were: the amount of heat generated that had to be dealt with; noise had to be reduced to a level acceptable for passengers and the general public; finally, the engine consumed diesel fuel at an alarming rate. The Austin Princess used for the trials happened to be the only vehicle in the company that could accommodate an engine of this size.

However, as a great deal of knowledge on gas turbines had been accumulated under Dr. Weaver's leadership, development was then turned to see if there was a commercial use for the engine as a stand-alone unit.

So a new single-shaft, axial flow engine was created to produce 250bhp at 29,000rev/min. Its specific fuel consumption was 1.05 pints per bhp.h. It was decided that if coupled to a generator it could be used to supply power for emergency power generation, as in hospitals in the event of power failure from the grid.

Another application proved to be to power a water pump, again for use in an emergency. In this case, by mounting the unit on a trailer it could be taken wherever it required.

A demonstration to publicise such applications of the engine was staged at the rear of East Works. The engine/gearbox attached to an inbuilt fuel tank was priced in 1961 at £2,500. Austin claimed this was cheaper by £500 than a diesel engine producing 250bhp.

The opening of the Electrical Engineers' Exhibition at Earls Court on the 21st March 1961 proved an important day for a small group working on gas turbines at Longbridge.

It was then, with the announcement of the 250 Austin Gas turbine engine, that years' of hard work by the research and development team in East Works was revealed to the world. The announcement culminated in 12 years' of team effort, during which many people worked on the project.

In early days, Dr. Weaving was joined by Mr. J Barton MSc. Another team member, Mr Bradley, was put in charge of engine testing, test housings, and equipment. He began his engineering apprentice with Austin in 1942 but later spent three and a half years at the NGTE before starting gas turbine work at

Longbridge in 1949. He drove the gas turbine-powered Austin Sheerline Princess at the Austin golden Jubilee in 1955 and demonstrated that vehicle to the Duke of Edinburgh.

A critical component of the engine proved to be the compressor. Dr. Tonks, who undertook his apprenticeship with the Nuffield Group, had responsibility for this item.

At the start, the work of producing the engineering drawings was done by three or four draughtsmen on an ad-hoc basis. But in 1957 a small design office was set up in East Research with six draughtsmen led by senior design engineer, Mr. D Rickman who had had experience of working on gas turbines since 1949.

Eventually the total team rose to 45 people in number. Perhaps not surprisingly the Austin gas turbine was expensive to produce and, while a 'fair number' were sold, in the end each engine produced lost money for the company so the project was stopped.

# 3 Bladon Jets

IF at first you don't succeed, try, try again. That could be the motto of Bladon Jets, the Coventry-based micro gas turbine specialist bidding to plough a new furrow for gas turbines in vehicle and power generation applications world-wide.

Told in the late 1980s by none other than Rolls-Royce that axial flow micro gas turbines just "could not be done", Chris and Paul Bladon – twin brothers, well-known for their racing motor cycle engineering skills – set about proving the elite aero engine maker wrong.

Paul Bladon sadly died from cancer in 2008, but before his death he pioneered machining minute axial flow compressor blade discs – blisks – from solid material using electro-discharge machines (EDM). Then, it took Paul two months to produce his first blisk; now with the latest machine, parts come at several per hour.

Bladon Jets has come far since its formation in 2007; now the worldwide application of micro gas turbine technology looks to be within its grasp, both as automotive range extenders and for electrical power generation sets of up to 100kW – and possibly even larger in the future.

After all it was in the 1960s that gas turbines first made a bid to power passenger cars and trucks when Rover and Leyland Gas Turbines developed their engines. But their Achilles Heel was poor heat exchangers and adverse fuel economy. Now, in range extender mode, a micro gas turbine need run for only a short time at constant speed to keep an electric vehicle's main battery charged.

Paul Barrett, Bladon Jets' chairman and technical director, admits serendipity has played a helpful part in the company's evolution.

'The way that Rolls-Royce helped was in challenging the Bladon brothers, who were very stubborn. Rolls told them they would not be able to do it,' said Barrett. 'They did not know that it was impossible so they went out and did it – using EDM to create the complex aerofoil blade shapes.'

'When we filed patents on the process in 2002, we had little idea of how – or even whether – we could ever turn it into a business '

'Then, when we set out to commercialise the technology in 2008, our primary interest was aero engines,' Barrett explained. 'But as we wrote the business plan and talked to investors, prospective customers and partners, we quickly shifted our focus to the automotive area, where we perceived a more immediate market opportunities.'

'Within a month of that decision, we were approached by Jaguar Land Rover (JLR) who were interested in the weight-saving potentials of a gas turbine range extender for electric vehicles,' he said.

'At that time, we were not on the map,' added Barrett. 'JLR tracked us down through our principal engineer Phill Heward, who had joined the company a few months previously.' Phill is the world's most prolific micro gas turbine designer. His business, Heward Microjets, had been making model engines based on turbocharger technology since 1988; and he was one of the first to crack the 'black art' of micro combustion technology. He built 15 or 16 combustors before finding one that worked. Since then he has made several hundred engines, of many different designs, from 5lb to 200lb of thrust.

'A lot of CFD work goes into developing combustors, but it can only help you once you have an established design,' said Barrett. 'It is getting the combustor to work in the first place that is crucial.'

'A collaborative project with Jaguar Land Rover (JLR), part funded by the Technology Strategy Board (TSB), led to the two 70kW micro gas turbine engines that featured in Jaguar's 2010 ground-breaking C-X75 concept electric sports car. Using Bladon Jets ultra-light weight range-extender technology, C-X75 has a top speed over 200mph and a range of 540 miles on a 60-litre tank of fuel, noted Barrett.

Bladon Jets' technology muscle was further enhanced by the arrival of three senior advisors, the 'three wise men'. Two had already examined the company for the Royal Society as Bladon Jets sought seed funding for new technology.

Professor Philip Ruffles CBE, former technical director of Rolls-Royce, and Professor John Denton, former director of Cambridge University's Whittle labs and a specialist in numerical methods for turbo-machinery, have recently been joined by Professor Geoff Kirk, until 2007 Rolls-Royce's chief design engineer and winner of numerous awards.

'Additional expertise is provided by partner PCA Engineering Ltd of Nettleham, Lincoln, specialists in aerodynamic design and CFD work,' said Barrett.

'PCA is one of the top turbo-machinery design companies in the world, working for the likes of Rolls-Royce, GEC and Pratt & Whitney. It creates the complex aerodynamic shapes that we then reproduce using our patented EDM process,' Barrett confirmed. 'Then there is Leuven Air Bearings of Leuven, Belgium. It develops advanced air bearings, crucial for high-speed rotating machines such as our micro turbines, which run at speeds up to 250,000 rpm.'

But it is the underscoring resourcefulness and support provided by India's Tata Group that may well prove crucial in enabling Bladon Jets to succeed where many others have failed. Tata's chairman, Ratan Tata, is a firm believer in gas turbines. He not only sees where the automotive market is heading, but believes that micro gas turbines will play a big part in addressing India's – and other emerging economies' – future power generation needs.

'Tata is a fantastic investor,' enthused Barrett. 'Besides the financial support, they give us access to a wealth of resources and expertise within their many group companies and, most importantly, with Tata we have a long-term partner who shares our vision of building the world's leading micro gas turbine business.' The field is currently wide open; and Bladon Jets have established themselves as a serious player in a very short space of time.

Bladon Jets' design is for a simple reverse-flow gas turbine of tiny proportions – 90mm is the internal diameter for Jaguar's 70kW machine with an entry hub/tip ratio of about 0.5. Overall thermal efficiency is around 25 per cent – achieved through high pressure ratios instead of the heavy (and expensive) recuperators found on most other commercial micro gas turbines. A multi-stage axial flow compressor is followed by a single-stage radial compressor, which feeds air into the reverse flow combustion chamber. The compressors share the same, 90,000rev/min shaft as a two-stage axial flow turbine. A third axial flow, 'free' turbine drives the generator at 60,000rev/min.

According to Barrett, 'shaft length is critical for rotor dynamics – especially as we use air bearings, which do not have much inherent damping. Hence our choice of reverse flow; it allows us to move the combustion chamber out in order not to take up space on the shaft.'

The absence of a recuperator has enabled Bladon Jets to keep the overall design compact and light: the target weight for the 70kW unit, including the generator, is less than 30kg. Turbine inlet temperature is close to 1,000°C. 'Currently, hot end components (turbines and combustors) are made from high temperature nickel alloys, but there could be a move to ceramics – and higher temperatures – in the future.

However, Bladon Jets' vision is for even smaller engines, down to 6kW or less for some applications. Here the miniaturisation of all the technologies will be pushed further – rotor dynamics, aerodynamics and thermodynamics. The gensets will require recuperators to enhance fuel economy; the constant-speed range-extender turbines are unrecuperated machines for simplicity, reliability and light weight.

'Due to the amazing power density of the gas turbine, we could fit a 6kW genset into a biscuit tin,' said Barrett. 'And, with its air bearings, air cooling and inherently clean combustion, it wouldn't need the bulky oil, water and exhaust treatment systems required by conventional internal combustion engines. Furthermore, gas turbines can run on pretty much any fuel – liquid or gas; and, with only one or two moving – that is rotating – parts, they are quieter and more reliable than their reciprocating counterparts. They really are the ultimate combustion engine'.

Bladon Jets has plans to build a complete range of engines. 'Although we started at 50kW and moved to 70kW, we will have 6, 12, 20, 35 and 100kW machines in time,' said Barrett. 'At present we are testing the engine we developed for Jaguar with funding, won in competition for low carbon vehicle initiatives, from the Technology Strategy Board.'

'We already have two customers for our first engine,' said Barrett. 'However, I am not convinced that 50kW is the right size for the high volume automotive market.'

'The sweet spot for Jaguar and other high end manufacturers might be in that range, but if you start with a small electric vehicle, such as the Nissan Leaf or Mitsubishi iMieV, then the range extender may be only need to be 5-10kW. So we are currently working on a smaller engine. This will have quite a different architecture from our 50kW machine,' Barrett told his audience.

Had it not been for the Bladon brothers' pioneering work in EDM, Bladon Jets would never have existed. And, thanks to the latest, £160,000, fully-automated Sodick AQ325L EDM machine, the company can produce micro turbine compressor and turbine blisks in less than an hour.

But does EDM lend itself to volumes of 100,000 a year? 'It is absolutely viable,' replied Barrett. 'We have developed the process for aluminium, titanium and nickel alloys and, using the latest CNC capabilities and our proprietary tooling, we can make multiple blisks in parallel in a 'lights-out' [un-manned] environment.'

'The number of blisks from one machine depends on blisk size, required surface finish and component complexity. We can make five, 10 or even 20 in one pass,' he added. 'The process is not power hungry, but depends on the amount of material removed – which is very little with such tiny blades. ' Bladon Jets has three patents granted and a further two in the pipeline.

And ceramics; will they come? 'For smaller engines ceramics are viable using clever design and production techniques,' said Barrett. 'They can be used for both static and rotating parts in the hot end. However, we don't believe they are ready just yet for rotating parts in larger units.'

Bladon Jets has won funding for a second TSB programme with Warwick University/Warwick Manufacturing Group (WMG), which has strong links with Tata and Jaguar. This is to develop gas turbine range extender validation programmes for automotive applications.

These programmes are important to customers. Bladon Jets has to prove to that its designs are robust enough for automotive applications.

'WMG was keen to work with us on this,' said Barrett. 'No one has developed validation standards for automotive gas turbines – or for range extenders. WMG has a lot of expertise in this area and good electric vehicle systems test facilities; it is a matter of applying the former and adapting the latter to the specific demands of this application.'

But Warwick is not the only university involved. Bladon Jets is working with Cambridge University's experimental combustion group to develop the CFD model, as well as others, including Cranfield University for performance testing and analysis.

What are the principal challenges? 'The biggest challenge is peoples' expectations, given our profile,' said Barrett. 'I think it is fair to say that our profile is somewhat ahead of our capability at present. Six months ago we were in a cowshed in Shropshire –albeit the best fitted-out cowshed you have ever seen. Now we have a custom facility in Coventry, but we are still two years away from production. People see the C-X75 and assume that the jet engines in it are fully developed.'

'Technology is the second challenge. Gas turbines maybe simple in concept, but there is a lot to learn about making them perform efficiently and reliably. However, at least the knowledge is out there. In respect of larger gas turbines anyway. We have to assimilate and apply that knowledge to micro engines. And third challenge is growing the business at the same time as doing development.

Added Barrett: 'We now have a clear understanding of the technology path and development project paths that we must take. We have 17 people – and everyone one who works here is an enthusiast – but we require more engineers. And it is very difficult to find anyone with micro gas turbine experience in the UK.'

'We are serious about making this succeed,' concluded Barrett, 'and we are very well positioned to do so. We have an incredible opportunity and the timing is right. We are extremely fortunate to have two major strategic partners in the form of Tata, with its resources and vision of the future, and Jaguar Land Rover, who are right on our doorstep at Whitley,

## Automotive gas turbines

Coventry and with whom we have developed strong working relationships – at all levels.'

He added: 'And we are also extremely fortunate to have the mentoring and guidance from our Advisors, who have seen and done it all before at Rolls Royce. But now we have to get our hands dirty. Our aim is to produce tens of thousands of micro gas turbines for the Indian power generation market and thereby reduce the unit cost to where we can compete with conventional engines in automotive applications. Our small genset is being designed from the outset for mass production. It is not rocket science, but we are applying automotive production techniques in an area where previous product runs have typically been measured in the hundreds. '

### Microturbines

When guest of honour Ratan Tata, chairman of TATA Group, officially declared 'open' the Bladon Jets engineering centre in Coventry on 2 February 2012 (below), he also unveiled a master class of manufacturing technology that Bladon Jets engineers have brought together to produce micro gas turbines destined for hybrid electric cars and vans, and generating sets.

TATA is one of several organisations giving financial support to Bladon Jets. JaguarLandRover, part of Tata Motors, is providing technical support; the microturbine also powers Jaguar's C-X75 hybrid electric concept car.

'What I have seen here vindicates the faith I had in what seemed like a tremendous idea,' Tata told *Professional Engineering*. 'What takes my mind is that it is far more sophisticated and impressive than I ever imagined. I think this company has a great future and I feel very proud to be a small part of it.'

The single-shaft microturbine Tata witnessed under test comprises only 49 parts. It has a four-stage axial compressor and a centrifugal compressor, an annular reverse-flow combustion chamber and a radial inflow turbine. Maximum speed is 100,000rev/min. At this level of technology, however, support comes in many guises and several businesses provide Bladon Jets with unique levels of technical expertise. Rolls-Royce has made very small gas turbines and been supportive.

One of the most remarkable components investors and supporters alike could inspect was the engine's nickel super alloy nozzle guide vane (NGV) assembly, together with the compressor/turbine, both central the micro gas turbine.

The highly complex, one-piece NGV is manufactured by Materials Solutions Ltd of Worcester, a small company focusing on additive layer manufacturing (ALM).

Materials Solutions, founded in 2006, is a particular specialist in metal powder bed ALM using EOS 200W argon laser machines to produce functional parts – so-called 'make to print'. It has five machines and has developed processes to deliver materials with equivalent mechanical properties to those of cast or forged parts. The Bladon Jets NGV demonstrates the fine art of ALM, namely high-value, thin-wall engineered components that might be extremely difficult to make, or if they can be made, would require heavy machining.

The size of component that Materials Solutions can comfortably handle is determined by the capacity of its EOS machines. The NGV is just within the limits of the machine, namely 250mm in diameter and 200mm high. It would be very difficult to manufacture this complex part by any other means.

The entire microturbine design is generated in CAD and this forms the basis of several manufacturing processes. In addition to the NGV's design features, which are transferred to Materials Solutions, the CAD model is a driver for parts inspection on the CMM, as well as offering data for A&M EDM Ltd of Smethwick. This company makes the tools that generate individual compressor and turbine blades on blisks (disks that contain either compressor or turbine blades). The CAD model is the driver for an electro discharge machine (EDM).

Materials Solutions can receive CAD files in multiple formats (it has five seats) but makes a virtue out of the fact that it is not an engineering design resource and has no interest in the function of the shapes it builds. It does, however, assist customers' designers with their designs for manufacturing process. All parts are explicitly or implicitly designed for a manufacturing process and as AML is a different manufacturing process it is rare that a part, as designed for machining, will be optimal for building by the ALM process.

'We frequently find break lines between components need to move and possibly we can make complete assemblies, obviating the need for machined faces and fastening systems,' said Carl Brancher, chief executive. 'We can make castings designs, but with far thinner walls. As such we have two deliverables: manufactured parts and technical capability.'

'We work in high grade stainless steels and specialise in high-temperature nickel based alloys such as Hastelloy and Inconels,' added Brancher. 'We focus on jet engine and gas turbine engine parts for the pragmatic reason that such engines contain many parts of the type we can build, using materials we can build in and with a major supportive customer in the form of Rolls-Royce.'

'So we have made jet engine parts our 'home' and expanded out of it into other related high-temperature applications, such as motor sport,' he noted. 'The aerospace industry accounts for

90 per cent of our business and exports run at over 50 per cent. Medical and dental sectors are already address by others.'

'We are an approved vendor for Rolls-Royce, Rolls-Royce's nuclear sector and Rolls-Royce Canada,' added Brancher. 'And we are approved by ITP in Spain and Mercedes-Benz AMG. Among customers in the last few months are: Sumitomo Precision Products (Japan), Tusas Engine Industries of Turkey, Alstom in Switzerland and Parker Aerospace in the US.'

Besides the NGV component, Materials Solutions makes other parts, including the microturbine's crucial central bearing housing.

A&M EDM Ltd is another young company; just nine years old it is one of the largest EDM firms in the UK with 20 wire/EDM machines. Among its parts (machine shown above with Ratan Tata - right) for Bladon Jets are: combustion housing, exhaust ducts, compressor casing and air bleed chamber, main shaft, bearing housing, centrifugal system housing and a deswirl diffuser. It has made parts sufficient for two development engines.

A&M EDM also makes the graphite electrodes that Bladon Jets needs for 'sparking' in its newly-delivered Sodick AQL325L EDM machine to generate the blade profiles on the crucial compressor and turbine blisks. A&M EDM produces these electrodes on five-axis milling machines. CAD data for the blade components are supplied by Bladon Jets and these are used in the EDM machine to generate the three-dimensional cutting strategy that creates blade profiles.

Surface finish, accuracy and precision are crucial to rotating components, including labyrinth seals, as these can affect the entire performance. This is especially so with compressor and turbine blades, where due to their very small size (20mm high), fillet radii and tip clearance are vital to hold within tolerance. Aerodynamic surfaces produce losses, especially end losses, and to reduce these requires excellent surface finish, accuracy and precision.

CFD – computational fluid dynamics – is just one tool Bladon Jets uses to optimise compressor and turbine. However, it is only through bench testing that engineers can verify their initial assumptions and then optimise further.

Blade tip and other clearances, acceptable at cold can quickly change as the engine reaches operational temperatures, creating 'rubbing'. These require an iterative development process to optimise the 'final' design. Typical turbine temperature is 1,050°C; exhaust temperature is 650/700°C.

Finally, Bladon Jets can see the day when microturbines serve as 100kW generator sets and these will require heat exchangers to improve thermal efficiency. To this end Bladon Jets is working with Brayton Energy Canada Inc. of Gatineau, Quebec. Its patented wire mesh heat exchangers, it claims, offer enhanced heat transfer and efficiencies better than a diesel engine.

Brayton expects to ship its first full-scale unit to Bladon in July with engine testing beginning in 2013. Meanwhile, Brayton has been asked to build a full-scale prototype intercooler for a hybrid truck development programme. This will take the form of a gas turbine with an intercooler and recuperator. Brayton claims the ultra-compact plate-fin counter-flow heat wire mesh exchangers will "dramatically" improve overall cycle efficiency.

Brayton expects to begin small-scale production in 2013 with ramp-up in 2014. Full-scale testing will take place in Canada, the US and the UK

# 4 Ceramics

PERFORMANCE goals for military engines and national needs to achieve fuel economy and environment quality at affordable costs were largely responsible in the US and elsewhere for the emphasis in the 1960s and 1970s on ceramics for heat engines. So, "mindful of the very high potential payoff to be obtained from successful exploitation of ceramics in heat engines," the US Defense Advanced Research Projects Agency (DARPA) initiated in 1971 a major study in Brittle Materials Design.

The aim of the wide ranging work was to encourage use of ceramics in heat engines by demonstrating 200-hour durability. Inevitably, the study attracted the attention of companies such as Ford, itself already examining such materials.

Art McLean and other researchers within Ford, and elsewhere in the US in the 1960s, 1970s and 1980s, presented a wide range of technical papers reflecting their work on the role of ceramics in heat engines, with the focus of the work in the area of gas turbines. But Britain was not in the slow lane.

In 1970 for example, *The Engineer* weekly journal (26 February) reported how "British expertise makes ceramic gas turbines practical, predicting the demise of the long-distance diesel engine". Materials under close study were silicon nitride and silicon carbide. And in its issue of 26 March 1970, Noel Penny outlined "why the gas turbine will be the engine of tomorrow".

*The Engineer* later reported its interview with Lord Stokes who said British Leyland had spent £10 million on gas turbine R&D, of which £1 million had been earmarked to develop heat exchangers. While in the same issue, 24 September 1970, the journal reported that "Leyland pins hopes on silicon nitride".

Two years on, *The Engineer* of 21/28 December 1972 reported Dr. Bertie Fogg, British Leyland's director of research, saying: "We have gone through a very difficult period with our turbines...but we believe we are in advance of most people".

One of the company's aims was to produce a silicon nitride heat exchanger disc in competition with that of Corning Glass in the US. Fogg added: "We have a substantial development in Solihull on silicon nitride aimed at products apart from heat exchangers". Fogg revealed that the idea of using Corning's Cercor heat exchangers as a catalyst by spraying on a special potassium salt coating was developed by BL.

A year later, *The Engineer*, in a feature dated 8 November 1973, noted that "performance of ceramic heat exchangers is poor and Ford of America must decide if it is to switch to a more forgiving silicate". The ceramic heat exchangers used lithium alumina silicate (LAS) which required "manufacturing to keep within very close tolerances if the heat exchanger will is to give the durability and reliability which engine makers are looking for".

But there were other materials such magnesium alumina silicate which was more forgiving.

While in its issue of 13 December 1972 *The Engineer*'s materials feature declared "as ceramics become even more vital, Britain drops behind". A year later, the same publication (7 November 1974) could report that a "Tough, heat resistant silicon/silicon carbide fibre composite material able to withstand 1,400°C could boost gas turbines".

Many of the technical reports of the period in the US were compiled for the Army Materials and Mechanics Research Center (AMMRC) in Ann Arbor, Michigan.

For example, one such report, by Art McLean and Eugene Fisher, both of Ford Motor Company, under the title *Brittle Materials Design, High Temperature Gas Turbine*, and written in August 1977, reflects the degree of the activity.

In their abstract, the authors note: "The demonstration of uncooled brittle materials in structural applications at 2500°F is the objective of the *Brittle Materials Design, High Temperature Gas Turbine* programme. Ford Motor Company, the contractor, is using a small vehicular gas turbine comprising an entire ceramic hot flow path including the highly stressed turbine rotors.

Westinghouse, the subcontractor, originally planned to evaluate ceramic first-stage stator vanes in an actual 30 MW test turbine engine; however, this objective was revised to demonstrate ceramic stator vanes in a static test rig. Both companies had in-house research programmes in this area prior to this contract.

At the time, a range of turbine engine components were under evaluation in ceramics of one type or another, including radial inflow turbine wheels and turbocharger components.

Much more recently as 2009, Art McLean was still busy writing on the subject of ceramics. In the *Naval Engineers Journal* (Vol. 87, Issue 2) he and Dr. E. R. Van Reuth and Dr. R. J. Blatter produced an article *Ceramic gas turbines for improved specific fuel consumption*, first published on 18 March 2009, suggesting further confirmation of the role that ceramics could play in advancing gas turbine technology.

Forty years earlier, and published in the proceedings of the Institution of Mechanical Engineers, Conference proceedings 1964-70, Vol. 178-184, McLean put forward an article titled *Case for single shaft vehicular gas turbine engines*. In this paper, given in 1969, McLean made one reference to heat exchangers, as mentioned. He concentrated on a single-shaft gas turbine arrangement and associated implications for transmissions, of which he put forward several alternatives.

He wrote: "Considering all of these infinitely variable transmissions, ones that would be obviously appealing to the turbine engineer would be the hydraulic torque converter and the aerodynamic torque converter, since these are members of the turbo-machinery family. Hydrostatic, split hydrostatic and electrical transmissions have special advantages in off-the-road vehicles where there are requirements for pulling or pushing heavy loads and for large accessory power take-off. For over-the-road vehicles, electric transmissions with a high-speed turbine-driven alternator seem appealing and would eliminate gear shifting; however, the cost of this system is a real problem."

As for ceramics he opined: "The use of ceramic materials for the sophisticated high temperature structure at the hot end of turbine engines is being explored. Although ceramics have attractive advantages of low cost, low expansion coefficient, and high temperature capability, one of the major disadvantages of using brittle materials is their very low strain tolerance.

"Simple component shapes, facilitated by the single-shaft turbine engine, are therefore favoured to equalize thermal gradients and minimize resultant thermal stresses, adding greater reliability to the structural components. Also, simple shapes are favoured from a fabrication and cost viewpoint, such as a turbine nozzle inlet duct made of silicon carbide and a turbine nozzle of silicon nitride as examples of ceramic components which could be used in a symmetrical flow path.

"The advantage of increasing turbine inlet temperature is well known and poses the challenge of developing relatively low cost techniques for air-cooling turbine blades. Apart from the detail of cooling each individual turbine blade, one of the biggest problems will be ducting the cooling air from the source, e.g. compressor discharge, to the blades. The single-shaft turbine engine simplifies this problem in that both turbine stages are attached to a common shaft; much greater simplification of blade cooling can be achieved if a suitable single-stage turbine can be developed.

No mention there of complex heat exchangers and the implications for engine development – a subject maybe too sensitive at the time – with the exception of two references, including one from John Lanning who fronted Corning Glass's entry into the heat exchanger stakes with Cercor.

Meanwhile, however, McLean was notable for taking out a number of US Patents, some associated with ceramics and their role in gas turbine technology.

For example, in addition to US patent 3623544, Gas turbine heat exchanging system, and US Patent 3999376, One-piece ceramic support for gas turbines, he is identified with US

## Automotive gas turbines

Patents 3485042 and 3486329 on control systems for gas turbines.

Notwithstanding these developments, the world still awaits Ford's supercharged gas turbine engines for superhighway trucks. It is not likely to happen, but the company did make an attempt to enter the heavy goods vehicle sector with its H-truck Transcontinental tractor (below) manufactured between 1975 and 1984, by Ford in Holland and Foden in Britain. A total of 8,735 were made – 8,231 in Amsterdam and 504 in Sandbach.

Assembled mostly from bought-in OEM components (cab shells from Berliet, engines from Cummins and transmissions from Eaton) Ford introduced it in April 1975 to fill a perceived gap in anticipation of relaxed HGV weight restrictions; as such it had a strong chassis and heavy duty suspension.

Although ahead of its time in almost all respects, 'Transconti' did not become a true success story for Ford, despite offering excellent standards of driver comfort and a high power outputs for its time using well-proven Cummins diesel engines with typical outputs of 290 to 350bhp. The Mk 1 (1975-78) carried the 'small cam' Cummins NTC engine while the Mk 2 (1978-84) had the 'big cam' Cummins NTE.

Sales failed to live up to expectations; its high duty construction was too heavy for UK 32-ton weight limit at the time. Today the 'Transconti' is a rare vehicle with about one per cent remaining. It is in favour with collectors of vintage commercials, although a few remain in revenue earning service throughout Europe.

# 5 Chrysler Corporation

THE CHRYSLER turbine car was the first and only consumer test ever conducted of gas turbine-powered cars. Of the total of 55 units built (five prototypes and 50 'production' cars), most were scrapped at the end of a trial period; only nine remain in museums and private collections. Chrysler terminated its turbine engine project in 1977, yet the turbine car was the high point of a three-decade project to perfect the engine for practical use.

The fourth-generation Chrysler gas turbine Chrysler engine rotated at speeds of 44,500rev/min, according to the owner's manual. It could operate using diesel fuel, unleaded gasoline, kerosene, JP-4 jet fuel and even vegetable oil. No air/fuel adjustments were required to switch from one fuel type to another and the only evidence of which fuel being used was the odour of the exhaust.

Chrysler claimed the engine had one-fifth of the moving parts of a traditional piston-based internal combustion engine (60 rather than 300). The turbine spun on simple sleeve bearings for vibration-free running. Its simplicity offered the potential for long life and, because no combustion contaminants entered engine oil, no oil changes were considered necessary.

The 1963 engine generated 130bhp (97kW; 132PS) and a near instant 425lbft (576Nm) of torque at stall speed, making it good for 0 to 60 mile/h (0 to 97km/h) in 12s at an ambient temperature of 85°F (29°C). The car could accelerate quicker if the air was cooler and thus denser.

The fewer number of moving parts and the lack of liquid coolant eased maintenance, while the exhaust did not contain carbon monoxide, unburned carbon, or raw hydrocarbons, it was claimed. Nevertheless, the turbine generated nitrogen oxides and the challenge of limiting them proved an ongoing problem throughout development.

# Automotive gas turbines

The power turbine was connected, without a torque converter, through a gear reduction unit to a moderately modified TorqueFlite automatic transmission. The flow of the combustion gases between the gas generator and free power turbine provided the same functionality as a torque converter but without using a conventional liquid medium. Twin rotating heat exchangers transferred exhaust heat to the inlet air, greatly improving fuel economy. Variable stator blades prevented excessive top end speeds and provided engine braking on deceleration.

Throttle lag and exhaust gas temperatures at idle plagued early models; Chrysler was able to remedy or mitigate these to some degree. Acceleration lag, however, remained a problem, and fuel consumption was excessive. Acceleration was outstanding provided the turbine was spun up (by applying power) prior to releasing the brakes. Otherwise it was mediocre. The turbine car (above) also featured a fully stainless steel exhaust system, the exits of which were flat in cross section. This was intended to spread the exhaust gases thinly and cool them further, in order to allow the vehicle to stand in traffic without risking damage to following traffic. The combustor, or burner, was somewhat primitive by the standards of modern turbojet engines. A single reverse-flow canister featuring a more-or-less standard spark plug for ignition was employed. Had the engine been further developed, annular combustion chambers and a second power turbine might have improved power and economy even more. The transmission had "idle" instead of "neutral".

The turbine car had some operational and aesthetic drawbacks. It did not make the traditional noise of an American V8 engine. High altitudes also caused problems for the combined starter-generator. Additionally, failing to follow the correct start-up procedure could cause the engine to stall; some consumers thought they could "warm" the engine up in a similar manner to the way they did with a gasoline engine. They would press the accelerator pedal to the floor before the engine had reached the correct temperature. Instead of warming the engine, the excess fuel slowed the turbine down and resulted in the opposite of the desired effect. Doing this, however, did not do any permanent damage to the engine. In fact, it was possible to apply full throttle immediately after starting the engine without much fear of excessive wear.

The engines proved remarkably durable considering how fragile turbine engines are when compared to internal combustion piston engines. Troubles were remarkably few for such a bold experiment. It is not known how many testers made the mistake of using the leaded pump gasoline of the era; the tetraethyl lead would leave debilitating deposits within the engine. It was the one flammable liquid Chrysler recommended not be used; it was also by far the easiest fuel to obtain. Even so, it is claimed that more than 1,100,000 test miles were accumulated by the 50 cars released to the public, and operational downtime stood at only four per cent.

The bodies and interiors were crafted by Ghia in Italy. The mostly completed bodies were shipped to Chrysler's Greenfield Avenue turbine research center in Detroit for final assembly. Outsourcing this procedure may have saved some

# Automotive gas turbines

money for Chrysler, but between the expensive Ghia bodies and the cost of the engine, each car may have cost as much as $50,000 to build. This would be over $350,000 in 2012 dollars.

A total of 50 "production" turbine cars were built between October 1963 and October 1964, plus five prototypes (three of which differed in roof/paint schemes). As each body was finished and shipped to Detroit, Chrysler employees installed the gas turbine engines, TorqueFlite transmissions, and electrical components to prepare the cars for use by the 203 motorists - 23 of them women - chosen to test them in the US. At least one car was brought to the UK for journalists to drive.

The turbine car was a two-door hardtop coupe with four individual bucket seats, power steering, power brakes, and power windows. Its most prominent design features were two large horizontal taillights and nozzles (back-up lights) mounted inside a heavy chrome-sculptured bumper. The single headlamps were mounted in chrome nacelles with a turbine styling theme to create a striking appearance. This theme carried through to the centre console and hub caps. Tyres carried small turbine vanes moulded into the white sidewalls.

The cars were finished in a reddish-brown "Frostfire Metallic" paint, later renamed "Turbine Bronze" and made available on production automobiles. The roof was covered in black vinyl, and the interior featured bronze-colored "English calfskin" leather upholstery with plush-cut pile bronze-colored carpet. Front suspension comprised upper and lower wishbones with coil springs and shock absorbers.

The dashboard could be illuminated using electro-luminescent panels in the gauge pods and on a call-out strip across the dashboard. This system did not use bulbs; instead, an inverter and transformer raised the battery voltage to over 100V AC and passed that high voltage through special plastic layers, causing the gauges to glow with a blue-green light. Instead of a water temperature gauge, Chrysler's turbine car had a turbine inlet temperature (TIT) gauge with the numbers 500, 1,000, 1,500, and 2,000°F.

The car was designed in the Chrysler studios under the direction of Elwood Engel, who had worked for Ford Motor Company before his move to Chrysler. The designer credited with the actual look of the car was Charles Mashigan, who designed a two-seat show car called the Typhoon displayed at the 1964 World's Fair in New York City. Engel used many older Ford styling themes. The rear taillight/bumper assembly was copied directly (with revisions) from a 1958 Ford styling study called the "La Galaxie". He used none of the themes associated with his 1964 Imperial. As Engel incorporated many of the design themes from the 1961 Thunderbird, and because the car had four seats of similar size and appointment, many enthusiasts call the Ghia turbine the "Englebird."

After Chrysler finished the user programme and other public displays of the cars, 46 of them were destroyed. Chrysler announced that this was necessary to avoid a stiff tariff, but that was only part of the story. The destruction of the cars was in line with the automobile industry's practice of not selling non-production or prototype cars to the public. This practice was also later used by General Motors with its electric car, the EV1.

Of the remaining nine cars, six had the engines de-activated and then donated to museums around the country. Chrysler retained three operational turbine cars for historic reasons; two of the three are owned by the WPC Museum. All of the turbine cars owned by the WPC Museum are in running condition at the archives of the museum. The last turbine car that is functional, owned by the Museum of Transportation in St. Louis, Missouri, appears at car shows around the US from time to time. One turbine car is on display at the Henry Ford Museum in Dearborn, Michigan, and formed part of the exhibition Driving America, opened in early 2012.

Only two Chrysler turbine cars are in the hands of private collectors: One was purchased by private automobile collector Frank Kleptz of Terre Haut, Indiana and is functional. Kleptz's turbine car was originally donated to the former Harrah Museum in Nevada. The second one is owned by comedian and television host Jay Leno, who purchased one of the three Chrysler turbine cars originally retained by Chrysler.

As mentioned, most of the turbine cars were destroyed by Chrysler for tax and liability reasons. This proved a great disappointment. Almost everyone riding in a gas turbine car would say: "Whoa, this feels like the future! You turn the key and there's a big woosh and a complete absence of vibration... I think it's the most collectible American car – it was so different. Most of all, the Chrysler gas turbine is a reminder that all the cool stuff used to be made in the US."

Chrysler's turbine engine program did not die completely with the destruction of the Ghia cars. A new coupe body, which was to become the 1966 Dodge Charger was considered for a new generation turbine engine. Nevertheless, Chrysler went on to develop a sixth-generation gas-turbine engine which did meet nitrogen oxide regulations. The company installed it in a 1966 Dodge Coronet, though never introduced it to the public.

A smaller, lighter-seventh generation engine was produced in the early 1970s, when the company received a grant from the United States Environmental Protection Agency (EPA) for further development. A specially-bodied turbine Chrysler LeBaron was built in 1977 as a prelude to a production run. By then the company was in dire financial straits and needed US government loan guarantees to avoid bankruptcy. A condition of that deal was that gas-turbine mass production be abandoned because it was "too risky" for a company of Chrysler's size.

A Chrysler turbine car, painted white with blue racing stripes, featured in the 1964 film *The Lively Set*. The car, loaned to the producers by Chrysler, was a test 'mule', and was returned to Chrysler after production. It was among the cars scrapped.

Chrysler's work with turbine engines did not pay off in the retail automobile sector, but the experiments proved fruitful with the incorporation of a Honeywell AGT1500 into the M1 Abrams Main Battle Tank, developed in the 1970s by Chrysler Defense, later sold to General Dynamics.

# 6 Daimler-Benz

ACCORDING to Eberhard Tiefenbacher, head of the technology department in Daimler-Benz AG, writing in a paper given to the International Symposium for Future Industries (22 – 25 March, 1988) in Kobe, Japan, his company had "started our work on automotive gas turbines at about 1972".

He added that when using metallic materials for turbine blades, gas temperatures of only about 1,000°C could be realised, which would give rise to high fuel consumption. The energy crisis of 1973-74 demonstrated the importance of fuel consumption.

On this basis a materials research programme was started, later was sponsored by the German ministry of research.

"Our goal from the beginning was to look for a turbine material which stands up to about 1,200°C at a minimum," said Tiefenbacher. "And this was without any cooling, because in the very small blades of vehicular gas turbines, you cannot bring in cooling channels as it is done in large aircraft engines."

The paper gave no indication of the engine's design details although reference was made of a vehicular gas turbine in the "100kW power class".

In 1978 the German government programme called for programmes to demonstrate research cars for the year 2000.

"And then a decision was made by our chief engineer," said Tiefenbacher. "He said: 'I want to have a gas turbine in this car.' That meant that we had a very tight schedule because the car had to run in 1981. We did not have the time to make a thorough finite element analysis, so we did it as good as we could with some engineering feeling. The engine should run at first with 1,250°C and later on with 1,350°C gas inlet temperatures."

The two shaft engine used a single rotating heat exchanger on top of the engine, with the two shafts mounted below.

# Automotive gas turbines

Air entered the engine (see cross section below) from the front into the compressor where it was heated to 200°C and the pressure raised to four atmospheres. The air then flowed into the heat exchanger. The heat exchanger extracted heat from the exhaust gas and passed it to the incoming air which is heated to 900°C. The air then travels to the combustion chamber where fuel is added.

The hot mixture, the result of continuous combustion, at about 1,250°C to 1,300°C then travels across the first turbine rotor (which ran at 60,000rev/min) before passing to the power turbine that ran at 55,000rev/min. The illustration on p. 357 shows the PWT 110 engine in 1984.

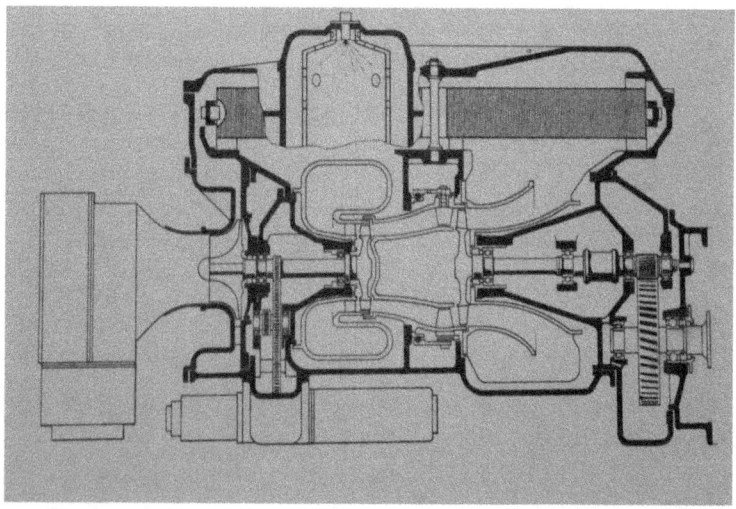

The high speed of both turbines meant high thermal stresses and centrifugal stresses had to be contained. Both turbines were of ceramics and had to be "connected" to a metallic shaft.

"That's another area where you can find a lot of difficulties," said Tiefenbacher. In fact, engineers used a mechanical bonding technique between the rotor and the shaft. The temperature of bonding was given as 800°C.

Tiefenbacher revealed that "we are just in the process of introducing sintered silicon nitride for the turbine rotors.

## Automotive gas turbines

Engineers made this by injection moulding. Typical rotor temperatures were in the region of 1,250°C to 1,300°C.

"We have 1,350°C but not yet in the car on the road," claimed Tiefenbacher at the time.

In his conclusion Tiefenbacher described the gas turbine as an engine which is simple in mechanics.

"It has good potential for high efficiency if it is possible to use high temperatures – 1,300°C.........I may say the vehicular gas turbine for automotive purposes will not be in mass production before 10 years (i.e. 1998). The ceramic and complicated parts made of ceramics must be improved for higher quality and must finally be produced with reasonable cost," said Tiefenbacher.

Tiefenbacher was proved right. Daimler-Benz has yet to produce a production gas turbine engine for passenger cars.

# 7 Detroit Diesel

IN LATE 1979, a programme was set up in the US to evaluate the application of gas turbines in transit buses. The transportation consulting division of Booz, Allen & Hamilton Inc. had the task of assessing the status of the programme.

In the first phase of the programme, reported in a 15 April 1980 brochure, five prototype GT-40404 gas turbine engines manufactured by the Detroit Diesel Allison division of General Motors Corporation were to be integrated into five RTS-11 model transit buses manufactured by General Motors and supplied by the Mass Transit Administration (MTA) of Baltimore, Maryland.

After acceptance testing and one of the coach's performance and mileage tested against a diesel-engined RTS-11 coach, plans called for the five coaches to be demonstrated in revenue service of the streets of Baltimore for one year, being driven by MTA drivers, and repaired and serviced by MTA mechanics.

By the early 1980s, the first (see above) of the diesel-to-turbine conversions had been completed at Booz, Allen & Hamilton's conversion facility (Modern Engineering Service of

Troy, Michigan). The decision to incorporate ceramic regenerators into the Phase 1 demonstration vehicles came too late for the material to be incorporated as original equipment "but they will be retrofitted into this engine following the shakedown tests" noted the consultants at the time.

In performance trials a number of problems emerged, most notably engine-generated noise levels from the air intake and exhaust. These were reduced by sound-absorbing materials with the result that in drive-by noise measurements the noise levels were 10dB below figures for the corresponding diesel engine.

Other problems to surface included poor standing-start acceleration; important in maintaining bus schedules. This was corrected to the point that acceleration was improved to within 0.25s of a comparable diesel coach.

The ability to demonstrate the multi-fuel capability of the gas turbine coach also required modification. In this respect two 125-gallon nylon tanks replaced the single 125-gallon metal tank allowing ethanol to be used as a fuel.

The demonstrations were due to run from the fourth quarter of 1980 through to the first quarter of 1982.

# 8 Fiat

IN NOVEMBER 1972, I met Giovanni Savonuzzi, director of research at Fiat, in Turin, Italy. He said he had a lot of faith in the gas turbine, which he called "a beautiful engine".

"If the automotive gas turbine is a success then I shall be happy," he said at the time. That production of the engine has not seen the light of day, might suggest that Savonuzzi turned out to be a very unhappy man.

Born in 1911, Savonuzzi worked in Fiat's aviation division before joining Cisitalia after WW2; there he developed a series of aerodynamic coupes with signature high rear fins.

Compared to the flashy styling trends across the Atlantic, Savonuzzi's research for Exner led to a purer wedge with rounded front and sides that curved under the body, while the

## Automotive gas turbines

fins ran the length of the sleek, smooth profile. But Savonuzzi was infatuated at the time with stunning Hollywood actress and redhead, Rita Hayworth; he tagged the prototype 'Gilda' after her 1946 movie.

A year later Savonuzzi followed the Gilda to the US, where he headed Chrysler's gas turbine division, graduating to director of automotive research before returning to Fiat in 1968.

It is reported that of the Chrysler gas turbine car only 55 were built. Because the Gilda was the first project developed in the Turin wind-tunnel, it seemed to make sense to him to fit a gas turbine engine. Not wanting to be branded as a man who destroyed the Gilda, he was adamant that the conversion would require minimal modifications: "As historian Michael Lamm put it, 'you've inherited the Mona Lisa, so don't bugger it up'".

In the late 1940s, Savonuzzi was involved with a special Cadillac built for Prince Aly Khan, married to Rita Hayworth.

When I met Savonuzzi, I reported that only a handful of companies were working seriously on gas turbine engines. They included: Ford, GM, British Leyland, Fiat, Volvo, Renault, Daimler-Benz, Chrysler, MTU and KHD.

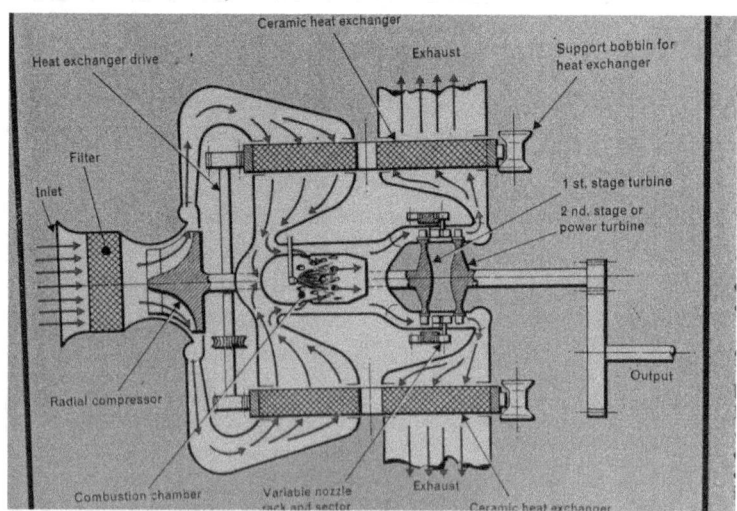

Fiat's design (see illustration above) for a truck gas turbine used two heat exchangers and variable nozzle guide vanes for the power turbine. The engine was designed to develop 400bhp

He called the gas turbine the ideal multi-fuel engine. He added that he did not spend his time at Chrysler for nothing.

"I brought the best ideas from Chrysler and put them with some more of my own," he told me. "And we have found that we can build an engine here quicker, even though Italy is not the US with all its technology."

Savonuzzi's engine used two types of heat exchanger: ceramic and steel.

"The ceramic ones we buy from Corning Glass in the US. They are very expensive. But our metal one has two advantages: it is cheap and we can make it ourselves," he said.

"The two 28in diameter Corning Glass heat exchangers were identical to those supplied to British Leyland and Ford," he said. Fiat's gas turbine engine was designed around these two heat exchangers.

Savonuzzi died in 1986 aged 75.

# 9 Ford Motor Company

A MAGAZINE, *Car Life*, in its July 1963 issue, quoted C. L. Bouchard, manager of Ford's gas Turbine Engine Department as saying: "Eight to 10 years". "Trucks such as the Ford Engineering test vehicle we drove will be the first commercial vehicles to be so powered," the magazine quoted Bouchard as saying. A bold prediction, if ever there was one.

The magazine wrote in its article titled *Taming the fan-jet in your future*: 'Ford has been investigating, developing and testing gas turbines since 1950. Research vehicles have included a Thunderbird, and a digital computer…..At Ford, the computer is a valuable and much-used research tool.'

The magazine noted that the Thunderbird was 'used to test various transmission configurations and for a long time was equipped with a Boeing 502 gas turbine for the basic power.

Unmuffled, (it had no regenerators) it singed ankles and shattered ears on the Dearborn test circuit.'

It gave the weight of the Boeing 520-6 engine as 316lb for a shaft horsepower of 400bhp. The Ford 704 weighed 650lb for a shaft horsepower of 300bhp, whereas the GMC Allison GMT-305 weighed the same as the Ford engine (650lb) for 225bhp.

The minimum specific fuel consumptions of the three engines were given as: 0.7, 0.56 and 0.56 lb/bhp.h respectively.

The Chrysler CR-2A of 140shp weighed 450lb and had a minimum specific fuel consumption of 0.51lb/bhp.h.

The magazine added: 'One interesting point: the g. t. needs very little in the way of transmission, having its own in-built torque converter.'

It added that an Army-Navy research programme 'has implemented research and resulted in Ford's Model 705, a 600bhp engine for a variety of military installations – from a stationary power plant to a hydrofoil sea-skimmer.'

'This project also fostered the Ford 704, a 300bhp commercial engine of surprisingly competitive characteristics.'

'The 704, in all likelihood, will be the prototype for a whole family of Ford turbines,' the magazine enthused and highlighted a photograph of the 600bhp Ford 705 turbine accompanied by two engineers: Ivan Swatman, design and development engineer, and C. L. Bouchard. Swatman went on to progress Ford's work in truck gas turbines.

That same year, journalists in the UK can recall visiting Germany and having a Ford commercial vehicle demonstrated to them on German autobahns; it was powered by a Boeing gas turbine engine, one journalist notes.

Two top engineers at Ford Motor Company's headquarters in Dearborn, Michigan in the US did more than most to proclaim their company's developments in gas turbines and ceramics.

# Automotive gas turbines

Ironically, it was Ivan Swatman and Art McLean respectively, through their various papers to the Society of Automotive Engineers and elsewhere that did most to offer glimpses of various Ford gas turbine programmes and associated work that otherwise might not have been seen for public scrutiny. For, in the 1960s and 1970s particularly, automotive companies were careful in the amount of information they revealed about their gas turbine programmes.

Ford in particular proved unusually protective of the various gas turbine engine data it generated, fearful that it could be seized upon by competitors and used to Ford's disadvantage.

Ford developed at least three generations of truck gas turbines – 704, 705 and 707: a 350bhp model in the late 1950s and early 1960s, a 'supercharged' version of 600bhp destined for the early 1970s, and a smaller version of around 250bhp. It is said that Swatman designed all of Ford's truck gas turbine engines.

Ivan Swatman, a naturalised American, was born in St. Albans, Hertfordshire, UK where he served an apprenticeship before taking a position in D. Napier & Son Ltd. He worked as a development engineer until moving away the US as an engineer working on industrial turbines. Later, he spent six years with Solar Aircraft's industrial turbines unit before joining Ford in Dearborn in 1956.

Among the early designs to be discussed in technical papers, Ford's 704 truck gas turbine engine was most prominent. In a paper in 1962, Swatman described the engine's cycle simply as that of a turbocharged gas turbine consisting of two stages of compression, resulting in an overall pressure ratio of 16:1, with air-to-air intercooling between stages to reduce the work, size, and tip speed of the high-pressure compressor. A plate-type recuperator was used between the high-pressure compressor outlet and the primary combustor of the high-pressure turbine. Reheat combustion was utilized between the high-pressure turbine exit and the power turbine. A unique feature of the cycle was sad to be the location of the power turbine between the high-pressure spool turbine and the super-charging or low-pressure spool turbines. Installation of the power turbine at this point in the cycle provided a 'very desirable speed relationship between the low-pressure and high-pressure compressors'.

According to Swatman this resulted in an extremely good part-load fuel economy of 0.47 lb/bhp.h at 55 per cent part-load. Near optimum load control was attained by varying the air flow, while turbine inlet temperatures were held almost constant. In the ideal case, cycle pressure ratio would be held constant.

Although the cycle concept was not new in the heavy industrial gas-turbine field, it soon became apparent that any attempt to stuff all of that machinery under the hood of an automobile would certainly be an ambitious project. The design phase began with this undertaking. As a design target, the engine compartment of a Ford passenger car was selected as the envelope dimension for the engine. While it was apparent at that time that the end-use of the engine would be in truck applications, packaging for a passenger car challenged the ability of the design team to compress the arrangement of the components to the utmost. The final design is shown below.

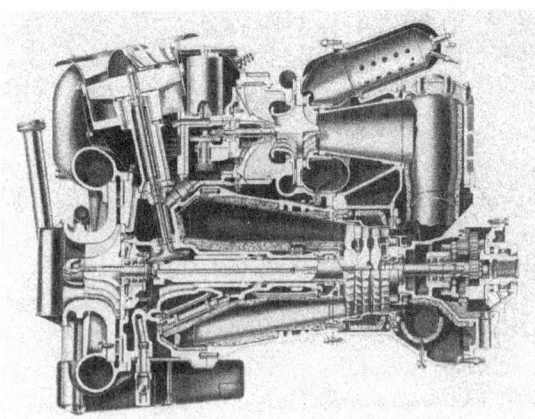

Fig. 6 — Final design of engine

Preliminary layouts for an engine with a 300bhp rating indicated that a symmetrical arrangement, using two intercoolers and two recuperators with the compressor spools located in the same vertical plane, was the most compact layout. The design also appeared to offer the greatest advantage from a mechanical and structural standpoint. Detail design followed the selection of this concept and resulted in the final engine. The final design package achieved the goal of fitting under the hood of a Ford passenger car - in fact, there was a little room to spare.

As could be expected, development of a bearing configuration for the high-pressure spool of the engine was a lengthy procedure. A test rig simulating the mass and inertia of the turbine wheel and driven by an air turbine was built for bearing development. In all, 91 bearing configurations were tested, and many thousands of hours of test rig operation accumulated. The problems associated with an overhung mass of this type are not entirely bearing development problems. A compromise in the damping characteristics of the bearing and the surrounding mount had to be made before a successful long-life bearing configuration could be developed. In the early period, development followed three separate paths: The cut-and-try approach, the more sophisticated approach of analysing oil whip phenomena and calculating the damping characteristics of various bearing configurations, and the computer approach.

For its second generation 'supercharged' engine, Ford engineers introduced a second-stage centrifugal compressor to raise pressure levels across the entire engine.

## Next generation

In the 1960s, anticipating completion of the 41,000-mile interstate superhighway network by 1972, Ford engineers devised a potential commercial vehicle able to take maximum advantage of this highway system.

Maximum utilization of such a network required a vehicle capable of hauling heavy payloads at maximum permissible speed which, in turn, dictated high horsepower requirements. And for this the engineers established their 'second generation' engine of 600bhp (560bhp of which had to be delivered to the drive-train) to meet vehicle performance objectives.

## Automotive gas turbines

A survey of existing and future powerplants for such service suggested nothing was available or even contemplated in the immediate future by major vehicle engine makers. Ford also concluded at the time that power levels of that magnitude for Big Red (below) would probably exclude gasoline and diesel piston engines because of packaging and weight.

A major factor in engine selection would be fuel consumption as such a vehicle must operate for relatively long, non-stop periods. Even lighter weight power plants, such as conventional gas turbines then available, would not be suitable due to their high specific air consumption, large specific volume, and poor part-load fuel economy.

Coincidently, at the time Ford received aid from the US government. It was developing jointly with the Military an engine to meet the general requirements for its 'super' vehicle.

This engine later became the Ford 705 supercharged gas turbine. And, according to Gerhard Peitscht and Ivan Swatman in their SAE Technical Paper published in 1965, it could deliver the necessary power in an acceptable package and weight suitable for mounting in the vehicle chassis. The engine was designed to compete with the then current diesel engines, both in fuel consumption and production cost.

To achieve the minimum specific fuel consumption objective of 0.4 lb/bhp.h, at approximately 50 per cent load required by the specification, a moderate increase in the 704 engine component performance had to be achieved. Typical increases in component efficiencies sought were: high pressure turbine – 0.86 to 0.87; recuperator – 0.74 to 0.8 and intercooler – 0.64 to 0.7.

According to Ford, 'the engine followed the historical trend, which is a steady but admittedly slow process of change from reciprocating to rotary motion in all modern-day stationary as well as transport equipment, namely pumps, compressors, blowers, and prime movers.'

That Ford should receive aid for its 'new' 600bhp gas turbine engine stemmed from the early 1960s, when agencies within the US Bureau of Ships and the US Army recognized the need for a 600bhp gas turbine engine. The requirements of each of these government agencies were gradually gelled into a basic specification, which was issued to industry for comment in August 1960, and, finally consolidated in a firm specification and request to bid in September 1960.

In addition to fulfilling the individual packaging and application requirements for a wide variety of Army and Navy applications with one engine, gas turbine engineers were expected to design an engine with a fuel consumption characteristic equivalent to that of a diesel engine and with comparable overhaul life characteristics and all for a cost of $12,000/engine, based on 1,000 engines a year production rate.

Coincident with the issue of the Bureau of Ships' specification, Ford was actively engaged in the development of its own, much smaller 300bhp gas turbine engine – the 704. This engine effectively became the prototype of the Ford supercharged gas turbine engine concept and provided information to support early engineering decisions in the selection of the engine cycle.

From this, two items were significant: the engine should have the fuel consumption characteristic of a piston engine, wherein minimum s.f.c. is achieved at approximately 50 per cent load was attainable with a gas turbine engine. Secondly, the greatest area of risk, which had been recognized when starting work on the supercharged-type engine, was controllability. However, on this score, using digital computer programming procedures and an analog model of the engine, the off-design, part-load, and transient characteristics of every component within the engine cycle were accurately defined.

Armed with this information in combination with actual engine hardware, Ford engineers made the control breakthrough, achieving successful steady-state and transient operation suitable for vehicle installation of an engine of this cycle. From these, together with data from the 704 engine development, Ford engineers reckoned a 600bhp version of the 704 concept would be the best design to follow to meet the Bureau of Ships' specification.

In early 1957, however, the 704 engine cycle concept and design had already been pressing the limits of the state-of-the-art gas turbines. By 1960 a number of design deficiencies and, in some cases, errors that could not be foreseen in a 'paper' engine were apparent. Therefore, in submitting a proposal to the Bureau of Ships, it was decided that the engine would follow the basic packaging arrangement of the 704 engine, incorporating proved 704 mechanical design features as well as all of the desirable design improvements that the 704 development had determined as essential to this type of engine.

To achieve the minimum specific fuel consumption objective of 0.4 lb/bhp.h at 50 per cent load required a moderate increase in the 704 engine component performance.

Although Military participation in the programme ceased in August 1963, it was the opinion of Ford that the original design and development programme was sound; especially the component rig test phase which was the major factor in the high degree of success of the initial engine build and test. A decision was therefore made that Ford would continue to fund, although on a reduced effort basis, engine development and component rig operation in order to provide a power plant for the super highway vehicle program which was already underway in Vehicle Concepts. In the flow path, air enters the inlet silencer which, in the case of the superhighway truck, is of a plenum-chamber type, serving both the low-pressure compressor and intercooler fan. Air is compressed in the low-pressure compressor to 4atm and then ducted through a stainless steel 'Zn-flow intercooler', where it enters the inlet to the high-pressure compressor. The intercooler cooling air is delivered by an axial fan driven by a bevel gear arrangement from the low-pressure spool shaft. The compressed air from the low-pressure compressor then enters the second stage or high-pressure compressor, where it is compressed to a final pressure ratio of 16:1. From the high-pressure compressor, the air runs in manifolds to the recuperators; after leaving the recuperators, the air enters the primary combustor where fuel is added to raise the temperature to 1750°F. The gas is expanded through a radial inflow turbine, driving the high-pressure compressor and the combined accessory and power take-off gearbox. From the radial in-flow high-pressure turbine, air enters the secondary combustor where fuel is added and the temperature again raised to 1750°F prior to expanding through the axial power turbine. This single-stage axial turbine is coupled to a planetary reduction gear, providing an output speed of 3,080rev/min into the vehicle transmission. The gas then passes through a two-stage, axial, low-pressure turbine which drives the low pressure, supercharging compressor and intercooler fan. Exhaust gases are then collected in a diffuser and directed through the gas side of the recuperator before being ducted through the exhaust silencer and out to atmosphere.

### Third generation

In 1968, Ford exhibited a gas turbine truck at the Earls Court, London, commercial vehicle exhibition alongside a gas turbine truck from Leyland. The Leyland truck was a joint development of Leyland and Leyland Gas Turbines. Leyland said at the time that it expected to go into limited production "late in 1970" and had plans for an annual output of 2,000 to 3,000. Initially, it was also said, the turbines would be limited to truck applications

## Automotive gas turbines

although long-distance coaches may prove a practical proposition.

The Ford gas turbine truck, then undergoing tests in the US, was also the product of a long development exercise of some 16 years, it was stated. That would take the start to 1952.

The engine, the 707, in the Ford truck was also rated at 375bhp like the Leyland truck but unlike Leyland, Ford said it anticipated the marketing of turbine trucks to begin in the US "in the early 1970s." If it had plans to market such trucks in the UK it kept those plans to itself.

The following year Ivan Swatman was in Britain from Detroit with his experimental gas turbine truck (see over), allowing UK journalists to drive the Ford 707 truck with its 375bhp gas turbine. It had also been brought over with the intention of undergoing demonstrations elsewhere in Europe.

It was in 1966, that Ford engineers fitted a production W-1000 6x4 tractor with a 375 bhp 707 gas turbine with the object of demonstrating the vehicle in North America and Europe – the UK and Germany. The double-drive tractor with its trailer weighed 12.5 tons – one fifth of the maximum train weight at which the tractor had been tested in the US. A realistic UK assessment of the vehicle fully laden was therefore out of the question. The low weight, if anything, offered journalists no more than an optimistic impression of the vehicle.

Commenting on the truck at the time, Swatman did not pretend that mass production of automotive gas turbines was nowhere near and it would be "several more years before all the problems were overcome".

In order to achieve comparable fuel consumption with that of the diesel engine, the heat output from the power turbine had to be harnessed as before to raise the temperature of the incoming air by means of a heat exchanger.

And it was in the development of the heat exchanger, or regenerator that Ford and Leyland had to break new ground. For Ford it meant doing away with the previous metal matrix

## Automotive gas turbines

recuperators and substitute then with Corning Glass ceramic matrices. Leyland used the same components. General Motors used a stainless steel regenerator which could distort far more than the ceramic units. But Ford had to find a way of joining ceramic discs to metal shafts – a technique it decided not to reveal at the time, much to journalists' annoyance at the time.

But more was to come. In July 1970 Ford announced a new automotive gas turbine in the 225bhp to 335bhp range featuring a new, lighter-weight design with fuel economy and durability to match present turbines. The company also stated that from August 1971 it would start production of gas turbines for industrial and marine duties in a new factory in Toledo, Ohio.

However, Ford closed the Toledo facility in 1973 after continuing issues of turbine overheating and a devastating flood that shuttered a single-source supplier's only plant. Ford got out of the gas turbines just as General Motors and Mack Trucks were beginning to use new ceramic materials to operate second generation gas turbines.

Developed from the 707, the new engine offered no radical change in design apart from detailed differences. The company said it had plans to produce 200 gas turbines for industrial and marine use from August 1971 to the end of the year with a steady increasing output thereafter.

It said most of the components would be machined in the factory in Ohio using the latest CNC processes. Ford also expected the workforce to reach 260 by the end of 1971 and 700 by 1975.

It also claimed the large and small gas turbines (the 707 engine was rated between 335bhp and 450bhp for a weight of 1,700lb) would be competitive with comparable diesel engines in price.

But, as already mentioned, those long-term plans did not come to fruition.

## 10 Garrett AiResearch/Ford

FORD Motor Company embarked on two gas turbine engine programmes: one with Garrett AiResearch for passenger cars and the other for heavy trucks. This latter programme it managed in-house. In the case of the former, Ford elected to act as a sub-contractor to Garrett through a teaming agreement for the AGT or Advanced Gas Turbine programme.

The two companies claimed they were 'committed to the accomplishment of the NASA/DOE AGT programme, with the team providing an 'outstanding blend of high technology aerospace gas turbine development experience with proven automotive production capability'.

Garrett claimed to be the foremost manufacturer of small gas turbine engines whereas Ford could post itself as the world's second largest producer of automobiles and 'has conducted in-depth evaluations of vehicular gas turbines'.

And, in a joint brochure (produced by Garrett Corporation) dated April 1979, Ford claimed it offered 'dedicated research and development in vehicular gas turbines since 1956' having worked with ARPA on a Ford Model 820 advanced ceramic gas turbine with high temperature ceramics in the hot gas flow path.

Garrett on the other hand could point to 'over 40,000 gas turbines manufactured' and 'over 30 different production models', as well as its own Garrett GT601 gas turbine truck engine – see Appendix 10 – and the ARPA/Navair/Garrett ceramic gas turbine demonstration programme.

## Automotive gas turbines

The AGT engine was indeed advanced, using a large number of complex ceramic parts, including a rotary triangular matrix ceramic heat exchanger.

As a precursor to the main AGT programme that began in October 1979, the two companies could point to a preliminary programme involving an 'Improved gas Turbine' or IGT with Phases 1 and 1A. Phase 1A looked at component definition, most notably a compressor, a ceramic turbine, ceramic combustion chamber and ceramic transition liners.

The AGT programme proper began in Fiscal Year 1980 and was due to run until the end of Fiscal Year 1985 when vehicle test would be completed. Half-way through the programme, in Fiscal Year 1983, the schedule showed the teams would be engaged in the most challenging aspect of the work, namely the development of the ceramic turbine, having already started development of the ceramic combustor and other ceramic structures.

At one point the two companies noted: "Garrett/Ford Advanced Gas Turbine (AGT) Technology Project, authorized under NASA Contract DEN3-167, is sponsored by and is part of the US Department of Energy Gas Turbine Highway Vehicle System Program.

'Program effort is oriented at providing the US automotive industry to high-risk long-range technology necessary to produce gas turbine powertrains for automobiles that will have reduced fuel consumption and reduced environmental impact. The AGT101 power section is a 100bhp, regenerated single-shaft gas turbine engine operating at a maximum turbine inlet temperature 2500°F. Maximum rotor speed is 100,000rev/min.'

'All high temperature components, including the turbine rotor, are ceramic.'

"Development has progressed through aero-thermodynamic testing of all components with compressor and turbine performance goals achieved. Some 170 hours of AGT101 (1600°F) testing has been accumulated on three metal engines.

# Automotive gas turbines

'Individual and collective ceramic components screening tests to 2100°F have been successfully accomplished. In addition, ceramic turbine rotors have been successfully cold spun to the required 115,000rev/min proof speed (15 per cent over-speed) and subjected to dynamic thermal shock tests simulating engine conditions.'

'Assembly of the engine with ceramic structures is underway. Engine testing of the ceramic structures and the ceramic turbine rotor is planned in the near future.'

Of all the automotive gas turbine developments this AGT 101 concept was perhaps the most challenging as it contained so many ceramic components that amounted to 'first timers' in any automotive application.

With the benefit of hindsight it is amazing that anyone at the time really believed that a largely ceramic engine – a crockery engine – was actually feasible using the production technology and the cost structure in place at the time.

But that did not stop the teams in their work – work which had been split into three Mods: Mod 1 engine first build with a ceramic heat exchanger, Mod 1 with ceramic combustor and turbine, and Mod 11 for engine final build.

A glimpse at the mountain teams had set themselves to climb can be found in the third stage or Mod 11 in which turbine inlet temperatures of 2,500°F are happily discussed. This equates to a temperature of 1,370°C.

The slip cast radial turbine, quite discounting the temperatures at which it was expected to operate, was equally mind-blowing as engineers evaluated silicon nitride and silicon carbide materials.

Engineers noted, perhaps ruefully, that 'successful demonstration of the ceramic turbine rotor materials development is considered key to the program goal assessment'.

Even today, in 2014, no less than 30 years after those words were written by no doubt over-optimistic engineers and marketing people, the ceramic turbine in gas turbines is as elusive as ever.

The AGT101 brochure (produced in at least three versions dated: 4/79, 1/80 and 3/80 with a black outside front and back cover – suggesting doom?) makes no mention of the materials selected for the ceramic combustor, nor are any details given for the ceramic regenerator which in illustrations is shown as a solid disc, no doubt of Corning Glass origin as this was the only company at the time making such complex components.

It does however mention that Garrett/Ford development experience with 'previous ceramic engine programmes will also be factored into the design' for ceramic structures, including 'reaction bonded silicon nitride, sintered silicon carbide and reaction bonded silicon carbide'. A photograph in the brochure showed 'injection moulded reaction bonded silicon nitride axial flow stators' or nozzle guide vanes.

In 1987, a final report ensued. It stated: 'This report is the final in a series of Technical Summary Reports for the Advanced Gas Turbine (AGT) Technology Development Project, authorized under NASA Contract DEN3-167 and sponsored by the DOE. The project was administered by NASA-Lewis Research Center of Cleveland, Ohio. Plans and progress are summarized for the period October 1979 through June 1987. This program aims to provide the US automotive industry the high risk, long range technology necessary to produce gas turbine engines for automobiles that will reduce fuel consumption and reduce environmental impact. The intent is that this technology will reach the marketplace by the 1990s. The Garrett/Ford automotive AGT was designated AGT101. The AGT101 is a 74.5kW (100shp) engine, capable of speeds to 100,000rev/min, and operates at turbine inlet temperatures to 1370°C (2500°F) with a specific fuel consumption level of 0.18 kg/kW.h (0.3lb/bhp.h) over most of the operating range. This final report summarizes the powertrain design, power section development and component/ceramic technology development.'

The report noted however that performance of the regenerator cores supplied by Corning Glass and NGK-Locke was hampered by poor performance of the seals which suffered due to distortion of the engine at high temperatures. Further development of these seals was required before goals could be met.

As with other automotive gas turbine programmes, AGT101 came to nothing in terms of high-volume production units, despite the expertise of Garrett's AiResearch Manufacturing Company of Arizona and Ford Motor Company.

## 11 Garrett AiResearch

ONE LITTLE-KNOWN truck gas turbine programme was that conducted by the Garrett Corporation's AiResearch Manufacturing Company of Arizona on behalf of a consortium of companies known as ITI – Industrial Turbines International. The three companies – Garrett Corporation of Phoenix, KHD of Germany and Mack Trucks of the US – had a programme to develop the GT601. KHD is the title of diesel engine maker Klockner Humboldt Deutz.

The go-ahead was given in 1972 with the manufacture of three development engines set for 1977 to be followed by five development engines in 1979. Ongoing development testing would start in 1982 and continue until at least 1986. At the time the GT601 brochure was written the company claimed "seven engines had been built with two installed in trucks now in operation". No date was given by the companies for the start of production of volume engines.

The brochure for the GT601 contained scanty details of the programme. The engine was rated at 550bhp or 410kW with an output drive of 2,600rev/min. The design weight was given as 988kg or 2,175lb and performance curves suggested that a best fuel consumption of 0.35lb/bhp.hr could be achieved between around 350bhp and 450bhp. The engine was intended for use in an 80,000lb gross vehicle weight truck.

Beyond that the company would say little more except: "High horsepower gas turbine meets both social and economic needs by maintaining high average speed with no fuel economy penalty and without violating existing highway statutes."

How the company could make such pronouncements without fully developing the engine is only something which marketing people will understand.

## Cummins Engine Company

SEEMINGLY predating this, on 29 December 1969, the *TriCity Herald*, in a business item noted: 'COLBUMBUS, Ind. (AP) Cummins Engine Co., a major manufacturer of diesel truck engines, Monday announced an agreement with Garrett Corp., manufacturer of gas turbine engines for the aerospace industry, of a joint program to explore the possibilities of gas turbine application to highway trucks and construction equipment. The announcement said that if a turbine truck engine should be developed in the program, Garrett would manufacture the turbine power sections. Cummins would build the rest of the engine and assembly, sell and service it.'

In other words, Cummins had noted efforts by Ford, Garrett and GM at the time to develop truck gas turbines that might threaten Cummins' future viability, and the Indiana company would dip a toe in the water to keep an eye on things.

In Britain, *Commercial Motor*, in its issue of 3 January, 1969, just days after the US announcement, reported: 'Cummins Engine Company and the Garrett Corporation, one of the Signal Companies, have signed an agreement for a joint programme to explore the feasibility of gas turbine applications to highway trucks and construction equipment.'

'The companies say that recent advances in turbine technology, coupled with the increased speed and weight predicted for future vehicles, along with requirements for less air pollution and reduced noise, may permit the gas turbine to take its place alongside the diesel and petrol (reciprocating engine) engine as an economical power plant for future trucks and construction equipment. For this project, the two companies will utilize their separate existing skills and technologies to develop commercially-successful gas turbine units for these applications.'

Cummins' president, E. Don Tall, and Garrett president, Harry Wetzel, noted "The two companies jointly have a unique competence to determine both the technical and the economic feasibility of the gas turbine for heavy-duty truck applications."

There appears to be no public record the joint undertaking went any further. But in 2001, Cummins did sign a three-year agreement with Capstone Turbine Corporation of Chetsworth, California, to develop and market Capstone MicroTurbine-driven stationary power systems of Cummins' design.

# 12 General Motors

AUTOMAKERS Chrysler and Ford were attracted to the gas turbine's long life, lower maintenance costs, low weight and smooth vibration-free running. It was hoped that development would get over the barriers of high cost and relatively high fuel consumption. General Motors (GM), one of the 'Big Three', was no exception.

Some of GM's most dramatic show cars, the Firebird cars of the 1950s Autoramas, had gas turbine engines and apparently were not just static show cars – www.thetruthaboutcars.com

GM began its research on turbines in the 1930s and its Allison division was building aircraft jet engines by the 1950s. The company put Emmett Conklin in charge of the project to build the powerplants for three of Harley Earl's most famous (and outrageous) show cars. Firebird I was a single-seater weighing 2,500lb; it had a 370bhp gas turbine driving the rear wheels. GM dubbed its in-house gas turbine the 'Whirlfire Turbo Power' engine. It had a two-speed transmission and the exhaust gas, at 1,250°F, could melt any plastic and blister paint that happened to be behind it in traffic. In addition to air cooled brake drums, air flaps were used to slow the car at speed.

For the 1956 Autorama, GM introduced a slightly more practical Firebird II. It was a family vehicle with four seats. Earl had one of the two Firebird II bodies fabricated out of unpainted titanium, which must have been outrageously expensive. Engine output was down to 200bhp but exhaust heat was less of a problem as a regenerative heat exchanger allowed the engine to run 1,000°F cooler and the engine could power accessories.

The 1959's Firebird III was another two-seater with the engine's name changed to the 'Whirlfire GT-305'. With 225bhp (168 kW) output, engineers augmented the gas turbine's output with a small 10bhp gasoline engine to run accessories.

According to an internet forum on GM *Inside News*, William Turunen headed GM's automotive turbine programme for most of its life. Albert Bell III's name also frequently appeared in accounts of GM turbine cars. Bell went to his grave believing the turbine could still be a practical automotive engine.

Two years after I started working on *The Engineer*, GM's research laboratories revealed details of its GT-309 at a Society of Automotive Engineers (SAE) meeting in Cleveland, Ohio on 18-21 October 1965. The GT-309 was the latest in the series of engines developed by GM that began with the GT-300, a non-regenerative gas turbine that powered the first Turbo-Cruiser. Both the GT-304 and GT-305 (below) used drum regenerators pioneered by GM. GM engineers argued that the heavy vehicle market, then dominated by the diesel engine, offered the greatest potential for gas turbine applications.

Engineers reckoned the best potential usage appeared in the 60,000 to 76,000lb class. Here, in that weight class they opted for a horsepower level range of 260 to 280bhp. This would provide maximum speeds on the level of 60 to 65mile/h, or 27 to 35mile/h on a three per cent incline.

The design specification was based on a 280bhp on an 80°F day. The chosen design, a two-shaft engine with a drum regenerator, returned a weight of 950lb within an overall length of 36in. The rated gasifier speed was 35,700rev/min with power turbine speed set at 30,000rev/min. GM engineers achieved a specific fuel consumption of 0.45lb/bhp.h at 80°F. But if the

# Automotive gas turbines

ambient temperature fell to 60°F, then power output could rise to 320bhp and the specific fuel consumption drop to 0.425 lb/bhp.h. For their design of drum regenerator (see below) GM engineers claimed an effectiveness at full power of 91 per cent, and a total pressure drop of less than 5 per cent.

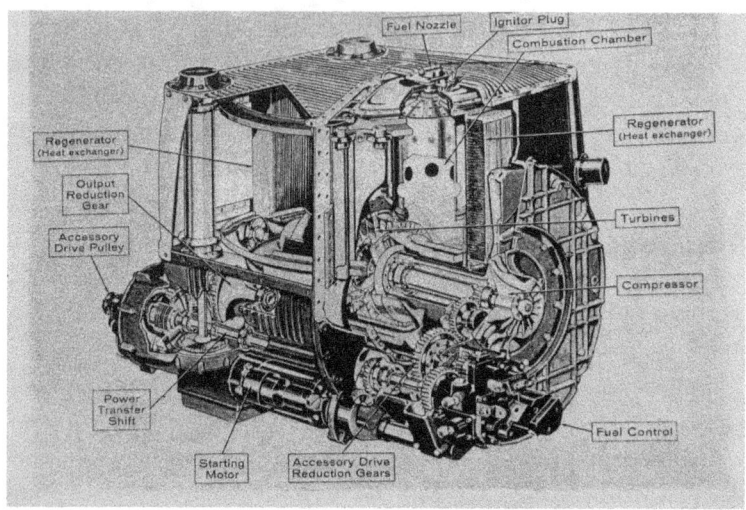

GM engineers installed their gas turbine in a Chevrolet Turbo Titan three-axle tractor. GM also produced the Bison - a low-slung turbine-powered concept tractor and trailer combination (below).

## Automotive gas turbines

It seems, according to some reports, that Albert Bell III was involved when GM recycled the turbine as a response to the oil shortages of the 1970s. Although gas turbines could be thirsty, they could also run on a variety of flammable fuels. GM even considered coal dust, which in the US could be cheaper per Btu than almost any other fuel, hence GM's interest in a coal burning engine in the late 1970s.

As a test mule, GM used the last full-sized Eldorado, the 1978 model. The huge Eldorado's cavernous engine compartment, designed to take a 500 cubic inch engine block with ease, had more than enough room for whatever machinery the prototype needed.

And the prototype indeed needed special machinery, according to retired GM engineer John Schult. He described the system to the *New York Times* in 2009:

"To keep the coal dust ready for delivery to the engine, it had to be continuously agitated. Then a small conveyor belt delivered the coal to the gasifier, the first section of GM's automotive turbine engine. When you stepped on the gas pedal, it actually moved a potentiometer that varied the speed of the coal conveyor belt. More fuel resulted in more power."

According to Schult, the fuel delivery system added an additional delay to the lag that automotive turbines already experienced. Schult said the car did accelerate fairly well due to turbine's high torque at low engine speeds. The experimental Eldorado's turbine, which turned over at a non-automotive 35,000rev/min, was geared down and that power was fed to a standard three-speed transmission.

The car used diesel/kerosene from a small tank in the boot to start the turbine. Once fired, it automatically changed over to the coal dust. Schult said that with the turbine whine and buzz of the coal dust agitator, plus compressed air blowing the dust into the gasifier, it did not sound much like a conventional Detroit V8 engine.

## Automotive gas turbines

GM made only a single prototype. Apparently, refuelling proved as messy as changing toner in a photocopier machine, with greasy coal dust everywhere.

However, that was not the end of GM's dream of gas turbine power for cars. In the April 1987 issue of *Popular Science*, (I knew well the UK editor of '*Pop Science* David Scott - Scotty) an article by Dan McCosh described the 'Chevy Express'. It was a 150mile/h gas turbine powered concept that from its styling appeared to have been part of GM's EV1 program.

The Express concept (below), which appeared in one of the *Back To The Future* films, was powered by an ACT-5 gas turbine with a regenerator, developed by a team headed by Albert Bell.

Some sources suggest GM continued its work on automotive gas turbines into the 1990s, when the programme was moved to GM's Allison division, which carried out the work on the prototype coal turbine. GM's automotive engineers shifted emphasis to fuel cells and other projects.

As part of the EV1 program, at the 1998 North American International Auto Show in Detroit GM revealed a number of alternative EV1 drivetrain concepts.

These included the EV1 Series Hybrid Concept. In range extended mode, the EV1 Series Hybrid used an auxiliary power unit powered by a small gas turbine developed with Williams International which, in one of those ironies of Detroit, was founded by Sam B. Williams.

Sam Williams who as a young engineer worked at Chrysler on its turbine project and would become famous in his own right, as Noel Penny found out.

# 13 George J. Huebner, Jr.

GEORGE J. HUEBNER Jr. is best known as the 'father of the automotive gas turbine engine'. The following is a speech he gave when he was director of research at Chrysler Corporation during a meeting with representatives of press, radio and television on 12 April 1966. George Huebner died 30 years later on 4 October 1996 in Ann Arbor, Michigan. He used the technical results of the 50-car programme as his subject.

Huebner received his B.S. in Engineering from the University of Michigan in 1932. He joined Chrysler Corporation in 1931 as a laboratory engineer in the mechanical laboratories of the engineering division. He continued his career with the company through to retirement in 1975. He served as chief engineer, researcher, in executive engineering, in the missile branch of engineering, and then as director of research for Chrysler until his retirement. He was president of the Chrysler Institute of Engineering from 1960 until his retirement.

Outside of his work at Chrysler, Huebner served as chair of the board for the Environmental Research Institute of Michigan and as president, fellow and member of the board of directors of the Society of Automotive Engineers.

Work on the Chrysler gas turbine engine began under his direction in 1945. He also organized a complete missile facility at Chrysler, including research, engineering and production. Author of numerous papers on automotive power plants, vehicles and components, he held more than 40 patents in the gas turbine automotive fields.

# Automotive gas turbines

"We in Detroit's automobile industry may feel that the automobile is the most important sociological and economic factor in our lives today. But, since the automobile is also a business for us, we are accustomed to appraising it from quite an objective standpoint. So, when we find that people in other parts of the country magnify our enthusiasm for something new, we are sometimes surprised. The story of the Chrysler automotive gas turbine (see Chrysler engine below) embodied in our consumer-evaluation 50-car program has been, to many industry people, one of those sunrises.

The vehicle itself did not represent a drastic departure from the normal configuration of an American automobile. It was a small, luxury car, equipped with power steering, power brakes, power windows and other customer conveniences. Standard instruments were to be found on the panel, with the addition of engine speed and temperature indicators.

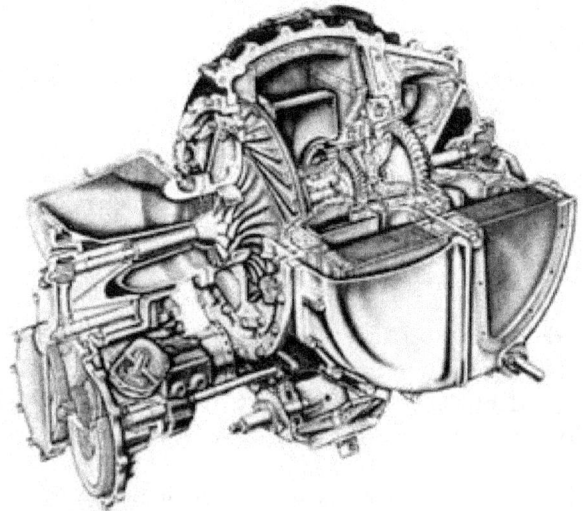

Most of you here today are very well aware of the Chrysler turbine research and development work which brought us to the 50-car consumer evaluation program. So, I will confine my remarks only to what we learned during this two-year period.

The turbine engine for the 50 cars was not considered to be a final production design. Most of the manufacturing techniques used for this limited quantity were necessarily those of the tool room and not the production plant. In many cases, this required a much different approach to the design of the parts than would be used for engines produced in larger quantity by highly automated engine plants of today. But in its basic concept, we felt that the power plant in the 50 cars appeared to have the potentiality of becoming an engine which could be manufactured in mass production volume.

During the course of the programme, we had our first opportunity to observe and judge the behavior of turbine engines under actual customer driving conditions. Thousands of hours of engine testing in laboratory test cells and many tens of thousands of miles of driving at our proving grounds and on the highways had not only given us the basis for rapid and continuous power plant development but had also proved reliability and endurance. However, the best 'controlled testing' in the world cannot completely replace the great experience accumulated through the usage and conditions of daily operation when the vehicle becomes an everyday tool of transportation.

For over two years our turbine cars were driven in cities and on highways, in deserts and in the mountains as well as even below sea level.

The wealth of information derived from this experiment is invaluable. Of principal interest to us were the life of parts and components of the 50-car engines, their performance, their reliability, the degree and nature of maintenance required, and the amount of training desirable for service people. The opportunity to test even more advanced turbine engine concepts by unexpected methods of operation also were revealed, indicating once again the value of use by non-experts.

Basically, all factors related to component life have been gratifying. We have had failures and disappointments, of course, but nothing we could even remotely consider unexplainable.

# Automotive gas turbines

The data available so far permit us to state that most parts have an endurance potential of over 50,000 miles. That is, as they were in the 50 cars, before the improvements the programme has taught us we could make. Furthermore, although only one car achieved 50,000 miles in the two-year test, our inspections of high-mileage cars indicate that most of those will enjoy far longer life than 50,000 miles and pass the 100,000-mile mark generally considered acceptable for passenger cars.

However, there were a few parts that gave us 'fits' because we could not readily duplicate field deterioration in the laboratory and consequently could not immediately pinpoint and solve the problem.

Regular inspections indicated that some engines had been subjected to temperatures very much higher than those normally allowed by the fuel control. Yet, a check of that component revealed no deficiency. It was finally noticed, however, that some drivers would initiate the automatic starting cycle with the ignition key and then very quickly shift the gear selector from start position before the engine had reached idle speed, thus by-passing the automatic start system. In a piston engine this is roughly analogous to over-choking, resulting in scuffed pistons, piston rings, and cylinder bores. In a gas turbine engine, the process is different, but the damage is still there, and we end up with scored regenerators and burned turbine blades. Once discovered, the trouble was an easy matter to cure, simply by modifying the automatic start system so that the driver could not over-ride and thus misuse it.

The most serious problems concerned the electrical system. Our use of a combined-function starter-generator caused a severe operational deficiency. Although high altitude performance testing to 13,000 feet had been carried out in the mountains near Denver, Colorado, it was not initially discovered that a combination of high altitude and low humidity caused rapid and catastrophic destruction of the starter-generator brushes of the otherwise excellent unit. Subsequent investigation indicated that the addition of barium salts to the graphite brush compound would reduce or eliminate the high

# Automotive gas turbines

altitude brush wear, but with this change a fundamental problem remained. Under cold-starting conditions the brushes were required to carry high current, thus requiring a soft, low electrical resistance short-mileage brush, whereas under generating conditions at commutator rotational speeds up to 20,000rev/min, a hard long-wearing brush with high resistance, was desirable. These mutually exclusive requirements were finally compromised in a brush which, under test, showed a life expectancy of less than 25,000 miles -- not considered satisfactory. This problem was corrected during the programme but convinced us that automotive turbines should be equipped with separate starters and separate alternators.

Early ignitors showed distress at the thirty-day inspection period required for all of the cars. This distress was indicated both by the appearance of rapid electrical erosion and by severe oxidation of the electrodes, despite the fact that the electrodes are supplied with cooling air discharged as excess from the fuel nozzle air pump. Since the ignitor must be inserted through the combustor sleeve in such a way that it is continuously exposed to some flame impingement and to radiation from the hottest part of the flame, modifications to the hollow electrodes and the means of discharge or the cooling air from them were incorporated. Test mileage on ignitors is now in excess of 20,000 miles. Although this may be satisfactory for a piston engine, it is not considered satisfactory for an automotive gas turbine and redesigned ignitors now under test will hopefully more than double this life.

On the plus side of the 50-car programme has been the overall performance of the experimental turbine engine. Fuel mileage on the 50-car test, although reasonable, was not quite as good as that of a comparable piston-powered car. However, consideration must be given to the mode of operation which for most users included an abnormal number of starts and a high proportion of stop-go driving while demonstrating the car to interested and curious people. We believe that our own tests of fuel mileage, which indicate fuel consumption comparable to piston powered cars, are more realistic.

Cold starting ability, claimed to be one of the greatest qualities of the turbine, proved to be just that. And of course, everyone appreciated getting instant heat in the passenger compartment on a cold winter day, or not having to think of anti-freeze. Power output of each individual engine as built remained consistently close to its original value, and even normal deterioration of power with usage ceased to bother us after the discovery of a highly efficient engine cleaner. In a piston engine, deterioration is corrected by a tune-up which, although not especially difficult, is costly and time consuming. In our turbine the lost power is recovered almost instantly by using a harmless compound which is simply introduced into the engine intake. It then removes accumulated deposits while on its way to the exhaust.

Chrysler Research was successful in developing families of low-cost, low-alloy content materials which are highly satisfactory for the purposes intended. A six percent aluminum-iron alloy, which we refer to as CRM-4 (Chrysler Research Material - Number 4), was used for much of the internal sheet metal in the engine. The strength required of most of these parts is not severe, and the hot strength of CRM-4 makes it a satisfactory low-cost substitute for the far more expensive CrNi stainless steels used in aircraft.

The most critical phase of the research program has been the development of both materials and fabrication techniques for turbine wheels and nozzles. The compressor turbine wheel is subjected to a metal temperature or 1500°F under full power. Although in passenger car applications this temperature is maintained for less than 10 per cent of the total operating life of the engine, this nevertheless represents a great number of hours and is further aggravated by the acceleration temperatures of the engine, which under some conditions exceeds the full power gas temperature of 1700°F by as much as 135°F. This problem was solved in the compressor turbine wheels in the 50 cars by employing cast, CRM6D, a member of the family of high-strength, high-temperature, low-cost turbine wheel alloys developed by Chrysler Research. The patent granted for this alloy shows that the material is principally iron and that alloying elements used are readily available domestically.

All but three of the engines in the 50-car program were operated with compressor turbine wheels integrally cast from CRM6D, and operating experience with this material has been highly satisfactory. The other three turbine wheels were made from an expensive, aircraft type, high-temperature alloy. They were of the same general design and the Chrysler alloy wheels and although they did not fail, their operation was not completely satisfactory in this design because they caused other engine problems which were not present with the Chrysler material. In addition to these materials tests, the opportunity was taken to test progressive design modifications and to explore various turbine wheel fabrication techniques.

A new and different, but still low-cost, version of CRM-6D alloy has proved satisfactory for the first-stage nozzle, which is subjected to metal temperature in excess of 1800°F under acceleration conditions, and a further modification of 6D has been proved successful for the variable nozzle vanes.

An extremely beneficial aspect of the programme has been the experience gained in turbine engine maintenance and in the training of service personnel. The five field service men and two supervisors kept close track of the days during which the engine could not be operated due to malfunction. During the early weeks of the programme, operating days lost to users were a little over 4 per cent but during later periods this had been reduced to slightly over 1 per cent. It should be pointed out that those days lost included travel time for the service representative to reach the vehicle and in many cases also included the time required to ship a part from the service stores in Detroit. In addition, some portion of this lost time should be charged to the installation of advanced experimental parts for the development testing in the field. We believe it is rather remarkable that all the required servicing on 50 cars scattered the length and breadth of the United States, was performed essentially by five men.

We were pleased to see a substantial reduction in the number of down-days as the programme progressed, and quite satisfied with the low frequency of required maintenance on an experimental engine out on its own for the first time. But, naturally, we engineers are never completely satisfied, and will not be until maintenance requirements had become practically nil over a reasonable engine life span.

Although the actual service arrangements employed would not be applicable to a possible large volume of vehicles, the experience of the 50-car programme indicates that training of personnel in the maintenance and repair of gas turbines presents no unusual problems. The principal difficulty, if there is any, is to make the trainee forget some of his piston engine knowledge. Mechanically, the power plant is less complex than most piston engines, automatic transmissions, and other current automotive components, so that the average conscientious mechanic should have no trouble performing any maintenance or repair operation which would normally be done in the field.

To this point we have seen the value of the 1,100,000 miles accumulated during the 50-car program as a direct source of information on the behaviour of gas turbine engines and components going through their baptism of fire. This was of great interest to us, but would remain purely academic unless the lessons learned could be translated into improvements in performance, reliability, life and manufacturing methods now incorporated in improved engines currently operating.

Vehicle response and acceleration were surprisingly good during the programme, when it is considered that the engine was rated at only 130bhp and the car weighed about the same as a Chrysler Newport. The time required to accelerate from 0 to 60miles/hour was generally around 12s, based on an outside temperature of 85°F.

On cooler days, greater performance could be achieved. Since then, vehicle acceleration had been substantially increased by means of a faster-acting variable nozzle actuator. In other words, the nozzle blades snap into their acceleration position

# Automotive gas turbines

much faster than they used to – something like three times faster. This is not only an obvious gain in actual response time but it makes the driver 'feel' the sensation of a snappy forward motion. The same is true of the engine braking, though at the other end of the line. The nozzle blades switch to their braking position in much less time than before and the vehicle slows down more suddenly and over a shorter distance. Additional engine braking has also been obtained by making it possible for the variable nozzle blades to go a little further than before into braking position without an increase in temperature which could cause damage.

Gas generator response – without which, incidentally, there is no vehicle response – has also been improved as a result of operating experience. In our first automotive turbines, back in the 1954-1959 period, it took 7s for the gas generator to accelerate from idle speed to full power.

In the 1962 turbine engines this was reduced to 3s, and response time in the 50-car program engines was down further in the 1.5 to 2s zone. Extensive operation established very clearly that we could increase acceleration temperature without hurting the hot parts of the engine. This, with a reduction in the inertia of the gas generator rotor, resulted in lopping off another half second in the time it takes for the gas generator to reach full speed.

Our turbine cars, located all over the nation, were exposed to all ranges of starting temperatures. Some were very low and required use of a 24V battery system, which was a purely temporary field expedient. Since then, we have reduced the accessory load and the bearing losses in the gas generator to the point where dependable starting is consistently and rapidly accomplished with a 12V system.

The layman associates a certain quality of sound with a turbine as used in aircraft jet engines. This sound is not entirely a characteristic of the turbine but is principally caused, in aircraft jets, by the accessory drive gears. It may have served as a means of product identification, but the initial attraction must inevitably wear off, except perhaps for the 'jet set', and it becomes more reasonable to follow the classic line of noise reduction. This was accomplished, particularly at low speeds, by using different types of gears, reducing the speed at which accessories run at idle, and modifying the intake filter-silencers.

In conclusion, we at Chrysler would like to take this opportunity to express once again our deepest appreciation for the encouragement received from the press and the public."

In 1974, John Hartley, a freelance journalist working for *The Engineer*, met George Huebner who, under the auspices of the US Environmental Protection Agency, had started working on its gas turbine programme again.

The first part of the contract called for Chrysler to build a sixth-generation gas turbine to establish the state-of-the-art, and to serve as the baseline. Moreover, tests to show the engine could survive a 3,500 hour durability test had to be carried out and which simulated 100,000 miles of operation.

Hartley found in discussions that Huebner's then hope was that he could deliver to the EPA two uprated engines that would be 40 to 50 per cent more efficient than the baseline engine by the summer of 1975. The contract was planned to be complete by 1976 and by 1977 hoped that Chrysler would be working on really advanced and efficient gas turbine engines.

However, as already mentioned, although engineers built a specially-bodied turbine Chrysler LeBaron in 1977 as a prelude to a production run, Chrysler Corporation was by then in dire financial straits and in need of US government loan guarantees to avoid bankruptcy. A condition of that deal was that gas turbine mass production be abandoned because it was 'too risky' for an automotive company of the size of Chrysler.

This brought to an end the work of the 'father of the automotive gas turbine' and his team.

# 14 Leyland Gas Turbines

FOR THE RECORD, Donald Stokes, later Lord Stokes, previously sales director, was appointed managing director of Leyland Motors Limited in September 1962. Originally a Leyland Motors student apprentice he grew up with the company and eventually in 1966 became chairman.

In 1968, Leyland Motor Corporation Limited merged with British Motor Holdings (BMH) to form the British Leyland Motor Corporation (BLMC).

BMH brought with it into the new organisation some more famous British commercial vehicle and bus and coach marques, including Daimler, Guy, BMC, Austin and Morris.

However, the BLMC group proved a difficult animal to manage because of the many companies under its control, often making similar products. This, and other reasons, led to financial difficulties and in December 1974 British Leyland had to receive a guarantee from the British government.

In 1975, after the publication of Lord Ryder's Ryder Report, BLMC was nationalised as British Leyland (BL) and split into four divisions with the bus and truck production becoming the Leyland Truck & Bus division within the Land Rover Leyland Group. This division was split into Leyland Bus and Leyland Trucks in 1981. In 1986, BL changed its name to Rover Group. The equity stake in *Ashok Leyland* was controlled by Land Rover Leyland International Holdings, and sold in 1987.

There was positive news for BL at the end of 1976 when its new Rover SD1 executive car was voted European Car of the Year, having gained plaudits for its innovative design. The SD1 was the first visible step that British Leyland took towards rationalising its passenger car ranges, as the SD1 was a single car replacing two cars competing in the same sector: the Rover P6 (Rover 2000) and Triumph 2000.

More positive news for the company came at the end of 1976 with the approval by Industry Minister Eric Varley of a £140 million investment of public money in refitting the Longbridge plant, Birmingham, for production of the

## Automotive gas turbines

company's ADO88 (Mini replacement) model, due for launch in 1979. However, the UK success of the Ford Fiesta, launched in 1976, redefined the small car class and ADO88 would soon be cancelled. Massive investment in the Longbridge plant would nevertheless take place in preparation for the introduction of the slightly larger LC8 subcompact hatchback, which would be launched as the Austin Mini Metro.

In 1977 Sir Michael Edwardes was appointed chief executive by the National Enterprise Board (NEB) and Leyland Cars was split up into Austin Morris (the volume car business) and Jaguar Rover Triumph (JRT) (the specialist or upmarket division). Austin Morris included MG.

Land Rover and Range Rover were later separated from JRT to form the Land Rover Group. JRT later split up into Rover-Triumph and Jaguar Car Holdings (which included Daimler).

From the gas turbine standpoint, in March 1967 the Rover team engaged in gas turbine work was regrouped under the company Leyland Gas Turbines Ltd. This led to a larger unit in the 260kW to 300kW range for commercial vehicles, to be known as the 2S/350R. The company quoted a minimum specific fuel consumption of 0.2373kg/kWh at 20°C.

One commercial vehicle journalist has recently commented (2014) that Roy Aston was the development engineer in charge of the Leyland gas turbine truck programmes (shown above) at Leyland, in Lancashire. As a gesture to Ray for his work,

Centurion Way at Leyland was renamed Aston Way following his death.

The journalist, John Dickson-Simpson, who is familiar with the work of Leyland Vehicles, commented: "Leyland was pushed over the edge with their gas turbine development. The sheer sense of competition was driving the programme forward from a political point of view. One of the problems with the gas turbine was that they couldn't get the seals for the heat exchanger to work properly. The fuel consumption of the truck was 'horrific'.

Someone asked Donald Stokes why he was spending so much money on gas turbines for trucks. He replied that it was worth every penny for the publicity they received. 'It tells the world that we are a progressive company,' said Stokes, 'and that we have great imagination.'"

Illustrated above is one of three gas turbine powered Leyland Comet tractor units used by oil companies to build up operating experience. This vehicle was operated by BP; the others were employed by Castrol and Shell.

# 15 Mack Trucks

THE following information is taken from bigmacktrucks.com

**Posted by – kscarbel. Pedigreed Bulldog**

# Automotive gas turbines

THE Mack gas turbine was the most powerful and promising of them all. And Mack chose two superb partners to work with.

The GT-601 gas turbine tested by Mack Trucks was designed and produced by Industrial Turbines International (ITI), a partnership between Garrett AiResearch, Mack Trucks and KHD (Klöckner Humboldt Deutz AG).

The joint venture launched a programme in 1972 to develop a 450-638bhp (335-476kW) power range of gas turbine engines specifically designed for heavy trucks. This effort resulted in the GT-601, a free-turbine all-metal engine with a fixed-boundary recuperator.

The GT-601 gas turbine eliminates winterizing requirements, operates without vibration, with less noise and favourable torque characteristics, and is air cooled. For density, weight and reliability, the GT-601 was far ahead of the diesel engine. However, in typical on-road applications, US truck weight (80,000lb/36.3 metric tons) and size regulations do not allow the efficient use of the 550bhp GT-601. Only the use of larger trucks (97,000lb/44 metric tons) would allow the GT-601 to yield economic savings.

Several engines were built during the programme and tested extensively.

More than 7,000 hours of operation were accumulated, including more than 2,000 hours in highway trucks.

These engines demonstrated the feasibility of fixed-boundary recuperators along with superior torque compared to diesels of similar rated power, easy cold starting, and the ability to survive limited foreign object ingestion.

In 1979, a GT601 was experimentally fitted in a Mack R-795S tractor, and tested throughout 1980.

In 1980, GT-601s were installed in a 1979 Hayward-production Cruise-Liner (WS-760LST) and Super-Liner (RWS-760LST).

# Automotive gas turbines

Mack installed the engines in Allentown and in Phoenix, Arizona, on the property of the local Mack distributor where the IMI joint venture rented a building. Phoenix was the home of Garrett AiResearch.

The two trucks made an appearance at the 1982 Gas Turbine Conference in Wembley, UK and operated throughout 1986.

**Posted by – kscarbel. Pedigreed Bulldog**

DEVELOPING a gas turbine for vehicular use posed unique challenges, as vehicular operation encompasses a wider operating envelope than aircraft or stationary gas turbine operation over a limited altitude range.

The GT-601 vehicular gas turbine incorporated variable turbine nozzles for engine braking and gear shifting, as well as its conventional use as a temperature trim device.

The GT-601 was designed to work with a manual clutch and transmission, or with automatic transmissions (with torque converters omitted).

To meet the challenges of vehicular operation, the GT-601 incorporated features not found in previously-designed gas turbine engines.

- Two-Stage Compressor - To meet the high specific power requirement not obtainable with a single stage compressor.

- Radial Inflow Gas-Generator Turbine – Minimum cost solution for equivalent performance/life trade-off. Cooled radial stators considerably cheaper to fabricate offered possibility of later transition to ceramic stators.

- Two-Stage Power Turbine with Two Stages of Variable Stators – Higher efficiency – Reduced diffuser loss-turbine /recuperator. Requirement to match with manual transmission dictated need to decelerate turbine during gear shifts in 2s. Only two stages of variable stators will suffice.

- Multi-Module Recuperator – High specific power dictated the pressure ratio. Pressure ratio mitigated against rotary regenerator – leakage incompatible.

# Automotive gas turbines

### Posted by – kscarbel. Pedigreed Bulldog

*Commercial Motor.* June 27, 1969

ANOTHER American vehicle manufacturer has entered the gas-turbine league. Mack Trucks unveiled a prototype turbine-powered truck to its executives earlier this month. The truck uses an industrial gas turbine engine made by The Garrett Corporation, a Mack associate; both Garrett and Mack are part of The Signal Companies Inc.

In announcing the development, Mack states that the prototype is intended to provide the firm's engineers with data needed before the introduction of pre-production test models. Data to be obtained includes acceleration rate, intake and exhaust silencing requirements and general performance information. No information is given on the turbine, such as output and type, but the unit is coupled to a modified Allison semi-automatic transmission in a standard Mack forward control chassis (F-model).

There is added interest in the Mack development given the agreement signed at the end of last year by Garrett and Cummins. This covers a joint program between these two companies to explore the feasibility of gas turbines for trucks and construction equipment. It was stated at the time that Cummins would handle marketing of such units.

### Posted by – kscarbel. Pedigreed Bulldog

A NEW gas turbine engine for trucks, the GT601, is being developed by Industrial Turbines International (ITI), a consortium of Mack Trucks, Inc. and Garrett Corp., American firms, and one German company, Klockner-Humboldt-Deutz AG. All have extensive backgrounds in worldwide truck engine manufacturing.

The GT601 is an all-metric design that ITI refers to as a recuperated cycle free-power-turbine engine in the 300kW to 560kW (402bhp to 750bhp) shaft power class.

The engine weighs 988kg (2178lb) and measures 1.492m (58 3/4-in) long, 1.038m (41in) wide, and 1.119m (44-in) high.

# Automotive gas turbines

All engine accessories are gear driven, and include the starter, lubrication pumps, fuel pump, governor, and an air pump used for starting only. Likewise, all vehicle accessories (brake air compressor, air-conditioning compressor, alternator, and power steering pump) are gear driven.

Aero-mechanically, the engine (below) consists of a gas compressor section, a recuperator section, a combustor section, and a power turbine section. An electronic computer oversees engine operation for minimum fuel consumption. At its commercial rating of 410kW (550bhp), the GT601 should have an overhaul life of 10,000h in over-the-road truck use.

The lack of belt-drive accessories and water-cooling system contribute to low maintenance, and there is easy access to engine components.

The variable stator, free power turbine design makes transmission requirements relatively simple. Though laboratory test-cell and in-vehicle evaluations of the GT601 have just begun, results to date look very good.

In combustion tests, the GT601 produced 3.7g/bhp.h nitrogen oxides and hydrocarbon; the carbon monoxide content of the exhaust was 0.076g/bhp.h.

Installed in an 80,000lb tractor-trailer combination (Mack R-795S), the engine accelerated the loaded rig from a dead stop to road speed using only top gear of a five-speed manual

# Automotive gas turbines

transmission. Initial fuel consumption results are in the area of 238g/kW.h (0.39Ib/bhp.h), which is within the diesel engine range.

If all goes well, production of the GT601 could begin as early as 1981. The truck is shown in Mack Museum below.

**Posted by – kscarbel. Pedigreed Bulldog**

*Commercial Motor*, May 1, 1982

TWO gas-turbine powered Mack trucks made an appearance at last week's Gas Turbine Conference at Wembley. The GT 601 turbine is produced by Industrial Turbines International, a consortium comprising Garrett and Mack of the USA, and Klockner Humboldt Deutz of West Germany.

The gas turbine is particularly sensitive to ambient air temperature, hence the qualification that it develops 410kW (550bhp) at 29°C (85°F). At sea level, with an ambient temperature of 15°C (59°F), the GT601 is rated at 475kW (638bhp). The turbine-powered Mack trucks are both 6x4 tractors, a forward-control Cruise-Liner and a bonneted Super-Liner.

They have been brought to Europe to demonstrate the potential of the turbine to military authorities as it is also produced in 520kW (700bhp) form for use in a light tank.

# Automotive gas turbines

To be compatible with automotive transmissions, the output drive speed of the GT 601 is 2,600rev/min compared with its generator and power turbine speeds of 37,000 and 26,000rev/min respectively. The highest temperature recorded in the engine is the 1,040°C (1,900°F) at the turbine inlet.

According to engine load, the temperature prior to the stainless steel heat exchanger (or 'recuperator' in Garrett terminology) is 650°C (1200°F) and 315°C. (600°F) after.

One advantage claimed for the gas turbine concept is that it is far lighter than equivalent high horsepower diesel engines. At the Wembley exhibition, Garrett was quoting a weight of 998kg (2,200lb) for the GT 601 compared with 1,360kg (3,000lb) for a 600bhp-plus diesel engine.

The GT 601 has been designed specifically for heavy trucks (30tons plus) where it is claimed to achieve 10,000 hours between major overhauls. At a gross weight of 80,000lb (approximately 36tons) in US tests, the Super-Liner achieved 56.5l/100km (5mile/gal) compared to 62.7l/100km (4.5 mile/gal) at 100,000lb (45tons) and 70.6 l/100 km (4mile/gal) at 120,000lb (54 tons).

The gas turbine conference was organized by the American Society of Mechanical Engineers and the Institution of Mechanical Engineers.

**Posted by – kscarbel. Pedigreed Bulldog**

Motor Vehicle Nitrogen Oxides Standard Committee,

Assembly of Engineering, National Research Council, 1981

ANOTHER promising alternative engine for heavy-duty applications is the GT-601 gas turbine being developed by a consortium of the Garrett Turbine Engine Company, the Mack Truck Company, and German engine manufacturer KHD (Klöckner Humboldt Deutz AG). This a very large engine rated at 550bhp in commercial applications.

Emissions of hydrocarbons, carbon monoxide and nitrogen (NOx) have been measured at 0.05, 1.89 and 3.13g/bhp.h respectively, on the steady-state test cycle.

At its most efficient operating point, it has a brake-specific fuel consumption of 0.393lb/bhp.h.

# Automotive gas turbines

Its particulate emissions have been reported as 0.33g/kg of fuel.

Using the best fuel economy figure of 0.393lb/bhp.h gives a minimum brake-specific particulate emission rate of 0.38g/bhp.h. (Of course, the actual emission rate over the test cycle would be greater than this, and no direct comparison with the 1986 particulate standard of 0.25g/bhp.h, on the transient test procedure, is possible.)

The developers of this engine are optimistic about its introduction in the latter half of the 1980s, but they recognize that this high-powered engine's uses will be limited. Initial applications are expected to be in on-and-off-road applications such as logging and mining. The first application in trucks would be in a Class 8 trucks operating on rugged terrain.

This particular engine is cited here because it is currently undergoing on-road evaluations in a truck. Future use of this engine will obviously depend on its ability to meet NOx and particulate standards, as well as on its fuel economy as compared to that of the diesel engine it would replace.

**Posted by – kscarbel. Pedigreed Bulldog**

THE GT-601 gas turbine was produced by Industrial Turbines International (ITI), a joint venture between Garrett AiResearch, Mack Trucks and KHD (Klöckner Humboldt Deutz AG) established in 1972 to produce purpose-designed gas turbines for heavy trucks.

Klöckner Humboldt Deutz had been producing gas turbines from 1956. In 1980, the KHD gas turbine unit was renamed KHD Luftfahrttechnik GmbH and sold to BMW Rolls-Royce in 1990, now known as Rolls-Royce LLC and a major producer of jet engines.

The GT 601 is a free-turbine all-metal engine with a fixed boundary recuperator. More than 7,000 hours of operation were accumulated, including over 2,000 hours in highway trucks (above, Mack RW). These engines demonstrated the feasibility of fixed boundary recuperators along with superior torque

# Automotive gas turbines

compared to diesels of similar rated power, easy cold starting, and the ability to survive limited foreign object ingestion.

While the initial chose power range was 450-638bhp (335-476kW) for heavy trucks, ITI later created a 550-750bhp (410-559kW) engine range to better target the tracked military vehicle segment.

Installed in the experimental XM723 mechanized infantry combat vehicle (the forerunner of the Bradley fighting vehicle), the GT-601 was mated to a Detroit Diesel Allison X-300 four-speed automatic transmission with the torque converter omitted.

GT-601 Development History

1974 – Initial design work begins

1976 – Final design completed.

1978 – Production of early prototypes. Results proved the GT-601 to be highly reliable.

1979 – The first GT-601 underwent endurance testing in a Mack R-795S 6x4 tractor. The U.S. Army Tank Automotive Research Development and Engineering Center (TARDEC) and ITI sign a $1 million contract to identify gas turbine power as a viable power plant for tracked military vehicles. Specific test programs were created for the US Army's XM2 infantry fighting vehicle and XM3 cavalry (scout

# Automotive gas turbines

reconnaissance) fighting vehicle, requiring a power increase to 750bhp.

1980 – Gas turbine truck testing (Mack Trucks) at the US Army Yuma Proving Ground in south-western Arizona.

1981 – A 638bhp GT-601 installed in a 45,000lb XM723 mechanized infantry combat vehicle (MICV) underwent TACOM Phase II demonstration testing at the General Motors proving grounds in Milford, Michigan from August 17 through to November 5 to demonstrate the characteristics of a Garrett GT601 gas turbine engine mated with a Detroit Diesel Allison X-300 transmission without torque converter installed. Virtually no engine-related problems were observed. Combined with additional US Army testing of the GT-601 in M2 infantry fighting vehicles, Mack 6x4 tractors, and M-48 main battle tanks, 6,200 miles of testing study were completed.

1982 – The GT-601 completed 1,000 hour qualification testing, and testing using multiple types of fuels.

1983 – The GT-601-powered XM723 underwent TACOM Phase III demonstration testing at the General Motors proving grounds in Milford, Michigan from January 4 through to 18, and from March 24 through to 30. Testing completed June 7 through to June 9, 1983.

Conclusions:

- Vane Braking - The US Army concluded that the distinct advantages of gas turbine "vane braking", a feature unique to the GT-601, provides enhanced drivability and the can maintain low speeds on downgrades without vehicle brake application.

- Inlet Blockage - The engine showed no adverse effects from a 55in-H20 inlet depression other than reduced power output

- Maximum Braking - Power turbine inertia does not have any adverse effect on the gear train during vehicle braking.

- Forward-Reverse-Forward Maneuvers - This manoeuver did not adversely affect engine or transmission integrity.

- Gradeability - The GT601-powered vehicle demonstrated it could maintain speed close to computer predictions on a 60% grade.

The positive experience prompted the US Army to arrange for additional testing at their Yuma, Arizona proving grounds to

# Automotive gas turbines

evaluate various inlet filter systems and demonstrate the GT-601's ability to operate in extreme sand and dust conditions.

The GT-601 was installed in additional vehicles for testing, including the General Dynamics Land Systems Division Electric Vehicle Test Bed (EVTB), a "Chieftain" main battle tank from the UK's Royal Armament Research and Development Establishment, a French AMX-30 main battle tank, an Israeli Russian T-55 main battle tank, an M-109 self-propelled howitzer (at Yuma and Fort Sill proving grounds), a Weasel air-transportable armoured fighting vehicle and lastly an M-48 main battle tank from the Federal Republic of Germany.

1987 – Sixteen prototypes had been completing logging over 60,000 miles. Development costs reached US$90 million. The GT-601 gas turbine has been extensively tested by the United States, Britain, France, Israel and Germany. ITI hoped for orders to refit M-47, M-48, M-109, AMX-30, and German main battle tanks, as well as orders related to power generation, industrial and marine power. ITI estimated the military spec GT-601 in mass production would cost US$250,000 per unit.

With reference to the XM723: From 1958, The US Army was imagining an infantry fighting vehicle that, with substantial armament and greater protection than the M113 armoured personnel carrier, would allow the armoured infantry squad to fight from the vehicle. In 1964, the Army initiated a development effort for a Mechanized Infantry Combat Vehicle (MICV) to include an interim vehicle, the MICV-65 (XM765), and an objective vehicle, the MICV-70.

After rejecting the XM765 AIFV derived from the M113, the Army gave FMC a contract to develop a superior version. In 1972, the Army awarded FMC $29.3 million to design, develop and fabricate three prototype MICV-70 vehicles, a ballistic vehicle, 12 pilot vehicles, and associated systems engineering, product assurance, and test support.

The MICV–70 project led to a purpose-built vehicle, the XM723 MICV, armed with light cannon and a machine gun in a one-man turret, and provided with vision devices and firing ports for the mechanized infantry squad it carried.

Prototype to the Bradley Fighting Vehicle, the XM723 operated with a crew of three and carried nine infantrymen. The turret was armed with a 20mm cannon and 7.62mm coaxial machine gun. The design had novel laminated steel/aluminium armour which was relatively light but gave improved protection against small arms fire up to 14.5mm.

The XM723 prototypes were completed in 1975 and owed some design heritage to the US Marine Corps Amtrak series of vehicles, rather than the M113. Although the infantry (XM2) and scout (XM3) variants of the MICV were still mechanically identical, they were envisioned as having different weapons stations. The infantry version was to continue with the one-man turret as planned. A dual turret version was planned also.

The scout version with its reconnaissance mission placed a premium on observation for the commander. The original MICV arrangement, with the commander stationed in the hull behind the driver and beside the turret, was unacceptable. Thus the scout version was to have a two-man turret so the commander could be stationed at the highest point in the vehicle with a 360° field of view. As well as the cannon and coaxial machine gun, the scout version included an optically-tracked wire-guided (TOW) heavy antitank missile launcher.

In 1975, the Army rejected two prototype designs for the scout version and began developing the prototype XM723 as a cavalry vehicle. The Army combined both cavalry and infantry fighting vehicle requirements under the MICV in 1976. At the same time, an army task force new programme called the Fighting Vehicle System of two vehicles was created: the XM2 Infantry Fighting Vehicle and XM3 Cavalry Fighting Vehicle.

The XM723 was designated the XM2 for the Infantry Fighting Vehicle requirement and the XM3 for the Cavalry Fighting Vehicle requirement. FMC was awarded a contract to produce the modified vehicles. Based on recommended changes to the XM723, the M2 Bradley Fighting Vehicles was manufactured and fielded in the early 1980s.

**Tom B**

A FEW amendments to the otherwise excellent information.

I was the project engineer for the GT-601 installation in the R model. My team at the Engineering Development and Test Center (ED&TC) in Allentown, Pennsylvania, installed the engine in 1979-80 and modified the vehicle accordingly. The technicians and machinists at the ED&TC were absolute masters of innovation and deserve huge credit for their work. The vehicle was eventually driven out to Phoenix where AiResearch was headquartered and extensively tested in desert and mountainous conditions (up to Flagstaff).

When idling, the air velocity of the gas turbine intake made horrendous noise at the front of the vehicle but there was no time to work on silencers so out it went. The vehicle was interesting to drive because it needed no clutch. Compared to Maxidyne engines, it was not very impressive and I had many good conversations on the subject with Win Pellizzoni the Maxidyne concept inventor.

The fuel economy was poor (it was a prototype), the engine management was buggy and black art stuff, plus the engine costs were enormous. It was reasoned that 500bhp and the costs would not be successful so the idea was dropped.

**Posted by – kscarbel. Pedigreed Bulldog**

Tom, thank you so much for adding to this discussion and correcting my memory where necessary.

Once upon a time, I found myself at the Mack Trucks distributor in Phoenix. As you recall, the distributor was leasing to Mack Trucks a building across the yard from their shop, for turbine truck work (owing to Garrett AiResearch being in Phoenix). I spent the day watching Mack engineers going about the installation of a turbine into a Hayward-production Value-Liner. And yet, I've never read any mention about that truck.

When you say that Win Pellizzoni was the developer of the game-changing Maxidyne high-torque rise engine, then would you describe Walter May as the project manager?

Tom, the work that was done by you and your colleagues at the Mack Engineering, Engineering Development and Test Center (R&D center) in Allentown, put the global truck industry on notice time and time again. In effect, it declared that Mack Truck engineering was second to none. Thank you for your many years of dedication to the greatest name in trucks. What Volvo has done, reduce a global icon to a mere shell of its former self, should be a crime.

**Tom B**
THANK you for the kind welcome. I really enjoyed working for Mack Trucks but things went downhill when AiResearch bought the company...development investment was drastically scaled back. RVI was purchased, and the rest is sad history. After Mack, I enjoyed positions at Mercedes-Benz, Britax Child Safety, and lastly BMW where I retired as VP of engineering.

Walter May eventually became chief engineer and then retired, followed by Bob Zalokar and then Steve Homcha, Win was the father of the Maxidyne concept...Walter May was the father of the 'Walter May' test. That was, for anything mounted on the truck with a bracket, smack it with your hand and if it moves it's not robust enough. It was a seat of the pants method that all young engineers could understand, and it forced hands-on development. He also helped establish the fuel tank 'bump test' (with Dr. Bill Geiger) where a filled tank and brackets on a frame were lifted and dropped at least 1 million times. If Mr. May had stayed the company would have continued to flourish.

I remember the big development projects for UPS trucks since they always got special features like mirrors and steps, etc. The head guy at UPS had worked for Mack at the Test Center so he was a big supporter of Mack. Eventually, some bean counters tightened the belt and UPS moved on.

There were a lot of very talented people that left Mack. The strikes and 'us vs. them' mentality were devastating. During one strike, I was warned not to look at strikers while driving into work, or risk getting my tires slashed. I was pelted with grommets from line workers on break at the Macungie plant when I was there to fix leaking fire trucks. Engineers could not pick up a screwdriver and use it even once or there would be a grievance filed. Sadly, management was also to blame. As a senior test engineer in 1979 with about seven big projects at once, I was paid the same as a plant floor sweeper. When engineers met with management and showed them the numbers their response was 'you don't have a union.' I left shortly after.

I still have the actual bulldog from the prototype MA model cab-over build that I supervised. Nice memories.

**Posted by – kscarbel. Pedigreed Bulldog**

Just so there's no confusion, Mack Trucks was never owned by Garrett AiResearch (Garrett Corporation). In 1964, Garrett allowed itself to become a subsidiary of The Signal Companies, the nation's largest west coast oil company, in order to avoid a hostile takeover by Curtiss-Wright Corporation.

In 1967, in order to raise capital to support Mack's rapid growth, Mack Trucks agreed to become a subsidiary of The Signal Companies on the condition that Mack was guaranteed complete autonomy (a requirement demanded by legendary Mack president and CEO Zenon C.R. Hansen). Thus, both Mack Trucks and Garrett AiResearch were subsidiaries of The Signal Companies.

Signal purchased Mack Trucks in 1967, sold 40 per cent to Renault in July 1983 and then sold 50 per cent in a public offering in August 1983. The remaining 10 percent (3.1 million shares) was sold in July 1986 by the Henley Group, a corporate spin-off resulting from the Signal-Allied merger.

In 1985, Signal merged with Allied Corporation, and became Allied-Signal. The company bought Honeywell in 1999 and adopted the Honeywell name.

# 6 MTU

IN THE 1970s, all but one of the world's gas turbine builders for aircraft turned their backs on the building of commercial vehicle gas turbine engines.

In 1969, Daimler-Benz and MAN – Maschinenfabrik Augsburg Nurnberg – formed aero engine maker MTU (Motoren-und Turbinen-Union) by taking equal shareholdings. The parent companies subsequently awarded MTU a contract to develop a truck gas turbine engine. The first engine ran in November 1971 and, at the time of writing my article for *The Engineer,* new engines were being prepared for handing over to both Daimler-Benz and MAN for road testing in trucks.

MTU believed at the time that spinoff from aero gas turbine could be cross-fertilised into truck engines. However, at the time there was little evidence of cross-fertilisation. My article in *The Engineer* noted that the 7042 engine (below) 'looks much the same as those under development by British Leyland, Ford, General Motors, Chrysler and Fiat'.

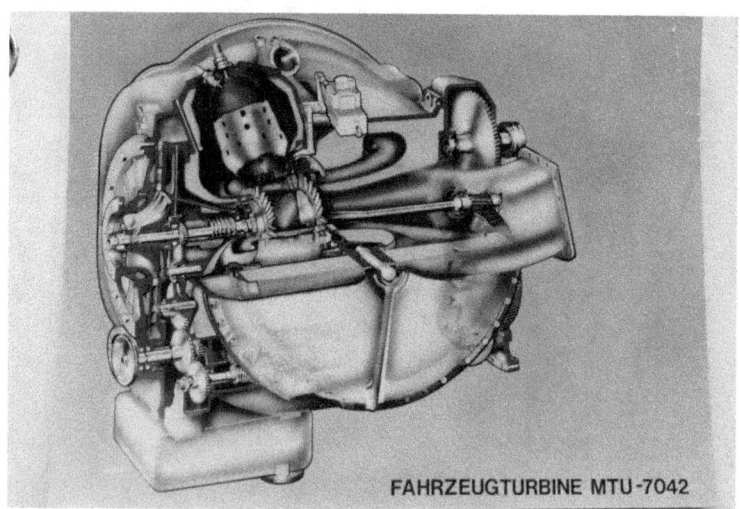

Dr. Wolfgang Heilmann, in charge of MTU truck gas turbine programme, had no doubts of the reality of a spinoff.

"We have many other programmes under way at MTU and there will be very extensive information exchange between the various groups," he said at the time.

According to Dr. Heilmann, under the contract all the previous development undertaken by Daimler-Benz and MAN was brought into the joint programme.

It seems Daimler-Benz started work on truck gas turbines in 1955. MAN started later but achieved useful results since it converted two 6022 helicopter engines and installed them in trucks, running extensive road-based test programmes extending into several hundred hours.

The engines were fitted with silencers but an engineer said these silencers would be eliminated in engines were fitted with heat exchangers.

It was in 1969 that MTU started with a clean sheet of paper to design a two-shaft regenerative gas turbine – the MTU 7042. The engine was designed for 350bhp and a weight of 680kg.

"We are trying to get more than 350bhp out of the engine but it is rather like the man who wears a belt," said Dr. Heilmann. "The belt will not let the man put on weight. So with the gas turbine the heat exchanger is the big factor restricting growth."

Dr. Heilmann conceded that it would be attractive to use only one heat exchanger. An appropriate power range for one heat exchanger would be 250bhp He did not think it sensible to design a 350bhp engine with only one heat exchanger, especially with regard to the low fuel consumption required.

Daimler-Benz had conducted extensive trial with various heat exchanger designs – including recuperative and regenerative metal matrix designs – but in the end resorted to the Corning Glass ceramic heat exchanger.

"I think we have a great chance at 350bhp and higher to compete with the diesel engine. But it needs more time and there are a lot of problems that have to be solved," said Dr. Heilmann. "And we need to carry out more work on components to be able to say that the gas turbine will beat the diesel engine."

"When you make the calculations on the gas turbine you very quickly come to the same conclusion as other truck gas turbine makers. With the thermodynamic cycle, a pressure ratio of 4 to 1, a maximum temperature of 1,273K and with the only available heat exchanger from Corning, gives the basic concept of the engine," he added.

He repeatedly noted that MTU was only a subcontractor. It was up to Daimler-Benz and MAN as how the engines would be used. But as Daimler-Benz and MAN could manufacture only the engine that MTU developed, both truck makers would be using identical engines but perhaps with different power ratings. And it was up to both companies to decide if further engines should be developed.

As it was, the gas turbine did not beat the diesel engine and neither company entered the truck gas turbine stakes.

# 17 Rolls-Royce Ltd

IN MARCH 1945, the experimental department of Rolls-Royce compiled an interesting report. *The possibilities of the gas turbine as a motive power for the automobile* reflected a detailed study conducted at the time by some of the company's engineers.

The author concluded: 'It is shown that the gas turbine is likely to be superior in all respects to the piston engine as a motive power for road vehicles. It is suggested that a turbine unit on the lines given in this report should be immediately put in hand if we are to enter the car field in the post-war era and successfully compete with other firms who will undoubtedly follow this line of development.'

Although the report gave a glowing prospect for the gas turbine in land-based vehicles, the directors of Rolls-Royce chose to ignore it. And presumably they never gave the matter a second thought. Even Rolls-Royce's adventure with the Cottage Loaf diesel – an attempt to put a Wankel engine – another novelty of the 1960s – in a Main Battle Tank failed to make production.

In the summary, the report (Reference Lov/FRB. 2/JF. 2/3/45) stated: 'A complete project is worked out for a gas turbine car engine of about 90bhp. A preliminary scheme is given and the performance has been calculated for the engine fitted into 4.25-litre Bentley. The performance figures show that for steady running on a level road fuel consumption varies from 25 miles/gallon between 20 and 50 M.P.H. to 17 M.P.G. at 80 M.P.H.'

'The turbine car shows many advantages when compared with the piston engined car, for instance the turbine unit should be cheaper, weigh less, have simpler auxiliaries, and require either no gears or possibly a simple two-speed gear and reverse.'

'With some development the gas turbine should be at least as reliable as the piston engine and probably as quiet or, even quieter. An outstanding advantage will be its smoothness.'

The engineer who compiled the report was clearly an enthusiast for a gas turbine engine installed in a car.

A scheme of the engine shows a radial compressor mechanically linked to an axial turbine. A power turbine extracts further energy from hot gases from the combustion chamber. Simple speed reduction transfers the power through a 25:1 reduction to the drive shaft.

The engine had a heat exchanger with a central combustion chamber.

## 18 Rover Gas Turbines Ltd

THE Rover Cycle Company Ltd., founded in 1877, was as a partnership between John Kemp Starley and William Sutton. The company originally manufactured cycles and motor cycles. By 1906 the company was making cars and changed its name to The Rover Company Limited. In 1933, it came under new management from the Wilks brothers, Maurice and Spencer.

# Automotive gas turbines

Mark Barnard, who worked closely with the development of gas turbines at Rover from 1953 when he joined the company as a graduate, has written a book, *Piston to Blades* (see references). He records Rover's early time with Sir Frank Whittle's gas turbine engine.

According to Barnard, the company's involvement with this novel engine began in 1940 when Maurice Wilks, then chief engineer of Rover, met Frank Whittle for the first time.

Whittle it seems had become frustrated with the progress of the work he was guiding to create his first jet engine at British Thomson Houston (BTH) in Rugby. He had been put in touch with Wilks through a mutual friend. Wilks was interested and it was suggested that Whittle's small firm Power Jets might arrange for the Air Ministry to place some sub-contract work with Rover. And in fact, direct contracts were awarded to the company. Things happened quickly from then onwards and a team of about six drawing office staff was set up at Rover's factory in Helen Street, Coventry.

But when Germany began bombing cities in England the company decided life was becoming too precarious and decided to move engineering staff to Chesford Grange Hotel, midway between Leamington Spa and Kenilworth. This transfer was largely completed by a Saturday in November 1940, with work planned to start the following Monday morning. The move proving to be of amazing timing: on that Sunday, in Coventry's blitz, the Helen Street facility was destroyed.

And so it was that Rover's production director, Geoffrey Savage, wisely instructed his buyer Bernard Smith to find premises in the northern half of England to carry out Whittle's turbine work, far from the dangerous environment of Coventry.

Having been told not to return to Coventry until he had found an available factory, Smith located an old cotton mill, Bankfield Shed in Barnoldswick, Lancashire. This provided enough space to accommodate the manufacture of components for a gas turbine. By the end of 1940, a further building, Waterloo Mill, had been located in Clitheroe. Both were requisition by the Ministry of Aircraft production (MAP) for Rover to develop, manufacture and assemble the components of the turbojet engine and proceed with testing them, in addition to the compete engines.

Rover sub-contracted some work, such as the fuelling system and combustion chamber to Lucas. By 1941 staff from both companies had moved to Clitheroe and by the end of that year Lucas staff and their manufacturing equipment had moved also to another disused textile mill, Wood Top Mill in Burnley, which had been requisitioned for their specialised work.

While everyone was working hard to produce the Whittle-designed reverse flow engine, Rover became interested, according to Barnard, on an alternative design of engine which could significantly reduce the temperature of the sheet–metal ducting carrying hot gases from the combustion chamber to the turbine. This was the ST or straight-through concept, but it required the turbine shaft to be lengthened, together with the distance between the compressor and turbine bearings.

Whittle, according to Barnard, had considered this but rejected it as it might incur serious shaft whirling problems. Perhaps more significantly, the ST concept required a new type of combustion chamber, a critical aspect of the entire design and one of which Whittle it seems was less confident. A young man, Adrian Lombard, devised a way of splitting the two shafts in half and re-joining them with a sleeve over a spherical joint to be later known as the Lombard coupling.

According to Barnard, MAP gave Rover an additional contact to design, build and test this engine, a task completed in nine months.

The two engines shared the same aerodynamic components and were developed in parallel, the designation being W2B for Whittle unit and W2B/26 for the Rover straight-through (ST) design. In the event, both designs were developed by Rolls-Royce: the Whittle design became the Welland and the Rover design became the Derwent 1.

## Quid pro quo

# Automotive gas turbines

There is another version of events, similar but different. Indeed, there may be many versions of events. But shortly after the start of World War, Spencer Wilks, managing director of the Rover Car Company, had been approached by government ministers and asked to give support to a radically new and very secret development - Frank Whittle's gas turbine 'Jet' engine project.

The concept was totally new and an entirely different venture for Rover. It was so top secret that the British Government, being terrified of security leaks, instructed Rover and all other people involved in the work to refer to Whittle's invention as a 'supercharger'. Whittle's small company, Power Jets Ltd, was based in a foundry in Lutterworth, 16 miles from Coventry. The company was the world leader in turbo-jet design, but the first Whittle jet engine flight would not take place until 1941.

Power Jets at Lutterworth was an experimental workshop with only a pilot-build plant at Whetstone, near Leicester, not set to be established until later in the war. Component parts for Whittle jets engine were sourced mainly from the British Thomson Houston Company in Rugby, a business more accustomed to building large industrial turbines, as well as from Joseph Lucas in Birmingham.

Rover's brief was to develop Whittle's design to the production stage, and later begin manufacture of jet engines for the RAF. Meanwhile, Gloster Aircraft was designing the first jet fighter, later to become known as the Meteor. Rover was already associated with Gloster Aircraft as the company was building Albemarle airframes for them.

Work started on the jet engine in early 1940 but before long there were serious personality clashes between Frank Whittle and Rover engineers. Although Rover was improving the initial design, Frank Whittle could not accept that his designs in both detail and basic concept should be altered.

By November 1940, Rover's work was considered of vital importance and from that point onwards there was no doubt that improvement and proving of the Whittle machine was of great national importance.

The German Luftwaffe's attempt to obliterate Coventry's industrial area on the night of 14th/15th November reinforced the decision to implement dispersal of production. Consequently, manufacture was transferred to Clitheroe and Barnoldswick in Lancashire. Bankfield Shed at Barnoldswick had been a serious player in the textile industry, but had been closed in the gloomy days of the 1930's textile trade depression. Later Barnoldswick would become even more famous in Rolls-Royce hands and today, modern Rolls-Royce engines such as the RB2ll, carry the designation 'RB' for Rolls-Royce Barnoldswick.

The first development jet engines were virtually unchanged from Whittle's 'W2' design, but used Rover's expertise for the accessory drives. There was serious trouble from surging and the failure of turbine blades, and the 'W2' proved to be seriously underpowered.

Rover was then asked to go ahead with a development based on Whittle's 'W2B' design. This meant Rover could undertake considerable mechanical design of its own. Misunderstandings and the difficult atmosphere between Whittle's firm Power Jets Ltd. and Rover deteriorated even further with Rover's development of the 'W2B'. Whittle became particularly furious when the first Rover-built 'W2B's were running in the Lancashire factories in October 1941. This was due to many design changes Rover had made to his firm's original design layout for the 'W2B' engine.

With the apparent success of its 'W2B', Rover was asked to plan for quantity production at Clitheroe and Barnoldswick, but Rover's engine designers had reached the conclusion that Whittle's WI' and W2' engines were aerodynamically inefficient because of the counter-flow arrangement of compressors, combustion canisters and turbines.

Due to this, Rover received permission to begin design of a new 'straight-through' engine having a different layout concept, to be known as the 'B26'. This new project was top secret; so secret that even Frank Whittle would not be informed about it. Rover's 'B26' straight-through design retained the best of the original centrifugal compressor layout. The prototype B26' ran for the first time in November 1942, and almost immediately showed a great improvement in thrust and reliability.

In August 1942, a Whittle-type 'W2B' had been installed in the tail of a Wellington Bomber and test-flown from the Rolls-Royce flight test field at Hucknall, near Nottingham.

At this stage more than 30 engines of this type, mainly developments of 'WB2's had been built and a government decision was about to be made on quantity production of gas turbine engines. Meanwhile, aircraft manufacturer Gloster Aircraft was ready to begin manufacture of its 'Meteor' fighter planes, and these would become the first 'Jet Fighter' aircraft.

At that juncture, Rolls-Royce began to take particular interest in the new jet engine technology. When first approached by the British Government, Rolls-Royce directors said they were far too busy to become involved in this new technology, instead becoming committed to producing Merlin V-12 aero engines. These legendary 'Merlin' engines would go on to become an essential part of Britain's air defence and would play a major role in the air war against the Luftwaffe, and the RAF's ultimate victory during the Battle of Britain.

However, even though Rolls-Royce was fully committed to the manufacture of piston engines, the company could not help but be intrigued by Rover's gas turbine jet developments, and decided to take a serious interest once the test flights of the 'Wellington Bomber' proved successful. Rolls-Royce directors quickly announced that as a matter of policy they would now like to become involved in gas turbine engine development. With their obvious commitment to successful aero engine development this new policy had to be taken seriously by the British Government.

At the same time the Rover management, while acknowledging their continued success with the new technology gas turbine engines, did not wish to move on to continued development of aircraft engines on a permanent basis.

The solution was a compromise that saw Rover hand over its gas turbine development work to Rolls-Royce, receiving in return from Rolls-Royce a large and vital tank engine contract. Rover took on board the entire Rolls-Royce Meteor engine project, itself an advanced piston engine design. The engine consisted of a much modified but unsupercharged version of the V-12 Merlin aircraft engine, and would be produced for the latest heavy allied tanks.

This engine became the most powerful piston engine with which Rover had ever been associated, developing around 700bhp. The engine was further developed and in Mark IV version, sprang to prominence in the world famous 'Centurion' tank. Later, fuel injected M120 versions powered the bigger and more impressive 'Conqueror' tanks.

Later a development from the 'Meteor' was the Meteorite engine range, which was essentially two thirds of a Meteor, being a V-8 while the Meteor was a V-12. Meteorites were developed during the late 1940s and were built in both petrol and diesel engine forms for vehicles, marine use and use as stationary power units. The mighty Antar tank transporter, built by Thornycroft, was powered by a Meteorite, and was often to be seen dragging Meteor engined tanks around the world.

According to Barnard, when Rover's involvement with gas turbines ended in 1943, Maurice Wilks and his brother Spencer Wilks, managing director of Rover, were planning the new business of Rover cars after the destruction of their pre-war Coventry base at Helen Street. The government had built a shadow factory on Lode Lane between Solihull and Sheldon where Rover people had been building, repairing and testing RAF aero engines; namely, 600bhp Armstrong Siddeley Cheetah, 1,200bhp Bristol Pegasus and 1,800bhp Hercules power units. All were radial piston designs.

At the end of WW2 the Lode Lane factory became the main Rover factory, although there were subsidiaries at Tyseley,

# Automotive gas turbines

Acocks Green, Ryeland Road and Springfield; these provided the main technical services needed for engineering work and which had all played their part in wartime. However, it was Lode Lane that became the headquarters of the company.

## Looking ahead

Because of the Wilks brothers' considerable interest in the gas turbine engine they had considered other applications of this new prime mover in fields outside those of aircraft, particularly in road vehicles. It was not long before they pioneered the world's first design of a smaller engine of more modest power. They also incorporated two major features: the two-shaft engine and the use of a heat exchanger.

In the two-shaft engine one shaft carries the compressor and its own turbine, but a second shaft uses surplus energy and is mechanically independent of the first shaft. This 'power turbine' could be geared to drive road wheels of a vehicle or a boat's propeller.

On leaving the power turbine the exhaust gas has lost most of its pressure energy but not all its heat energy. So the second feature was the need to design a heat exchanger that would remove as much heat energy as possible and transfer it to the incoming airflow as it leaves the compressor and before combustion.

In 1945, a nephew of Maurice Wilks joined Rover having served an apprenticeship in gas turbines at Rolls-Royce Derby. He was Charles Spencer King, hitherto known as Spen King. Spen King brought with him to Rover a colleague from Derby, Frank Bell, hitherto known as 'Tinker' Bell. Another early member of the team was Harry Knowles, a bachelor with an unusual personality. He was meticulous with his calculations, whether it be stressing or form development. He would always throw away his calculations when the task was complete.

The team was sited originally is a corner of the Rover car design office and every morning a half-hour discussion would take place between Frank and Harry on any topical subject, not necessarily concerned with gas turbines.

Barnard, who joined Rover in 1953, writing to me in a private letter dated 5 September 2011 stated: "I am pretty sure the first heat exchanger was Spen King's idea, and an engineer Phil Gardiner designed and made the first. Another bloke, Mark Barnard became heat exchange engineer No. 2 and I made our first primary surface and also adapted the Cercor ceramic discs."

Spen King died in 2010 as a result of injuries sustained in a collision between his bicycle and a van. He was 85 and best known as the driving force behind the 1970 Range Rover. In July 2010 Noel Penny sent me the following email following the death of Spen King: "Myra (Spen's wife) died this year; I attended the funeral. Spen was a great friend, he worked hard for NPT. It's sad. He will join Myra."

From the end of WW2, Rover's Acocks Green factory had become designated as the home of 'Fighting Vehicle Engine Research', and this association with the defence ministries carried on for 21 continuous years into 1964.

Rover's wartime activities were large and although the gas turbine engine projects were a crucial development, it must be remembered that the aircraft construction work was also vital to the defence of Britain. Hundreds of radial piston engines and Meteor engines poured out of Shadow Factories, and thousands of airframe parts were built in the damaged Helen Street works. Nor should be forgotten the army's webbing contracts, completed at Hinkley, the overhaul and repair of aircraft magnetos at Lutterworth, and vehicle body manufacture and the building of aero engine test stands at the London Seagrave Road service buildings.

By 1944 it became obvious that an end to the hostilities was in sight, and the company was able to look ahead towards car production once again. The question was which model and where would it be built? There were no new models on the drawing board and the Helen Street premises remained badly damaged.

Rover's staff members were dispersed throughout several factories and the company was a much larger concern than ever before. Post-war conditions proved to be much different to the 1930s and the big question now was just what kind of demand would there be for Rover's type of quality, middle-class and reliable cars? Maurice and Spencer Wilks were not at all sure just what would await them in peacetime, but as usual, they were looking forward to the challenge.

A full and detailed description of Rover's involvement with Sir Frank Whittle's jet engine can be found in *Vikings at Waterloo* by David Brooks, see References, while a full description of Rover's involvement in small gas turbines can be found in Mark C. S. Barnard's assessment of the work in *Pistons to blades*, and again see References.

# 19 Volvo Corporation

THE HISTORY of gas turbines within the Volvo group goes back to the first jet engines made for the Swedish Air Force in 1946. Since then, various jet engines have been produced up to today's RM12 for the Saab Gripen Fighter. For ground vehicles, the development of the power plant for the Swedish S-tank was the first achievement. S. O. Kronogård and his team at Volvo worked with a regenerated gas turbine for this combined diesel and gas turbine unit.

A ceramic regenerator was developed, but time was running out and it was decided to go for a simple cycle gas turbine instead. A Boeing 553 engine (initially for helicopters) was chosen in spite of its high fuel consumption. This was seen as a minor problem as the tank was supposed to work with gas turbine power only for short periods.

With respect to trucks, Volvo was for some time a partner in the ITI GT 601 engine. This engine employed a stainless steel recuperator to avoid the many problems encountered with the ceramic regenerators. The engine was installed in various vehicles and there remains a Mack truck with a gas turbine at the Mack Museum in Allentown.

## Automotive gas turbines

The three-shaft gas turbine patented by S.O. Kronogård was offered to Volvo in 1974. This engine employs three turbines, the first driving the compressor, the second (the power turbine) driving the output shaft and the third, the auxiliary turbine, splitting its power between the compressor and the output shaft through a planetary gear. The advantages claimed are very high low-speed torque and good gas generator acceleration.

United Turbine (later part of Volvo Aero Corporation) was formed to develop this idea into a working passenger car gas turbine, namely the KTT 150. Up to 1983 six KTT 150 engines were built and demonstrated both in rig form as well as being installed in a couple of cars. An interesting test during this period was a trip with a car from Malmö to Monaco.

The interest for ceramic components during the 1980s was high and much work was done in this area. It was decided to design a high-temperature engine using ceramic components, but based on the KTT 150, which should also have a low emission combustor. This engine, named GT 110, used one regenerator instead of two in an effort to minimize the length of the leakage path. Also, the seals were improved. A two-shaft arrangement was employed to avoid the added complexity of the three-shaft KTT 150.

The ceramic components used in this GT 110 engine were, aside from the regenerator, the compressor turbine, the turbine shroud ring and the turbine nozzle.

The combustion chamber of the LPP-type (Lean Pre-mix Pre-vaporization) was developed in-house. A ceramic version was made and successfully rig tested. In the engine a version made in O.D.S. material (Oxide Dispersion Strengthened) was used, allowing TIT (Turbine Inlet Temperatures) up to 1,250°C. Emission characteristics were very good.

The GT 110 was used to verify gas turbine emission characteristics in tests at Volvo Truck in Gothenburg and at United Turbine in Malmö before the start of the VT300-project. It was also used to evaluate transient performance. The engine suffered from failures of the ceramic turbine, cracking of

## Automotive gas turbines

the regenerator (a problem which was solved at the end of the project) and unreliable heat exchanger seals (although the seals were improved compared to KTT 150) The three-way catalytic converter for passenger car gasoline piston engines took care of the emissions and the project ended in 1987.

Volvo became involved in the investigations on high-speed generators being conducted at The Royal Institute of Technology in Sweden, by designing a test set-up consisting of a turbocharger with a combustion chamber driving the generator. This research work subsequently resulted in the 40kW HSG-engine (High Speed Generator) used in the Volvo Cars ECC-project. (Environmental Concept Car). This engine was also using a ceramic regenerator.

It was followed by the 100kW unit for the Volvo ECT (Environmental Concept Truck, and the ECB (Environmental Concept Bus), where a stainless steel recuperator was used. These HSG-units all employed low-emission combustors, with the VT40 running on diesel fuel (as had the earlier United Turbine engines) but with the VT100 running on ethanol.

Two city buses with hybrid power plants using the ethanol-fuelled VT100 were tested in passenger service in Gothenburg. The exhaust emission values were very low as could be expected. Drivability and performance of the buses proved well within specified values, and the response from drivers and passengers was very good. The fuel consumption was high, but possible to improve by at least 8 to 10per cent by choosing an improved battery system. Further reduction could have been reached by turbine improvement and upgrading the power electronics.

The HSG project has resulted in the formation of Turbec, a company set up by Volvo Aero and ABB to develop the Turbec T100 for use as a CHP (Combined Heat and Power) power plant. By 2005, approximately 100 units had been sold. Natural gas is the fuel that is normally used, but to cope with customer requirements the combustor has been tested on various fuels such as LPG, diesel fuel, kerosene, ethanol and methanol.

# Automotive gas turbines

Essentially no changes were required on the basic flame tube and swirler design, and the necessary changes were limited to the injectors. Emissions were very low for all versions, according to Volvo.

Prior to fully developing a heavy truck gas turbine, during 1990 a decision was taken to investigate what was thought at the time as probably the best alternative to the diesel engine in terms of low emissions and good fuel economy, namely the advanced heat exchanged gas turbine. A pre-study was performed in which several configurations were examined. Altogether six different gas turbines were simulated and compared to the ITI GT-601 and to a diesel engine. Four gas turbines were intercooled with two-stage centrifugal compressors, and two were non-intercooled. The TIT was 1,360°K in all cases. Component efficiencies were state-of-the-art (and as such conservative), while pressure losses, heat losses and mechanical losses were based on in-house experience from earlier gas turbines. Design point fuel economy, truck driving cycles, response and torque characteristics as well as achievable emission levels were studied. A production cost study based on preliminary designs was made and critical components identified.

Some conclusions:

• The best alternative was the three-shaft, intercooled, recuperated engine.

• The two-shaft, intercooled, recuperated engine was very close to the three-shaft engine.

• The gas turbine using today's technology (metallic components) could be competitive in terms of fuel consumption and had very low emissions.

• Production costs were estimated to be higher than for the diesel engine.

Based on the pre-study, a decision was taken to evaluate the gas turbine alternative further by designing and building two engines for rig testing as well as for installation in a truck. The chosen concept was:

- Two-shaft, non-ceramic (1,360°K TIT)
- Two-stage compression with intercooling
- Recuperator type heat exchanger
- Lean Premix Prevaporizing combustor with variable geometry
- Radial or axial gas generator turbine
- One or two axial power turbines

The two-shaft alternative was chosen in favour of the three-shaft to avoid the added complexity of the three-shaft arrangement.

The main issues tackled in the concept study and specification stages concerned the thermodynamic layout of the new engine, and a reworked engine model was created. The engine configuration became clearer:

- Two-stage centrifugal compressor
- First stage compressor inlet guide vanes (IGV)
- Intercooler
- Recuperator
- Lean Premix Prevaporization (LPP) combustor
- Radial compressor turbine
- Two-stage power turbine with first stage variable nozzle

The inlet guide vanes to the first stage compressor were added in spite of a more complicated design for reasons of multiple engine brake capacity, improved acceleration, and improved idle fuel consumption.

The first stage compressor is a critical component as the surge margin is very small in the low power area. It was suggested that this component should be tested very early in the programme. The final pressure ratio was not frozen, the choice being between 7:1 and 9:1. The higher pressure ratio had the advantage since engine volume flow would be lower, resulting in smaller components like the intercooler and recuperator. It was decided therefore to investigate this more closely, especially the influence on the recuperator performance and life.

The radial compressor turbine seemed to have advantages over the axial one. However, it was decided to further study the radial turbine (stress situation) for the low-pressure alternative.

Work on the preliminary engine design made during the concept study resulted in an engine layout not far removed from the final one. The main difference was the arrangement of the compressors 'in series' (inlet section in the same direction) and the arrangement of the accessory gearbox which was driven by a bevel gear and 'tower shaft' from the gas generator. The IGV was of the axial type. The engine was built up in modules separated by walls to avoid internal leakage and a power turbine arrangement that allowed assembly as a balanced unit without separating the rotor during final assembly.

During the second phase of the project the thermodynamic specification was finalized as well as engine layout. Final choice of pressure ratio was selected at 9:1. The compressor turbine is of radial type. Maximum turbine inlet temperature is 1,350°K. A condensed summary of the engine specification is shown below.

**Table 1. Engine specification:**
ECE R49 fuel consumption 219.4g/kWh
Power rating between -20°C and 30°C, 350kW
Data at design point (400 kW ISO)
Cycle pressure ratio 9
Engine efficiency 42.2%
Turbine inlet temperature 1,350 K
Air flow 1.5kg/s
Efficiency (t-s) compressor 1 80.3%
Efficiency (t-s) compressor 2 83.8%
Efficiency (t-t) HP turbine 90.1%
Efficiency (t-s) Power turbine 86.4%
Recuperator, efficiency 91.1%
Intercooler, efficiency 80.3%

# Automotive gas turbines

Inlet pressure losses 3.3%
Intercooler pressure losses 3.9%
Recuperator pressure losses between 3.0 and 3.6%
Combustor pressure losses 3.9%
IDD pressure losses 1.0%
Exhaust duct pressure losses 2.8 %
Heat losses 7.3kW
Mechanical losses between 4 and 13kW

Following this, the engine layout work was finalized and detail drawings completed. The layout was changed significantly in two areas, with the following remarks made during the preliminary design review.

Compressor arrangement: First, the compressors were arranged in a 'back-to-back' arrangement (instead of the 'series' arrangement studied during the concept phase) with radial IGV for the first stage. This reduced engine length and made for very good control of compressor tip clearance as the gas generator thrust bearing is situated between the compressors. This bearing is a double-row pre-loaded angular contact unit of special design.

Accessory gearbox layout: Secondly, the accessory gearbox drive had cylindrical gears and the gear numbers were reduced from 21 to 19 (including the reduction gears in the power turbine section and those necessary for the power transfer)

From start it was decided not to use ceramic materials in the engine for two reasons: use of the materials would have increased the design effort considerably and the opinion was that ceramic materials were not yet reliable enough for engine usage (at least not for rotor components) Thus, conventional gas turbine materials were used. A simple cooling arrangement was used for the compressor turbine nozzle, and the compressor turbine had some cooling flow on the rear face.

• Compressors – machined from aluminum forgings

• Combustion chamber – Hastelloy X

**Automotive gas turbines**

- Compressor turbine nozzle – IN939
- Compressor turbine – MAR-M-247
- Variable nozzle, power turbine – Hastelloy X housing with HS 31 nozzle vanes
- Second stage nozzle, power turbine – IN 738 LC
- Power turbine, both stages – IN 738 LC
- Recuperator – AISI 347 SS

In conclusion, the three Volvo engineers who compiled their report for the SAE noted that the truck gas turbine project has been very valuable and although all goals were not reached it has resulted in much experience with this type of power plant. However, they concluded also that the all-important fuel consumption remained as the major disadvantage for gas turbines. Their view was such that the current standpoint was that a diesel engine would have lower fuel consumption also for US10 and Euro 6 emission levels compared to a gas turbine with metallic construction materials. Future developments in ceramic components, even more strict emission legislation or other things can change the situation. Volvo will follow the future developments closely and the VT300 engine can, if needed, be used as a good test bench for further developments toward high performance multi-fuel heavy trucks. The engineers added that project had been a continuation of Volvo's long traditions in exploring alternative powertrains and especially gas turbines (see References).

# 20 Williams International

WILLIAMS International was founded on the vision of Dr. Sam B. Williams. In the early 1950s, Williams foresaw extensive markets for small gas turbines. Such engines, he believed, would provide the performance efficiencies of very large turbine engines in very small sizes by using revolutionary design. Innovative manufacturing technologies would also help keep the process simple and relatively low cost.

This vision prompted Williams to leave the security of Chrysler Corporation and, with his very own limited funds, start his own company, Williams Research Corporation, in 1955. The company name changed in 1981 to Williams International. The first contract was for an experimental gas turbine for a marine outboard application, a trait that would soon characterize the engineering, manufacturing and management personality of the company. The Williams International employee team has taken the risks of using new ideas and has boldly introduced products that have changed the aerospace industry.

Most noteworthy of these products are the miniature turbofans that enabled the creation of cruise missiles; the X-Jet flying platform; turbojets that are so small and inexpensive that navies around the world use drones powered by them to train their gunners; and finally the simple, rugged FJ44/FJ33 family of turbofans that enabled the development of new categories of very light business jets.

The company claims its strong focus on customer service has earned it the No. 1 rating in ProPilot's Survey Customer Survey every year since 1998.

The company declares that the culture of continuous improvement in all aspects of the business is paramount to the success of each person the company.

Williams International is privately owned and because of this, according to the company, the vision stays focused, communication is straightforward, decisions are made quickly, and efforts are concentrated efficiently. Resources, both funds and manpower, are assigned where and when they are needed most, allowing the company to adjust quickly and seamlessly.

Williams International's product development, product support and administrative headquarters are located in Commerce Township, Michigan. A second facility in Ogden, Utah, is the most modern and efficient design-to-production facility in the world, it is claimed. Through a combination of equipment improvements and management innovations, the company claims to have the development, manufacturing, product support capability and capacity to meet growth objectives in aviation, industrial and military markets.

Williams International claims to be the leader in small gas turbines engine development, manufacture and field support. It notes that it took the lead through customer focus, vision, innovation, determination and perspiration.

# 21 Walmart

In 2014, the US supermarket giant Walmart revealed its futuristic Walmart Advanced Vehicle Experience, (Wave) based on an advanced turbine-powered range-extending series-hybrid powertrain mounted below its bullet-shaped cab.

As part of its plan to double the efficiency of its 6,000-truck fleet, Walmart Corporation of Bentonville, Arkansas, worked with several companies to build the Walmart Advanced Vehicle Experience. It is a test bed of cutting-edge technologies that might show up in the next generation of trucks and trailers used to deliver goods across the country.

The futuristic Wave truck includes several improvements, thanks to its partners including Capstone Turbine Corporation of Chatsworth, California, Great Dane Trailers, Peterbilt, and Roush Engineering. On the outside, the truck sports LEDs for most lighting needs. This reduces weight and power requirements. And the truck is assembled using advanced high-strength adhesives rather than rivets. This simplifies assembly and cuts inventory costs. Other features include:

**Aerodynamics:** Engineers redesigned the cab and trailer (illustrated right) using CFD to cut aerodynamic drag by 20 per cent compared to the newest conventional trucks on the road. The wheelbase is shortened by mounting the cab over the engine. This reduces the truck's weight and improves its manoeuvrability.

**Hybrid powertrain:** The truck's engine, a range-extending series-hybrid, uses much smaller and lighter batteries than pure electric trucks. The engines are well suited to the shorter trips

# Automotive gas turbines

and lower speeds that characterize most Walmart deliveries (Walmart distribution centres are now closer to metropolitan areas so final-delivery trucks make shorter trips.) These shorter trips also create more opportunities to recover energy through regenerative braking. The generators and energy-storage devices on the truck are scalable to the needed range.

**Microturbine power:** A small turbine on the hybrid engine improves fuel economy and gives the truck a greater range. The hybrid powertrain lets the turbine spin at its most-efficient speed, while the electric motor/energy-storage device handles acceleration and deceleration. Trucks scheduled for longer trips would be outfitted with larger turbines and smaller energy-storage devices.

**Fuel neutrality:** Turbines can use a wide variety of fuels, including diesel, natural gas, and biodiesel, while generating few emissions without the need for after-treatment. Turbines also have few moving parts which reduces maintenance requirements and weight.

**Charge mode:** In this mode, sensors can determine the batteries' state of charge and begin charging using the turbine if needed. Drivers can select charge mode if they want to "top off" the batteries prior to shutting down.

**Electric-vehicle mode:** The truck can run on electric power alone on urban streets, until the battery charge drops to below 50 per cent. Then the turbine starts and charges the batteries.

**Hybrid-electric mode:** For maximum range, the turbine runs continuously, only shutting down if the batteries run down.

**Component electrification:** The truck uses scaled-up versions of automotive subsystems powered by electric motors instead of hydraulics. They include power steering and air conditioning.

**Trailer:** The vehicle's trailer is built almost exclusively from carbon fibre, including one-piece panels for the roof and sidewalls, saving nearly 4,000 lb. compared to traditional designs. The trailer's convex nose also improves aerodynamics while providing storage space inside the trailer. Other features include a one-piece, fiberglass-reinforced floor panel with a 16,000-lb forklift rating.

**Cab:** The driver's compartment has a lightweight sliding door and a simplified dashboard with few controls and displays. In addition, customizable LCDs allow drivers to keep clutter to a minimum. There is also a full-sized sleeper with a fold-down bed.

# 22 Wrightspeed

Wrightspeed and The Ratto Group have formed a partnership to convert North Bay refuse collection and recycling vehicles from clean diesel to electric drive. The partnership was announced three weeks after Wrightspeed, of San Jose, California, publicized its heavy-duty electric driveline product, the route HD. The route HD is a plug-in electric truck powertrain that uses an on-board turbine generator to charge the battery, as needed, on the road. It uses CNG, LNG, diesel, or landfill gases and burns cleaner per kilowatt-hour than the average mix of US electrical power plants, making it "cleaner than an EV."

Surpassing California Air Resources Board's (CARB) ever-tightening emissions standards by 1,000 per cent, Wrightspeed's Powertrains are not only hyper-clean, they are commercially future-proof.

"We're always looking for ways to reduce greenhouse gases in our pursuit of an environmentally sustainable economy," says chief operating officer, Lou Ratto. "Wrightspeed's very efficient and super clean powertrains are a great fit for our fleet." The Ratto Group of Companies, based in Santa Rosa, provides refuse and recycling services to cities and other areas in Sonoma and Marin Counties.

Ian Wright, Wrightspeed's founder and chief executive officer, agrees, "The route HD was engineered for the refuse and recycling truck application, where it can reduce fuel spend by $35k per year and dramatically reduce noise pollution."

Wrightspeed claims it is "passionate" about designing and delivering products that really work. Started by one of Tesla Motor's co-founders and headquartered in Silicon Valley, Wrightspeed uses electric drive together with an on-board turbine generator for exceptional efficiency, optimal performance, and unlimited range. Built on a tradition of quality systems engineering, Wrightspeed's powertrains are the next step in the evolution of vehicle propulsion.

# 23 Miscellaneous

The Internet can be a happy hunting ground for information. In the case of gas turbine trucks various sites offer glimpses of attempts (mostly US) to put gas turbines in trucks. Two sites in particular offer a variety of vehicles. These are: www.pinterest.com and www.trucksplanet.com. Among the illustrations shown on these websites are:

• 1950 Kenworth rig with a Boeing gas turbine engine of 175bhp output. The picture caption noted that even though the Boeing engine weighed in at only 200lb, acceleration and fuel economy posed a problem. The picture caption reads that Boeing's Kenworth gas turbine-powered truck has logged over 40,000 test miles lugging loads of up to 68,000lb. Another illustration shows the truck nearing the summit of Snoqualmic Pass on the route from

# Automotive gas turbines

Seattle to Yakima, Washington. Clearly Boeing made serious attempts to enter this sector.

- Boeing also had a gas turbine-powered US Army ordnance truck which it claimed achieved 70mile/h. Boeing's engine was the 502-10 gas turbine.

- 1964. GMC Bison gas turbine concept truck which was displayed at that year's New York World Trade Fair as part of GM's Futurama exhibit.

- 1964. Ford's experimental Big Red gas turbine-powered Superhighway truck.

- 1965 Freightliner Turboliner tilt-cab tractor unit powered by a Boeing 553 300bhp gas turbine engine mounted in a rig operated by Consolidated Freightways.

- 1968 International Turbostar rig, powered by a gas turbine engine of unknown origin.

- Chevrolet Turbo Titan Jet Turbine Truck concept.

- Ford gas turbine truck powered by Ford's own design of gas turbine and illustrated during its 1970s UK tour.

- GMC Astro forward control tractor unit

- Mack WS760LST gas turbine tractor unit which can be found in Mack Trucks Museum in Allentown, Pennsylvania.

- Mack RW bonneted tractor unit powered by a gas turbine of unspecified manufacture.

# Automotive gas turbines

# Illustrations

In 1950 the world's first gas turbine car, Jet 1, powered by a T8 engine mounted at the rear of the car appeared. It can be seen at the Science Museum, London.

In 1953, Rover's Jet 1 with Spen King at the wheel, set the first world speed record for a gas turbine car of 151.965 mile/h on the Jabbeke Highway, in West Flanders, Belgium.

# Automotive gas turbines

The 1949 Rover T8 twin-shaft automotive gas turbine engine of 200bhp used a single combustion chamber

By 1950 the 200bhp T8 automotive gas turbine engine had been equipped with a single combustion chamber

## Automotive gas turbines

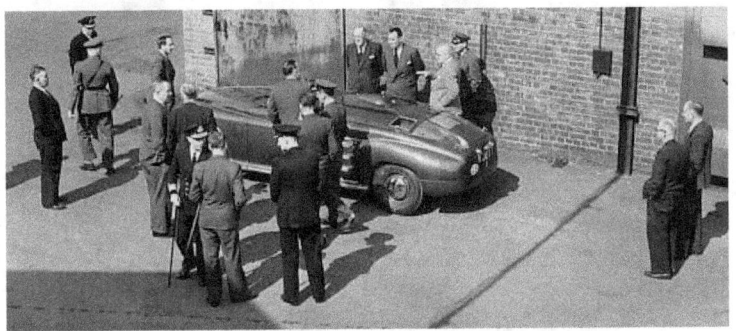

Rover's Jet 1 gas turbine car being studied by officials.

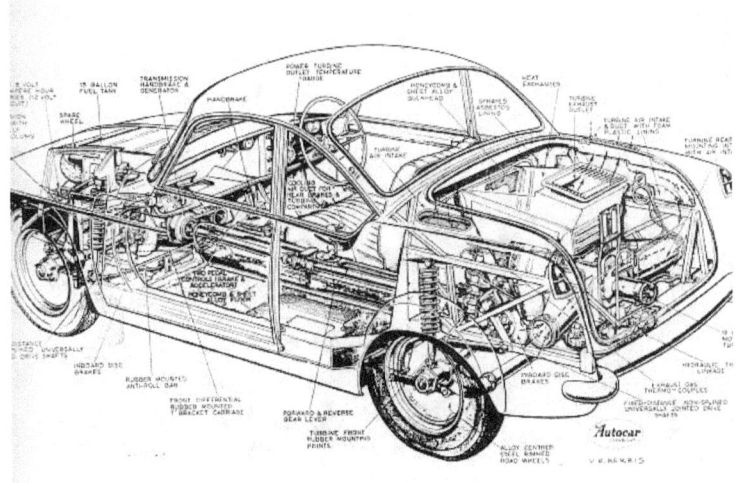

A 1956 cutaway drawing published in Autocar of the T3. According to Noel Penny, the T3 could take a hairpin bend at 80mile/h and was glued to the road. "It was a remarkable little car. It was my favourite vehicle," he wrote.

## Automotive gas turbines

The members of the team who worked on the T3 car, both in its polished state, and on the test track.

Noel Penny wrote that the 1956 trials of the T3 car proved the car an "immense success", while the fuel consumption was even better than predicted, so it was decided to exhibit it at the Motor Show at Earl's Court. The picture shows some of the participants of T3 programme.

## Automotive gas turbines

The Rover T3, the second prototype gas turbine car, had four-wheel drive with the 2S/100 100bhp engine mounted over the rear axle, as shown below. The car covered "thousands of miles" under normal traffic conditions.

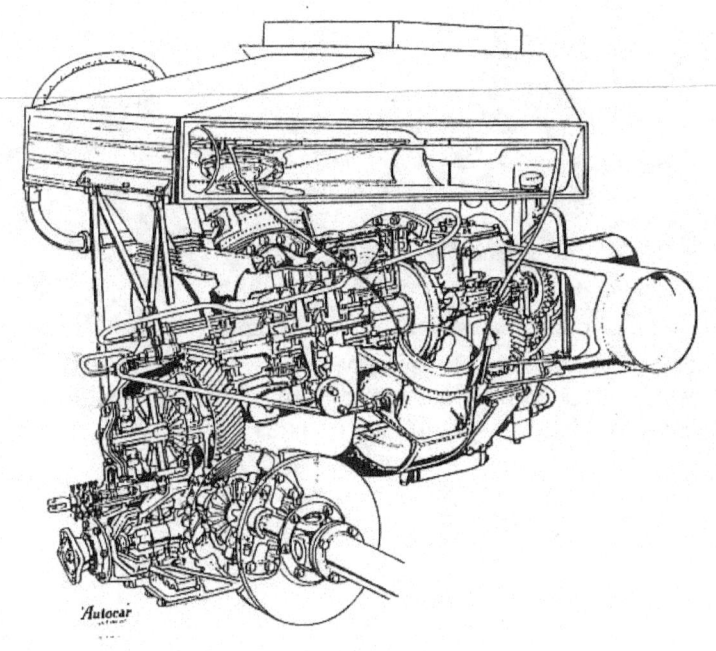

## Automotive gas turbines

Illustrated here is the team who worked on gas turbine engines at Rover. Included (seated fourth from the left) is Charles Spencer King, known as Spen King, and Noel Penny, with pipe. Also included is Frank 'Tinker' Bell, who left the team.

From left to right: the Rover/BRM racing car; the T3 and T4 passenger cars, and the Leyland Super Comet tractor unit, all powered by Rover's gas turbine engines.

# Automotive gas turbines

The first twin-shaft gas turbine produced by Rover developed 100bhp and was intended for automtive applications. Below: a marinised version of the same 2S/100 engine for the Royal Navy with Leyland-Thornycroft being responsible for installation. Both engines date from 1956.

## Automotive gas turbines

Noel Penny's caption for this illustration noted the "T4 embodied everything up-to-date. The picture is of the team who spent most of time inventing the T4."

In 1960, Rover produced a turboprop version of its 1S/60 engine intended for light aircraft operation.

## Automotive gas turbines

The Currie Wot single-seat aircraft powered by the Rover 1S/60 gas turbine

Noel Penny with the Rover-BRM which was at Noel Penny Turbines Ltd. for servicing.

# Automotive gas turbines

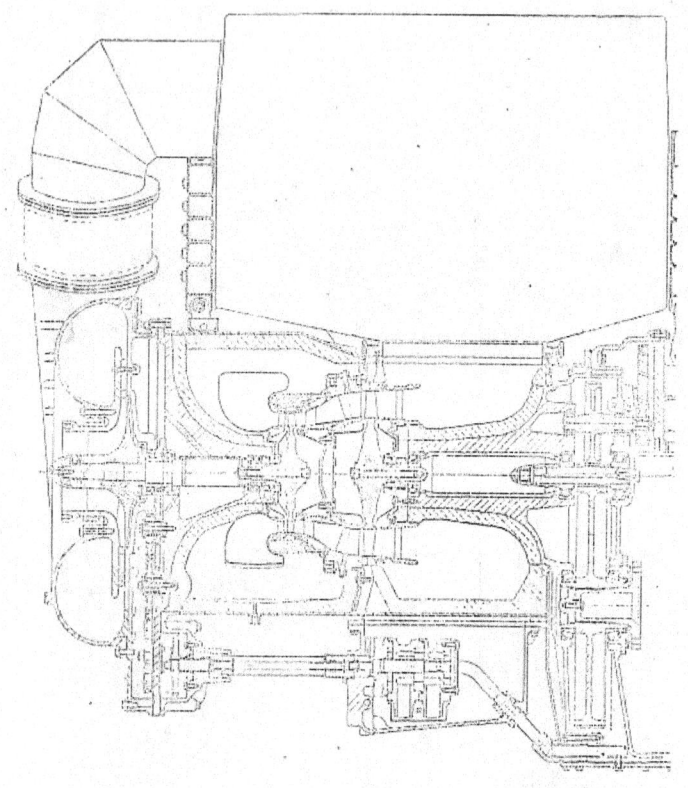

Cross-section through Project 201, the two-shaft 450bhp gas turbine engine Noel Penny Turbines Ltd designed for Caterpillar Inc. of Peoria, Illinois, USA.

## Automotive gas turbines

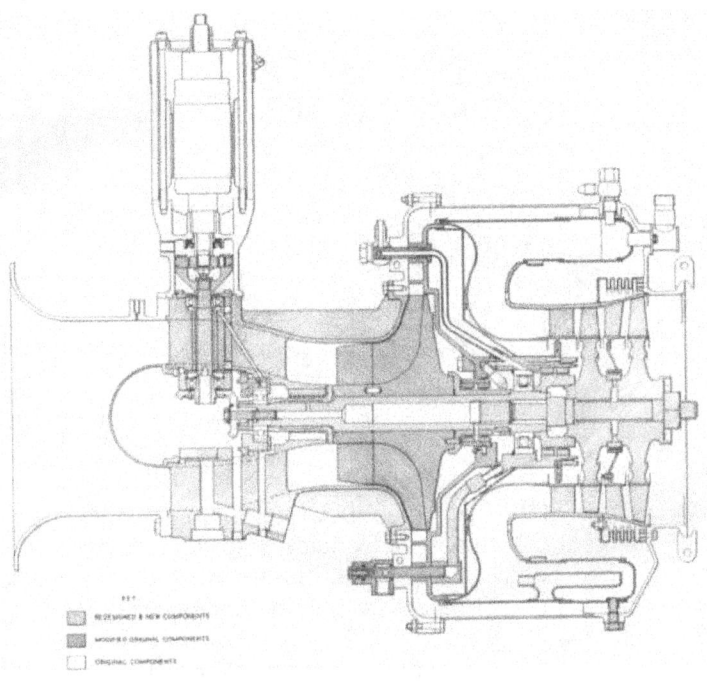

The NPT 100 Budworth engine comprised a single stage axial flow compressor, a single stage radial compressor and a two-stage axial flow turbine

## Automotive gas turbines

Perhaps one of the UK's most powerful turbojets in its class, the Noel Penny Turbines 301 engine was the last design to be produced before the company folded in 1990. This unit carries a number of similarities to its predecessor, the similar but much less powerful NPT109. This unit was developed to power recoverable drone type un-manned aircraft and UAVs. Two units were also fitted to a remarkable prototype four-seat business jet (bizjet) in the late 1980s. A simple but thirsty engine now replaced by much more elaborate units such as the Williams FJX33 turbofan.

The engine has an almost classic mechanical layout for a small turbojet engine. The NPT301 can be likened to a very large aircraft power plant (see illustration p.450).

Noel Penny Turbines Ltd. planned to produce a whole family of small turbojet engines as well as a turbo-shaft unit. Prototype NPT151 and NPT171 units were produced, a very small NPT051, a larger NPT401 and an even an axial flow NPT901 were on the drawing board.

Had the company not gone bust, the UK today could have had a very innovative supplier of small gas turbines. Competition from Williams and Microturbo was probably to blame for the down-fall of NPT.

## Automotive gas turbines

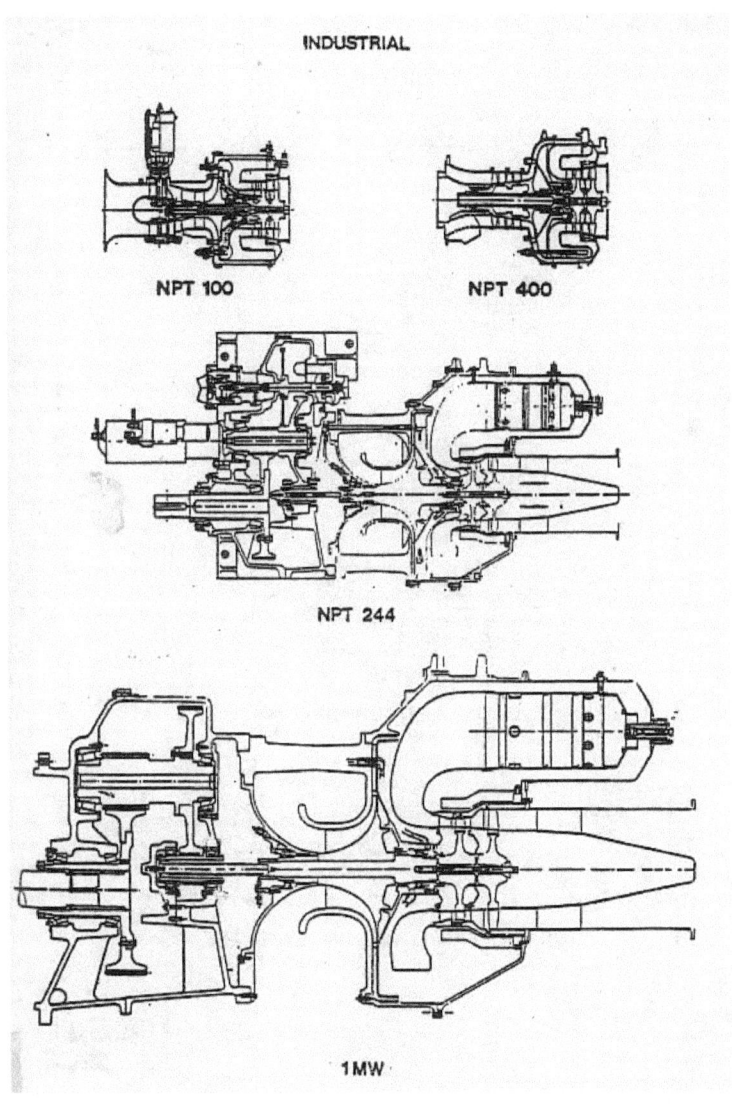

A range of turbines schemed by Noel Penny Turbines Ltd.

## Automotive gas turbines

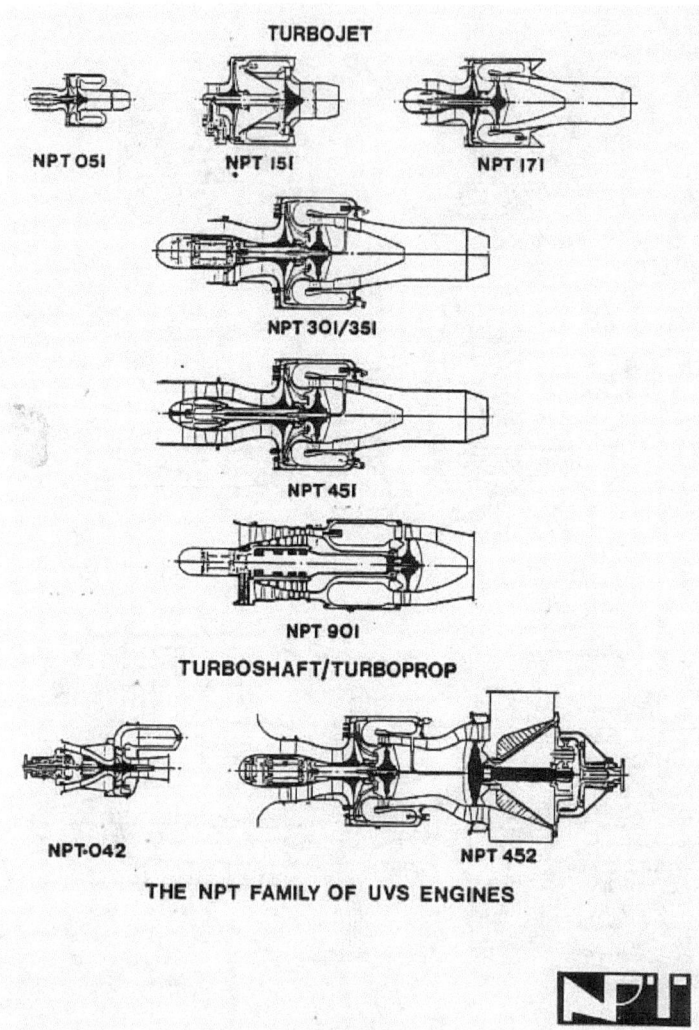

A family of turbojet and turboshaft/turboprop engines designed by Noel Penny Turbines Ltd.

# Automotive gas turbines

Rover JET 1 (1952)  NPT  Rover/BRM (1965)

Chrysler LeBaron (1978)  Reserved for production gas turbine car (1990?)

This 49pp technical brochure produced by Noel Penny Turbines Ltd gave acknowledgement to *The Engineer,* which had used it on its front cover for the 27 March 1980 issue. The title: *The ingredients for the commercial acceptance of gas turbine power,* was prepared by Noel Penny and delivered at the 5th International Symposium on Automotive Propulsion, 14-18 April, 1980. In it, Penny referred to the NPT G45 truck gas turbine which he said had 'proved its ability to meet all mandatory requirements for truck engines of the 1980s' with over 6,000 hours running achieved and fuel consumption better than 0.38lb/bhp.h. He forecast two-shaft truck engines of 220bhp.s/lb specific power for a 7:1 pressure ratio; and two-shaft car engines of 190bhp.s/lb specific power. And it gave details of a battle tank engine of

## Automotive gas turbines

1,450bhp, a specific power of 192bhp.s/lb and a specific weight of 40lb/lb/s to give a weight of 300lb.

Among advanced concepts illustrated was a schematic of a compound drive – in effect a turbocharged single-shaft engine with a low pressure spool and a high pressure spool, each feeding through fixed ratio primary reduction gearing to a Perbury CVT to the differential output. Among useful references quoted: No. 10. 'Gas turbines for future transport', by Noel Penny. *The Engineer*, April 1970.

Display of parts for Stage 3 of Caterpillar Inc.'s 450bhp truck gas turbine. Jill Wilks is just visible, right.

## Automotive gas turbines

The Yanmar engine designed by Noel Penny Turbines Ltd., developed 650bhp and was directly coupled to a 500kVA generator.

# References

Eight

**Gas turbines power trucks**

*On the road with an all-British turbine wagon*, by Alan Bunting. Commercial vehicles, November 1968

Nine

**Noel Penny Turbines Ltd**

*British genius backs his own judgement in gas turbine stakes*, by John Mortimer. The Engineer, 27 July 1972, pp 60-1, 98

*Small gas turbines make sense – But will they succeed?* by John Mortimer. *The Engineer*, 21 September 1972, pp 44-5, 47-9, 51

Ten

**Downfall**

*US sets pace in gas turbine car projects*, by Dr. Paul Butler. *The Engineer*, 27 March 1980, pp 36-8

**Appendix 3**

*Microturbines seek macro market*, by John Mortimer. *Professional Engineering*, December 2011

*Adding value, layer by layer*, by John Mortimer. *Professional Engineering*, March 2012

**Appendix 4**

*Ceramic turbines: Why Britain is leading the race*, by Duncan Peters and John Mortimer. *The Engineer*, 26 February 1970, pp 29-33

*Ceramics durability troubles get gas turbine men guessing*, by John Mortimer. *The Engineer*, 8 November 1973, pp 60-2

*As ceramics become even more vital Britain drops behind*, by John Mortimer. *The Engineer*, 13 December 1973, pp 52-4

*Tough, heat-resistant composite should boost gas turbines*, by Jeremy Sumner. *The Engineer*, 7 November 1974,

*Brittle Materials Design, High temperature Gas Turbine rotor blade development and turbine modifications*, December 1976, by D. G. Miller, Westinghouse

Electric Corporation for the Army Materials and Mechanics Research Center, Watertown, Ma. AMMRC CTR 76-32

*Brittle Materials Design, High Temperature Gas Turbines*, August 1977, by Arthur. F. McLean and Eugene A. Fisher, ford Motor Company, for the Army Materials and Mechanics Research Center, Watertown, Ma. AMMRC CTR 77-20

*Reliability of Ceramics for Heat Engine Applications*, by National Research Council, 1980

*Process Development for Silicon Carbide-based Structural Ceramics*, by Edward E. Hucke, University of Michigan, Ann Arbor, Mi., prepared for the Army Materials and Mechanics Research Center, Watertown, Ma. AMMRC

*Low-cost Net Shape Ceramic Radial Turbine Program*, February 1981, by D. W. Richerson and J, R. Smyth of Garrett Turbine Company for the Army Materials and Mechanics Research Center, Watertown, Ma. AMMRC TR 85-20

*Ceramic Technology for Advanced Heat Engines Project, (Oct 1984-Mar 1985)*, September 1985, by D. E. Johnson, Oak Ridge National Laboratory for the US Department of Energy.

## Appendix 5

*Chrysler hopes for gas turbine may be accomplished soon*, by John Hartley. *The Engineer*, 16 May 1974, pp 50-2

## Appendix 8

*I am designing for people, says Italy's gas turbine king*, by John Mortimer, *The Engineer*, 7 December 1972, pp 50-53

## Appendix 9

*Development of the Ford 704 Gas Turbine Engine*, by I. M. Swatman, Ford Motor Company. SAE Technical Paper 620516

*European debut of Ford's gas turbine, Commercial Motor*, 16 August 1968

*Unladen turbine truck holds no driving bogeys*, by Alan Bunting, *Commercial Vehicles*, November 1969, pp 67-69.

*Ford second generation hots up gas turbine war*, The Engineer, 2 July 1970, p 9.

## Appendix 10

Brochure: *AGT101. Advanced gas Turbine Powertrain; System development project. Proposal Summary. April 1979* Produced by The Garret Corporation with AiResearch manufacturing Company of Arizona, Phoenix, Arizona.

## Appendix 11

Brochure: *GT601. 550HP Gas Turbine Truck Engine.* Produced by The Garrett Corporation for Industrial Turbines International.

## Appendix 12

*The General Motors Research GT-309 Gas Turbine Engine*, by W. A. Turner and J. S. Collman. SAE Combined National Transportation and Power Plant Meeting, Cleveland, Ohio, 18-21 October 1965. *The Engineer*, 5 November 1965, pp 781-783

## Appendix 13

*Chrysler hopes for gas turbine may be accomplished soon*, by John Hartley. *The Engineer*, 16 May 1974, pp 50-2

## Appendix 14

*Both commerce and environment need gas turbines*, by John Mortimer. *The Engineer*, 21 December 1972, pp 28-9, 31

## Appendix 16

*The gas turbine man who came in from the cold*, by John Mortimer, *The Engineer*, 29 March 1973, pp 31-33, 56

## Appendix 17

*Vikings at Waterloo*, by David Brooks, Rolls-Royce Heritage Trust, Historical Series No. 22.

*Pistons to blades*, by Mark C. S. Barnard, Rolls-Royce Heritage Trust, Historical Series No 34.

## Appendix 19

SAE Technical Paper 2005-1-3504. *The Volvo Heavy Truck Gas Turbine VT300.* By Arne Olsson of Volvo Powertrain Sweden, and Bertil Jönsson and Lars Sundin of Volvo Aero Corporation.

# Index of names

Alvis, 5, 133, 141-2, 204, 208, 212-3, 215, 230, 240, 242, 328,

Austin, 5, 127, 131, 275, 293, 329, 330-2, 393-4,

Barnard, Mark, 162, 178, 203, 414, 415, 419, 421-2, 456,

Bell, Albert, 379, 381-2,

Bell, Frank, 125, 131, 420,

Bell, Frank (Tinker), 125, 127, 131, 132, 420, 422,

Benn, Tony, Minister of Technology, 204-5, 215, 224-6, 230,

Bladon Jets, 5, 332-339, 340, 342-3, 462,

Boeing, 281, 283, 361-2, 422, 435, 460,

Caterpillar Inc., 9, 200, 217, 232, 244-5, 247-9, 251-3, 257-8, 264, 269, 271, 277-8, 282, 294, 315, 446, 452, 461-3,

Cercor, 345, 347, 421,

Chevrolet Turbo Titan, 380, 435,

Chrysler Corporation, 5, 171, 174, 176, 199, 202, 313, 315, 327, 349-354, 360-2, 378, 383-4, 388-9, 390, 392, 410, 430, 455-6, 460,

Corning Glass, 145, 175-6, 181, 184, 187, 191, 194, 196, 198, 200, 222, 224, 227, 229, 242, 257-8, 264, 289, 345, 347, 361, 371, 375-6, 411, 460,

Cummins Engine Company, 348, 377-8, 398,

Cox, Harry, 141, 162, 178,

Currie Wot, 164-5, 273, 445,

Daimler-Benz, 5, 355, 357, 360, 410-2,

Detroit Diesel, 5, 358, 403-4,

Edinburgh, Duke of, 160, 332,

Ferguson, Harry, 248-9,

Fiat, 5, 216, 302, 359, 360-1, 410,

Fogg, Dr. Bertie, 137, 205-8, 210, 212-4, 222, 225, 228, 238, 239, 240-1, 243-5, 345, 461,

Ford Motor Company, 5, 7, 176, 257, 345, 353, 361-2, 372, 376, 435, 460,

Garrett AiResearch, 5, 372, 376, 396-7, 402, 407, 409,

General Motors, 5, 176, 244, 258, 353, 358, 371, 378, 404, 410, 456, 460-1,

Graham Hill, 178, 180, 187, 194, 461,

## Automotive gas turbines

Hartley, John, 392, 455-6,

Huebner, George J., 5, 174, 176, 199, 327, 383, 392,

Issigonis. Alec, 142, 230,

Jackie Stewart, 187, 194, 197, 461,

Jet 1, 126, 129, 130-1, 170, 174, 178, 216, 329, 437, 439, 461,

King, Charles Spencer, 420, 442,

Knowles, Harry, 127, 420,

Lanning, John, 191, 198, 229, 258, 347,

Le Mans, 174-5, 177-9, 180-7, 191, 194-6, 202, 207, 461, 463,

Leyland Gas Turbines Ltd., 5, 142, 212-3, 217, 236, 240, 261, 327, 333, 369, 394, 460-1, 463,

Leyland Comet, 210, 395, 460,

Leyland Super Comet, 207, 442,

McLean, A. F., 7, 8, 344-7, 363, 455,

Mack Trucks, 5, 371, 395-6, 398-9, 400-4, 407-9, 422, 431, 435,

Merlin, 127, 329, 418-9.

MTU, 5, 360, 410-2,

NGK-Locke, 376,

NGTE, 154, 204, 215, 270-1, 331,

Noel Penny Turbines Ltd, 4, 232, 249, 251, 253, 256, 282, 296, 445-6, 448-9, 450-1, 453, 454, 461-3,

P4, Rover, 129, 131, 173, 211, 329,

*Popular Science*, 382,

Rolls-Royce Heritage Trust, 456,

Rolls-Royce Ltd, 5, 7, 25, 33, 124-6, 137, 144-5, 163, 164, 172, 180, 192, 203-4, 217-9, 227, 228, 229, 233, 237, 242, 249, 252, 262, 274, 291, 306, 313-4, 320-4, 341-2, 402, 412, 415, 417-8, 419, 420, 456, 464,

Rover-BRM, 2, 175, 177-9, 185-6, 193, 197, 202, 223, 307, 445, 462,

Rover T3, 4, 141, 147, 151-2, 170, 178, 429, 439, 440-2, 448, 461,

Rover T4, 164, 168-9, 170, 174, 176, 178, 204, 309, 442, 444, 461,

Rover T8, 126-7, 129, 130-1, 134, 137-8, 437-8,

Rover Company, 7, 52, 56, 61, 123-6, 130, 138, 144, 175, 177, 183-4, 199, 202, 204, 206, 208, 216, 218, 282, 413, 460-1, 463,

**Automotive gas turbines**

Rover Gas Turbines Ltd., 5, 134, 139, 142, 144, 158, 160, 164-5, 173, 180, 183, 202, 204, 208, 212-3, 229, 230, 240, 242, 273, 306, 413, 461, 463,

Sam Williams, Dr., 176, 200, 245, 26, 309, 311, 315, 319, 383, 429, 430,

Science Museum, London, 129, 437,

Spen King, (see also King, Charles Spencer) 127, 154, 178, 225, 296, 420-1, 437, 442,

Stokes, Lord, 204, 206-7, 212-3, 217, 221, 223-6, 230, 238, 241-4, 344, 393, 395, 460, 461,

Swatman, Ivan, 225, 362-4, 366, 370, 455, 460,

Tata Group. 335, 337-9, 340,

Tata, Ratan, 335, 339, 342,

'Tinker' Bell, Frank – see Bell, Frank and Bell, Frank 'Tinker',

*The Engineer*, 10, 251, 327, 344-5, 379, 392, 410, 451-2, 454-6, 460, 464,

Volvo, 5, 216, 292-4, 296, 299, 300-2, 306, 360, 408, 422-5, 429, 456,

Walmart, 5, 431-2,

Wrightspeed, 5, 433-4

Wedgwood-Benn, Tony; see Benn, Tony,

Williams International, 5, 200-1, 288, 298, 304, 309, 310-9, 320-3, 383, 429, 430-1,

Whittle, Sir Frank, 9, 126, 162, 172-3, 218, 281, 291, 322-3, 334, 414-8, 422, 463,

Wilks, Maurice, 125, 160, 171-2, 175, 197, 414, 419, 420,

Wilks, Spencer, 126, 416, 419, 422,

Yanmar, 9, 256, 275, 283, 453,

# Obituaries

## *Professional Engineering*, May 2014

'IN JUNE 1950 in north London, a group of engineers were testing a new kind of automobile. It had no pistons, cylinders, carburettor and crankshaft. But it did have performance. It could accelerate from zero to 60mile/h in 14 seconds and reach 90 mile/h without effort.

By coincidence, at about the same time, 6,000 miles away in Seattle, Washington, USA, a 10-ton truck cruised the highways under the thrust of another kind of engine, brand new to truck wranglers.

# Automotive gas turbines

The truck, like the car in London, had the faint whiff of kerosene about it. And so it was that while the car in London was designed and built by the Rover Company of Solihull, the truck in Seattle likewise was powered by a gas turbine engine designed and built experimentally by the Boeing Aeroplane Company. Nothing came of Boeing's effort; the same could not be said of Rover's endeavours.

The British public first heard of Rover's mysterious silent Rover when the *News Chronicle* of Friday May 14, 1948, carried an item "Car firm is testing gas turbine engine". And so began to unfold the story of the Rover gas turbine engine and how it developed over the years to become the power unit of several Leyland trucks, the first of which was the Leyland Comet.

The architect of this engine was Noel Penny, who died aged 87 on Friday 17 May. Penny was managing director of Leyland Gas Turbines Ltd, (LGT) formed by Leyland boss Lord Stokes to design, develop and manufacture gas turbines for commercial vehicles.

I had the great fortune in the 1960s, as editor of *The Engineer*, to be as close to Noel Penny as any journalist was allowed to be. For Penny was a private man and did not like journalists following the development of his engines, especially when testing work did not proceed according to plan.

Not the least of LGT's challenges was the use of a highly innovative Corning Glass ceramic heat exchanger to extract heat from the exhaust and transpose it to the incoming air. When mysterious white powder appeared in the engine during testing, journalists were the last people Penny wished to get wind of any such 'technical hiccup'! In later life, when we were both retired, we became good friends, even to the extent that he allowed me to read his biography, "My Story", which he compiled several years before his death.

Penny and his team at LGT were pioneers, treading a path few others had taken. In the US, mighty companies like Chrysler, Ford and General Motors wrestled with a similar mountain of technical issues to bring gas turbine cars and trucks to fruition. By comparison, LGT was doing its work on a shoestring.

Penny was probably less that happy when rival Ford Motor Company brought its gas turbine truck (developed under the guidance of another technical genius, Watford-born Ivan Swatman) to show it off to journalists in Britain.

# Automotive gas turbines

That LGT achieved so much was attributable to the unwavering passion and commitment of Noel Penny who was not only a brilliant scientist and engineer, but an extremely competent and shrewd business man whose energy and enthusiasm for the subject of gas turbines was infectious.

The LGT engine was a much more powerful version of the engine fitted to the Rover-BRM Graham Hill and Jackie Stewart drove so successfully in the 1965 Le Mans 24 hour race. Smaller versions of these engines powered Rover's prototype cars in the 1950s and 1960s – the T3 and the T4. Perhaps the most famous Rover gas turbine car was JET 1, which in 1950 reached a top speed of 151 mile/h on the Jabbeke motorway in Belgium

Penny arrived at the Rover Company in May 1952 from the Atomic Energy Authority where he worked as a scientist. He told me that he began working at Rover in a department called "Water Pumps". The name disguised gas turbine development. Penny was technical assistant, responsible for high-speed bearings and fuel systems.

Penny worked his way up through the ranks until one day he was approached by Dr. Bertie Fogg, director of engineering at Leyland Corporation. Dr. Fogg asked Penny, then chief engineer of Rover Gas Turbines Ltd in Solihull, to set up a company to design, develop and manufacture gas turbine engines for trucks.

Although engine development proved successful, several external factors led to a change of course for Penny. Lord stokes wished that Leyland Gas Turbines Ltd in Solihull be moved to Leyland's headquarters in Lancashire. Penny opposed the move and offered his resignation to Dr. Fogg in 1971. On 22 July 1972 Penny formed his business, Noel Penny Turbines Ltd, based in Siskin Drive, Coventry.

Penny's immense knowledge of gas turbine development inevitably brought him unrivalled respect on both sides of the Atlantic. Such was the level of this worldwide respect that both General Motors and Caterpillar Tractor were anxious to have Penny on their payroll to work on their behalf in the US. Penny was not keen to move to America, just as he has not been keen to move from Solihull to Leyland.

In the end, the mighty Caterpillar of Peoria, ever anxious to make use of Penny's talents in whatever manner they could, awarded his newly-founded company NPT with a research contract to design and develop a two-shaft regenerative 350bhp gas turbine engine for off-highway application. The engine would be fitted to an International Tri-Star truck for highway development.

I visited NPT's offices in Coventry on many occasions; it was like stepping into another world where talented engineers and scientists worked with unbridled enthusiasm to develop Caterpillar's engines.

So secret was their activities that Penny was always reluctant to discuss any aspect of Caterpillar's work; he would not even openly admit he was working for Caterpillar, so secret were his endeavours. Certainly I, as a humble journalist, was never allowed to see this 'magic beast' nor touch any of the components associated with it. Penny took company confidentiality very seriously.

Caterpillar decided not to put with the engine into production. Compared with a diesel engine, the gas turbine was a thirsty animal. Notwithstanding this, Penny and his team focused their attention on other applications for their product and devised a range of gas turbine engines the company went on to manufacture for use in business jets.

Although the gas turbine has so far failed to find a niche in the automotive world, Penny never lost his enthusiasm for a power unit which he believed offered so much potential.

It is perhaps ironic that Coventry remains the home of Bladon Jets Ltd which is working with Jaguar Land Rover to introduce small gas turbine engines into passenger cars. Noel Penny, for one, would be pleased if that work has a successful outcome.' JM

## *Coventry Telegraph*, 10 June 2013

'Pioneering Coventry engineer Noel Penny has died (17 May) after a long illness aged 87. A survivor of the Coventry Blitz while growing up in Canley, he went on to set up Noel Penny Turbines Ltd which employed hundreds of engineers in the city for two decades.

He designed 3,000 gas turbine engines for automotive, aerospace, marine and industrial machinery still used around the world today.

One of ten siblings and a former Coventry Technical College student, Noel stayed in the city all his life and still holds 34 British patents as well as 258 worldwide.

Daughter Deborah Neville said "He was a proud Coventry boy and he brought a lot of jobs to the city. He picked up where Frank

## Automotive gas turbines

Whittle left off designing jet engines and became internationally renowned for his gas turbines."

After the war Noel worked as a scientist at the Atomic Energy Authority before moving to the Rover Company in 1952. He worked his way up to become chief engineer at Solihull-based Rover Gas Turbines in 1958 where he was responsible for the gas turbine racing cars which competed at Le Mans in 1963 and 1965.

He was head-hunted by the Leyland Corporation in 1966 to set up a new company called Leyland Gas Turbines. But when the parent company unveiled plans in 1971 to relocate Leyland Gas Turbines out of the area - from Solihull to Lancashire - Noel tendered his resignation. A year later he set up Noel Penny Turbines in Siskin Drive, Tollbar End.

Motoring journalist John Mortimer described the ground breaking Coventry company.

He said: "I visited NPT's offices in Coventry on many occasions – it was like stepping into another world where talented engineers and scientists collaborated with unbridled enthusiasm to develop Caterpillar's engines. So secret were their activities that Penny was always reluctant to discuss any aspect of Caterpillar's work. He would not even openly admit he was working for Caterpillar."

Noel Penny Turbines eventually went into receivership in 1991 and the rights to his key turbine technologies sold off to the highest bidder.

Noel went on to mentor many promising engineering students as well as publishing 34 international papers and lecturing in 12 countries.

In his final years Noel was battling the cruel impact of progressive supranuclear palsy, which sees sufferers lose movement and the power of speech. Daughter Deborah said that despite suffering from this "horrendous disease" his mind was still active until the end and he lived at his home in Canley until moving to a care home six months ago.

Noel Penny is survived by three children and five grandchildren.'

# Author

John Mortimer has an MSc in aircraft propulsion from Cranfield University, having worked for a number of aero engine companies in the early 1960s, including D. Napier & Son, de Havilland Engine Company, Bristol Siddeley Engines and Rolls-Royce. His main areas of work then were liquid propellant turbopumps, small gas turbine engines, and torpedo engines, and gearbox design.

For 11 years, John Mortimer served as editor and publishing director of *The Engineer*, a London (England) weekly engineering newspaper; he acted also as pro-tem editor of *Civil Engineering* and *Tunnels & Tunnelling* during acquisitions; also editor of *Career Choice*. He followed this with five years as managing director of IFS Publications Ltd, a company specializing in books and quarterly journals devoted to robotics, sensors and factory automation. During this time he authored *The FMS Report* and *The Ingersoll Report*, and co-wrote several books on robotics and automated assembly, as well as copy editing the English translation from German of *Automated Guided Vehicles*.

In 1985, John Mortimer established Industrial Newsletters Ltd, a company publishing monthly automotive industry newsletters and executive reports on benchmarking, simultaneous engineering, electronic data interchange and best practice product design. Each month he generated text required for the 32-page editorial-only *Auto Industry Newsletter*, which became a world-wide industry benchmark for authoritative news about the world's automotive industry – passenger cars, trucks and buses. Ten years later, he sold this successful activity to FTSE 100 enterprise United Business Media. He subsequently wrote news and features on diesel engines and vehicle manufacture for www.automotiveworld.com and the Institution of Mechanical Engineer's *Professional Engineering* and *Automotive Engineer*. Recently he launched his blog: www.autoindustrynewsletter.blogspot.co.uk

He is a Fellow of the Institution of Mechanical Engineers and a Member of the Guild of Motoring Writers. He has won numerous awards including the John Player Management Writer of the Year Award, the Blue Circle Award for Technical Journalism, The Delphi Award for Automotive Journalism and the Literati Award for Excellence 2003.

Following many years of detailed research, John Mortimer completed the biography: *Zerah Colburn: The Spirit of Darkness*, published in 2005 by Arima Publishing. The biography concentrates entirely on Colburn's engineering and writing career.

His first novel, *Angel in the house*, is set in New York, Philadelphia and Victorian London between 1860 and 1870. The principals are drawn from true-to-life characters of the period.

John Mortimer's third book traces the development of gas turbines for commercial vehicles, using Noel Penny's life story as its backbone.